T0318589

INTRODUCTION TO BIOMEDICAL INSTRUMENTATION AND ITS APPLICATIONS

INTRODUCTION TO BIOMEDICAL INSTRUMENTATION AND ITS APPLICATIONS

SUDIP PAUL

ANGANA SAIKIA

VINAYAK MAJHI

VINAY KUMAR PANDEY

ELSEVIER

ACADEMIC PRESS

An imprint of Elsevier

Academic Press is an imprint of Elsevier
125 London Wall, London EC2Y 5AS, United Kingdom
525 B Street, Suite 1650, San Diego, CA 92101, United States
50 Hampshire Street, 5th Floor, Cambridge, MA 02139, United States
The Boulevard, Langford Lane, Kidlington, Oxford OX5 1GB, United Kingdom

Notices
Knowledge and best practice in this field are constantly changing. As new research
and experience broaden our understanding, changes in research methods, professional
practices, or medical treatment may become necessary.

Practitioners and researchers must always rely on their own experience and knowledge
in evaluating and using any information, methods, compounds, or experiments
described herein. In using such information or methods they should be mindful of
their own safety and the safety of others, including parties for whom they have a
professional responsibility.

To the fullest extent of the law, neither the Publisher nor the authors, contributors, or
editors, assume any liability for any injury and/or damage to persons or property as a
matter of products liability, negligence or otherwise, or from any use or operation of
any methods, products, instructions, or ideas contained in the material herein.

Library of Congress Cataloging-in-Publication Data
A catalog record for this book is available from the Library of Congress

British Library Cataloguing-in-Publication Data
A catalogue record for this book is available from the British Library

ISBN: 978-0-12-821674-3

For information on all Academic Press publications visit our
website at https://www.elsevier.com/books-and-journals

Publisher: Mara Conner
Acquisitions Editor: Fiona Geraghty
Editorial Project Manager: Devlin Person
Production Project Manager: Sojan P. Pazhayattil
Cover Designer: Matthew Limbert

Typeset by TNQ Technologies

Contents

4. Radiological devices   169

Sudip Paul, Angana Saikia, Vinayak Majhi and Vinay Kumar Pandey

5. Analytical instruments   213

Sudip Paul, Angana Saikia, Vinayak Majhi and Vinay Kumar Pandey

6. Cardiac pacemakers and defibrillators   251

Sudip Paul, Angana Saikia, Vinayak Majhi and Vinay Kumar Pandey

10. Patient safety   **399**

Sudip Paul, Angana Saikia, Vinayak Majhi and Vinay Kumar Pandey

Contributors

Vinayak Majhi

Department of Biomedical Engineering, School of Technology, North-Eastern Hill University, Shillong, Meghalaya, India

Vinay Kumar Pandey

Department of Biomedical Engineering, School of Technology, North-Eastern Hill University, Shillong, Meghalaya, India

Sudip Paul

Department of Biomedical Engineering, School of Technology, North-Eastern Hill University, Shillong, Meghalaya, India

Angana Saikia

Department of Biomedical Engineering, School of Technology, North-Eastern Hill University, Shillong, Meghalaya, India; Mody University of Science and Technology, Laxmangarh, Rajasthan, India

Preface

We live in an age of modernization where electronics have flourished and been applied in many fields. It has revolutionized most fields: communication, aerospace, transportation, film and television, healthcare technologies, and many more. Applying electronics in healthcare has improved precision and made diagnoses more accurate. These factors have motivated many engineers to work in medical and healthcare technologies, but they have encountered difficulty with the physiological aspects. To overcome this, we have designed this book to provide essential knowledge about physiology and an overview of basic electronics that should prove helpful as a textbook for undergraduate and diploma-level students.

This edition of *Introduction to Biomedical Instrumentation and Its Application* comprises 10 units covering the crucial aspects of biomedical instrumentation in a straightforward, precise, and systematic way. They start with an overview of biomedical instrumentation and detailed explanations of medical and instrumentation terminology related to patient safety. The final unit is intended to provide a better understanding of the modern systems incorporated to provide best-in-class patient safety.

Unit 1 describes various medical terms and their physiological locations concerning the human body. Students overview various physiological parameters, such as heart rate, pulse rate, and SpO_2 level.

Unit 2 deals with various biosignals and the bioelectrodes used to acquire them. Our body produces many signals to communicate and transfer information from one place to another. These signals are captured using bioelectrodes as an interface.

Unit 3 relates to the transducers and amplifiers used to convert one form of energy to another. Amplification is important, as we know that biosignals are weak signals—without amplification, it is challenging to process such signals. This chapter describes various types of sensors and their application in medical electronics.

Unit 4 explains radiological devices, their functions, and their applications in diagnosing and treating various diseases. The topics covered include diagnostic terminologies, such as X-ray, CT-scan, MRI, and PET. The topics are explained with a brief system description, the principle of application, how it works, and the precautions to be taken when employing these procedures.

Unit 5 covers analytical instruments, which are integral to medical electronics and find application in analysis and parametric changes at molecular and genomic levels. Techniques like HPLC, chromatography, and DNA analysis are discussed with appropriate figurative explanations and analyses.

Unit 6 explains cardiac pacemakers and defibrillators used in inpatient treatment and design and considers issues related to healthcare technology. It covers various aspects of internal and external pacemakers, from design to application, with a great concern for patient safety.

Unit 7 describes ventilators and anesthesia systems, from a background history to the modern changes made in the present systems. This unit includes the breathing circuit, scavenger systems, generators, related components, and design considerations.

Unit 8 deals with physiotherapy equipment used in inpatient care in modern-day therapeutic clinics. It describes various forms of electrical stimulation, like transcutaneous electrical nerve stimulation, direct current stimulation, transelectrical nerve stimulation, and other nerve stimulation techniques used for various interventions and treatments.

Unit 9 extends the work of the previous unit dealing with therapeutic equipment. Various therapeutic products are explained in detail, addressing diagnostic purposes, prevention, conception, influences on and modification of physiological processes, design regimes, etc.

Unit 10, patient safety, starts with a background history and perceptions about patient safety. It explains various medical errors and their adverse effects. It also describes issues related to patient safety and its culture in primary care. The principle of clinical risk management is discussed, and the current government policies adopted by various developed nations are compared with those of developing nations.

CHAPTER 1

Overview of biomedical instrumentation

Sudip Paul[1], Angana Saikia[1,2], Vinayak Majhi[1] and Vinay Kumar Pandey[1]

[1]Department of Biomedical Engineering, School of Technology, North-Eastern Hill University, Shillong, Meghalaya, India; [2]Mody University of Science and Technology, Laxmangarh, Rajasthan, India

Contents

Biosignals are defined as the electrical signals acquired from the human body to measure various bodily parameters. Some common biosignals are measured using

(1) Electroencephalography (EEG): brain signal recordings
(2) Electromyography (EMG): muscle signal recordings
(3) Electrocardiography (ECG): heart signal recordings

Introduction to Biomedical Instrumentation and Its Applications
ISBN 978-0-12-821674-3
https://doi.org/10.1016/B978-0-12-821674-3.00010-3

1

(4) Electrooculography (EOG): eye signal recordings

Biomedical instrumentation is the application of knowledge and technologies to solve problems related to living biological systems. It involves diagnosis, treatment, and prevention of disease in humans. It involves biosignal measurement using ECG or EMG or measuring other electrical signals generated in the human body. Biomedical instrumentation helps physicians diagnose problems and provide treatment. To measure biological signals and design medical instruments, an understanding of electronics and measurement concepts and techniques is required.

Any medical instrument consists of the basic functional parts described below.

Measurand

The measurand is a physical quantity. The human body is a source for measurands, which generate biosignals such as those at the body surface or blood pressure in the heart. Some design criteria of a measurand are signal, environmental, medical, and economic factors. The different types of biomedical measurands are

➤ Internal: blood pressure
➤ Body surface: ECG or EEG potentials
➤ Peripheral: infrared radiation
➤ Offline: tissue sample extraction, blood analysis, or biopsy.

The various biomedical measurands are biopotential, pressure, flow, dimensions (imaging), displacement (velocity, acceleration, and force), impedance, temperature, and chemical concentration.

Transducer

The transducer converts one form of energy to another, which is mostly electrical. The transducer produces a usable output depending on the measurand. The sensor is used to sense the signal from the source. It is used to interface the signal with the human.

Signal conditioner

Signal conditioning circuits convert the output from the transducer into a digital value sent to the display unit. Signal conditioning comprises amplification, filtering, and analog-to-digital and digital-to-analog conversions.

Display

The display provides a visual representation of measured quantities, such as chart recorder data, cathode ray oscilloscope data, and signals generated in the Doppler ultrasound scanner used for fetal monitoring.

Data storage and transmission

Data storage is used to store data for future use. Electronic health records have recently been utilized in hospitals. Data transmission is used in telemetric systems, allowing data to be transmitted remotely from one location to another.

Fig. 1.1 shows a biomedical instrumentation system.

The study of biomedical instruments can be approached from

1. Biomedical measurement techniques can be grouped according to the quantity sensed, such as pressure, flow, or temperature. One advantage of this classification is that it makes it easy to compare different methods for measuring quantities.

2. A second technique uses the principle of transduction, such as resistive, inductive, capacitive, ultrasonic, or electrochemical. Different applications of each principle can be used to strengthen understanding of each concept.

3. Measurement techniques can be studied separately for each organ system, such as the cardiovascular, pulmonary, nervous, and endocrine systems. This approach isolates all important measurements for specialists

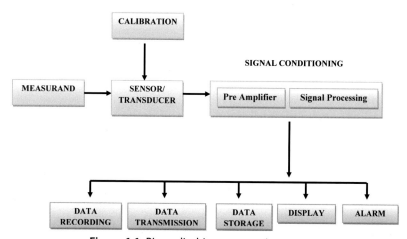

Figure 1.1 Biomedical instrumentation system.

who only need to know about a specific area but results in considerable overlap of quantities sensed and transduction principles.

4. Finally, biomedical instruments can be classified according to clinical medicine specialties, such as pediatrics, obstetrics, cardiology, or radiology. This approach is valuable for medical personnel interested in specialized instruments.

Some advantages of biomedical instrumentation are

➤ Augmented reality
➤ Cures for old problems
➤ Replacement of organs
➤ Extended lifetime
➤ Progressive knowledge

Some of its disadvantages are

➤ Introduction of foreign substances into the body
➤ Making humans more dependent on technology
➤ Some treatments that are difficult to access by people unfamiliar with them
➤ Treatments that may not fix the root cause of a problem or its long-term effects

Bioinstrumentation applies physics and activity principles and techniques to extensively studied natural perishable chemical compound formulations for drug delivery, wound dressing, and tissue engineering. So far, a restricted range of bioactivities is known from isolated compounds and research areas developing strategies for using them in human health promotion.

Biomedical engineering and research are presently challenged by the higher pro oom infrastructure of medical specialty engineering and its multidisciplinary nature. Medical specialty engineers can become quite familiar with applicable laws that set efficaciousness and safety in the long run.

Biomedical instrumentation can be classified into the following subdivisions:

1. **Cardiovascular technology** includes all drugs, biologics, and devices related to diagnostics and therapeutics of cardiovascular systems.
2. **Neural technology** includes all drugs, biologics, and devices related to diagnostics and therapeutics of the brain and nervous systems.
3. **Orthopedic technology** includes all drugs, biologics, and devices related to diagnostics and therapeutics of skeletal systems.
4. **Cancer technology** includes all drugs, biologics, and devices related to diagnostics and therapeutics of cancer.

Some characteristics of biomedical instrumentation are as follows:

1. **Accuracy** is defined as the difference between the true value and measured value.
2. **Precision** is defined as the number of distinguishable alternatives where the result has been selected.
3. **Resolution** is defined as the least incremental value that is measured with certainty.
4. **Reproducibility** is defined as the ability of any device to produce the same output for equal inputs given at a particular time interval.

Some commonly used biosignals and their instrumentation are described in Table 1.1.

Electroencephalography

An electroencephalogram (often abbreviated EEG, as is electroencephalography) is a test that detects electrical activity in the brain using small, metal disks (electrodes) attached to the scalp. Our brain cells communicate via electrical impulses and are active all the time, even when we sleep. This activity shows up as wavy lines on an electroencephalogram recording. EEG is a technique for recording and interpreting electrical activity of the brain. The nerve cells of the brain generate electrical impulses that fluctuate rhythmically in distinct patterns. In 1929, German scientist Hans Berger published the first study to employ an electroencephalograph, an instrument that measures and records these brain wave patterns and produces electroencephalograms.

Table 1.1 Some commonly used biosignals and their instrumentation.

Parameter	Range	Frequency (Hz)	Sensor
Blood flow	1−300 mL/s	DC-20	Flowmeter
Arterial blood pressure	25−400 mm Hg	DC-50	Cuff, strain gauge
Electrocardiography	0.5−4 mV	0.01−250	Skin electrodes
Electroencephalography	5−300 microV	DC-150	Scalp electrodes
Electromyography	0.1−5 mV	DC-10,000	Needle electrodes
Respiratory rate	2−50 breaths/min	0.1−10	Nasal thermistor

Table 1.2 Electroencephalography frequency bands.

Type	Frequency (Hz)	Location	Normally
Delta	Up to 4	Frontal part in adults, posterior half of a children's brains, high–amplitude waves	Slow-wave sleep in adults and during any continuous attention task in babies
Theta	4—8	Locations other than the human hand task	Drowsiness or arousal conditions in older children and adults
Alpha	8—13	Both sides of the posterior part of the human head	During relaxed states, such as with the eyes closed; mainly associated with inhibitory controls
Beta	>13—30	Symmetrical distribution on both sides, mostly frontal region with low-amplitude waves	During alert or working states when the person is active or busy
Gamma	30—100+	Somatosensory cortex	During any sensory processing; also seen during short-term memory matching of recognized objects, sounds, or tactile sensations
Mu	8—13	Sensorimotor cortex	Shows rest state motor neuron

The frequency bands in electroencephalography are described in Table 1.2.

The various applications of electroencephalography are as follows:

(a) Control of rehabilitation devices

(b) Early detection of various neurological disorders

(c) Therapeutic and recreational use

(d) Electroencephalography-based sensors used in medical assessment techniques

(e) Neuronal behavior of the human brain

The EEG test generally records cerebral activity in the brain. Recorded electrical activity not originated cerebrally are known as artifacts and can be divided primarily into physiologic and nonphysiologic artifacts. Artifacts are considered disturbances in measured brain signals. They are considered unwanted signals or signal interference. Physiologic artifacts are generated

Table 1.3 Electroencephalography artifacts.

Physiological	Nonphysiological
Electromyography	60 Hz interference
Eye movement	Electroencephalography electrodes
Electrocardiography	Environmental sources
Respiration	

from regions other than the brain (i.e., body), whereas nonphysiologic artifacts arise outside the body. Artifacts can affect many EEG features like mean, median, distribution, standard deviation, and signal-to-noise ratio. Some ways to minimize artifacts are the proper design of the EEG test, proper instruction before the test, response electromyography (EMG) or force grips, eye tracker, electrode localization equipment, comfortable environment, and good response device.

Different types of artifacts are shown in Table 1.3.

Electrode placement system of 10—20 electrodes

The 10—20 system or International 10—20 system is an internationally accepted EEG recording method. It mainly describes the location of scalp electrodes on the brain's surface in the context of an EEG experiment. The "10" and "20" refer to the actual distances between adjacent electrodes, which are either 10% or 20% of the total front—back or right—left distance of the skull. Each site has a letter to identify the lobe and a number to identify the hemisphere location. F, T, C, P, and O stand for the frontal, temporal, central, parietal, and occipital lobes. Even numbers (2, 4, 6, and 8) refer to electrode positions on the right hemisphere, whereas odd numbers (1, 3, 5, and 7) refer to those on the left hemisphere. A "z" (zero) refers to an electrode placed at the midline. In addition to these combinations, the letter codes A, Pg, and Fp identify the earlobes and nasopharyngeal and frontal polar sites, respectively. Fig. 1.2 shows the placement of the electrodes in the skull.

Encephalography uses a recording system. Fig. 1.3 shows a schematic of an EEG recording system. In an EEG recording system, the electrodes read the signal from the surface of the head. The proper function of these electrodes is critical to the acquisition of accurate and high-quality signals.

The various types of electrodes used are

(a) Disposable electrodes

(b) Reusable disk electrodes

(c) Headbands and electrode caps

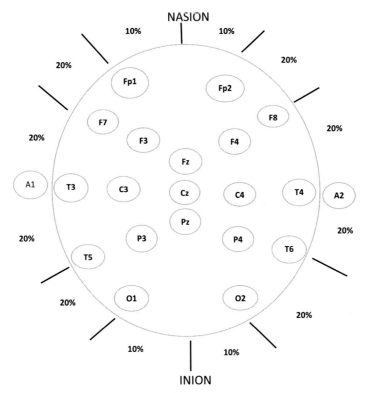

Figure 1.2 Electrode placement system for 10–20 electrodes.

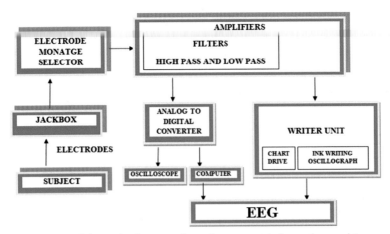

Figure 1.3 Schematic diagram of an electroencephalography machine.

(d) Saline-based electrodes

(e) Needle electrodes

The amplifiers amplify the recorded signal, i.e., convert microvolt signals to higher amplitudes. The converter changes the analog signal to digital form, and finally, the computer stores and displays the required data.

EEG measures potential differences between signal (active) and reference electrodes over time. The ground electrode is an extra electrode used to obtain differential voltage. A single-channel EEG system consists of one active electrode, one reference, and one ground electrode, whereas multichannel configurations can comprise as many as 128 or 256 active electrodes.

Types of brain disorders:

Irregular brain signals lead to various brain disorders such as

➤ Alzheimer's disease (AD), a brain disorder, is a quite prevalent and immensely growing type of dementia in aged groups of society. A person loses memory and thinking abilities, and with progression, the ability to carry out day-to-day tasks becomes slower and weakens. Treatment may lead to the management of disease symptoms, but a permanent cure is not possible at this time. AD generally starts deep inside the brain, where healthy neurons become inactive with less function and slowly die. Over time, this spreads to the hippocampus, which is the learning and memory unit of the brain. In addition, some other areas begin to shrink. Gradually, beta—amyloid plagues and neurofibrillary tangles begin to spread in the brain. According to scientists, these changes start 10—20 years before the onset of disease symptoms. The National Institute on Aging and the Alzheimer's Association have developed three diagnostic criteria for AD, comprising three stages: preclinical AD, mild cognitive impairment due to AD, and final-stage dementia due to AD. These stages can be distinguished by particular biological events such as glucose metabolism, accumulation of beta—amyloid plaques, and brain atrophy. These biomarkers can be detected through blood and cerebrospinal fluid testing and radiological techniques such as positron emission tomography (PET) and magnetic resonance imaging (MRI).

➤ Dementias lead to severe memory loss and decreased intellectual ability that affect a person's ability to carry out day-to-day work. A person with dementia suffers from defects in two or more thinking and mental areas such as forgetfulness, planning, organization, decision-making, and confusion. A decline in earlier activities also affects day-to-day life tasks like work duties, driving, preparing meals, and organizing tasks. Some symptoms of dementia are changes in memory, defect in solving problems, slowness completing similar tasks, time and place confusion, difficulty recognizing visual images and spatial relations, difficulty with words that leads to reduced writing and speaking skills, misplacing things and difficulty finding them, loss of judgment, less interest in social and cultural activities, mood swings, and personality fluctuations.

➤ Brain cancer is a primary brain tumor wherein cancer originates and contributes to 50% of brain tumors. Secondary brain cancer, wherein cells spread throughout the brain from a primary tumor situated at any body part, leads to various other cases. Peoples with metastatic cancer develop secondary cancer. The molecular imaging technique is quite useful for brain tumor management. A doctor uses PET studies to define the degree of malignancy, determine the severity of the disease, and detect its recurrence. New techniques such as magnetic resonance (MR) spectroscopy will enable clinicians to increasingly personalize treatment by providing information about tumor metabolism, differentiating recurring tumors from diagnosis-related changes, and accurately targeting radiation in recurring brain tumors.

➤ Epilepsy is a central nervous system dysfunction wherein abnormal brain activities lead to seizures, unusual behaviors or sensations, and loss of awareness. Epilepsy is a chronic disorder. The two basic types of seizures are generalized seizures that affect the whole brain and partial seizures that affect particular parts of the brain. Mild seizures are difficult to detect and last for only a few seconds. Stronger seizures lead to spasms and uncontrollable muscle twitches that remain for a few seconds to several minutes. Stronger seizures may result in loss of consciousness. Seizures occur from high fever, head trauma, low blood sugar levels, and alcohol/drug withdrawal. Some causes of epilepsy are

  a. Brain injury
  b. Brain injury scares
  c. High fever
  d. Stroke
  e. Vascular disease
  f. Decrease in oxygen level in the brain
  g. Brain tumor
  h. Dementia or AD
  i. Use of drugs and alcohol
  j. Infectious diseases
  k. Genetic and neurological disorders

➢ Parkinson's and other movement disorders: Parkinson's disease (PD) is a neurodegenerative disorder caused by reduced dopamine production in the brain. Dopamine is a neurotransmitter that coordinates brain and muscle activity. In a person with PD, the signal is disturbed and leads to irregular limb movements. Some PD symptoms are bradykinesia, rigidity, postural instability, tremor, sleep disorder, and depression. PD occurs mostly in the population greater than 50 years of age. Various environmental and genetic factors also lead to PD. People working in mines and exposed to harmful chemicals, and farmers in continuous contact with insecticides and pesticides, are more prone to PD.

Movement disorders are clinical syndromes indicating difficulty in voluntary and involuntary movements. Some movement disorders are

a. Ataxia: loss of coordination between muscles

b. Dystonia: involuntary contractions of muscles leading to twisting and repetitive movements

c. Huntington's disease: impairment of nerve cells in certain parts of the brain that affects voluntary movements

d. Tourette syndrome, leading to sudden twitches, movements, or sounds

e. Essential tremor, which causes involuntary trembling or shaking movements

Some causes of movement disorders are genetics, infections, medicine, brain and spinal cord damage, metabolic disorders, stroke and vascular disease, and toxins.

➢ Stroke occurs when blood flow to the brain is disturbed by blockage or bursting of blood vessels. Brain cells with less blood and oxygen supply can die, leading to permanent damage, including limb paralysis or deformations while speaking. There are three types of stroke:

a. Ischemic stroke is the common type of stroke and accounts for almost 87% of cases. Clotting of blood blocks the flow of blood and oxygen to the brain.

b. Hemorrhagic stroke is caused by rupturing of blood vessels that leads to hemorrhage due to aneurysms or arteriovenous malformations.

c. Transient ischemic attack is a type of stroke wherein blood flow to the brain is reduced for a brief period. The flow of blood returns to normal after a short period. This type of stroke is also known as a ministroke.

Recovery from stroke requires various therapies:

a. Speech therapy recovers issues related to producing or understanding speech. Communication skills can be practiced to achieve earlier recovery.

b. Physical therapy helps the person learn movement and coordination. Staying active is critical to treating stroke.

c. Occupational therapy helps a people improve the ability to carry out day-to-day activities such as bathing, cooking, and dressing.

d. Support from friends and family, including practical support and cooperation, is necessary for recovering from a stroke.

Some causes and risk factors are

a. Obesity

b. Above 55 years of age

c. Family history of stroke

d. High blood pressure

e. Diabetes

f. High cholesterol

g. Heart or vascular disorder

h. Excessive alcohol use

i. Smoking

j. Drug use

Electromyography

Electromyography (EMG) is a diagnostic technique that detects the health of muscles and the nerve cells controlling them, which are known as motor neurons. They transmit electrical signals that cause muscles to contract and relax. EMG translates these signals into graphs or numbers. A clinician will usually order an EMG test when someone shows symptoms of a muscle or nerve disorder. These symptoms may include tingling, numbness, or unexplained weakness in the limbs.

Some symptoms leading to an EMG test are

1. Tingling

2. Numbness

3. Muscle weakness

4. Muscle pain or cramping

5. Paralysis

6. Involuntary muscle twitching (or tics)

Muscle defects can be divided based on

a. Primary disorders due to direct muscle abnormalities and secondary disorders due to other conditions that may cause muscle damage
b. Genetic
c. Neuromuscular disorders or myopathies
Some causes of muscle disorders are
   a. Age
   b. Heredity
   c. Injuries such as sprains and cramps
   d. Cancers
   e. Inflammation
   f. Nerve diseases
   g. Infectious diseases
   h. Medicines
   i. Metabolic and endocrinal causes
   j. Autoimmune disorders
Some disorders are
1. Muscular dystrophy
2. Disorders that affect the ability of the motor neuron to send electrical signals to the muscle, such as myasthenia gravis
3. Radiculopathies
4. Peripheral nerve disorders that affect the nerves outside the spinal cord, such as carpal tunnel syndrome
5. Nerve disorders, such as amyotrophic lateral sclerosis
6. Sarcopenia: loss of muscle mass
7. Cramps: long, painful, involuntary muscle contractions
8. Rhabdomyolysis
9. Cardiac myopathies
An EMG test has two components:
1. The **nerve conduction study** is the first part of the procedure. It involves placing small sensors called surface electrodes on the skin to assess the ability of the motor neurons to send electrical signals.
2. **Needle EMG** also uses sensors to evaluate electrical signals. The sensors are called needle electrodes and are inserted directly into muscle tissue to evaluate muscle activity at rest and when contracted.

The nerve conduction study is performed first. During this portion of the procedure, the doctor applies several electrodes to the skin's surface, usually in the area where the symptom is experienced. These electrodes evaluate how well the motor neurons communicate with the muscles. Once the test is complete, the electrodes are removed from the skin. After the nerve conduction study, the doctor performs the needle EMG. The doctor will first clean the affected area with an antiseptic. A needle is then used to insert electrodes into the muscle tissue. One feels slight discomfort or pain while the needle is being inserted. The needle electrodes evaluate the electrical activity of the muscles when contracted and at rest. These electrodes are removed when the test is over. During both parts of the EMG procedure, the electrodes deliver tiny electrical signals to the nerves. A computer translates these signals into graphs or numerical values that the doctor can interpret. The entire procedure should take 30 to 60 min.

EMG electrodes and types

The electrical activities of human body muscles can be acquired using EMG electrodes. Most EMG electrodes can be classified as one of two varieties:

a. Surface electrodes: These electrodes are placed above the skin surface to detect the signal. They are mostly passive or active electrodes. Passive electrodes are connected to an external circuit using wires and leads. These are disposable as well as reusable and mostly gelled electrodes. Active electrodes have a preamplifier. These are mostly dry surface electrodes.

b. Inserted electrodes: These electrodes are inserted inside the body muscle to acquire the signal. These are mostly needle and fine-wire electrodes. Needle electrodes are used to detect neuromuscular diseases. The tip acts as the surface for detection and consists of a cannula. A needle electrode has a small acquisition area that can detect individual motor unit action potentials (aka MUAPs) during lesser force contraction events. Fine-wire electrodes have very small diameters and are highly nonoxidizing. The materials used are alloys of platinum, silver, nickel, and chromium. Wired electrodes are painless compared with needle electrodes.

Muscle architecture

The distribution of muscle fibers for the axis of force generation is known as the skeletal muscle architecture. A muscle's properties depend on its architecture. Muscles that go parallel to the muscle force-generating axis are parallel, fusiform, or longitudinally structured. Some of these types of muscles are biceps brachii and sartorius. Muscles oriented at a single angle to the force-generating axis are known as unipennate muscles. One of these types of muscles is the extensor digitorum longus. The measured angle between the fiber and force-generating axis fluctuates between 0 to

30 degrees. Muscles are mostly oriented at more than a single angle. Most skeletal muscles are within this particular category, which is known as the multipennate muscles. These include the rectus femoris, which is bipennate, and the deltoid, which is multipennate. Muscles that surround an opening to form a closed structure are termed circular muscles. One such muscle is the orbicularis oris (muscles in the mouth). The muscles where fibers converge at the insertion to increase the force of contraction are termed convergent muscles. An example of this type of muscle is the pectoralis major.

Skin preparation before electromyography procedure

During EMG acquisition, the subject's skin should be cleaned properly before application and insertion of the electrodes. Skin impedance must be greatly reduced to obtain a high-quality EMG signal, which can be achieved by properly removing dead skin from the skin surface, such as hair. The dead skin must be completely removed from the area from which the EMG signal is acquired. An abrasive gel is necessary for decreasing the dry skin surface. Moisture content on the skin should be avoided. The body surface should be cleaned properly with spirit to reduce wetness, moisture, and sweat from the skin.

Electromyography electrode placement

Muscle mechanics are essential to electrode placement. Mostly, two detecting surfaces are placed on the skin surface in a bipolar configuration. For obtaining a proper and correct signal, the electrodes must be placed exactly, and their orientation across the muscles is important. The EMG electrode must be placed properly across the motor units and tendinous insertion of the muscle along the longitudinal midline. The maximum distance across the center of each electrode must be 1–2 cm. The axis across both detecting surfaces must be parallel to the length of the fibers of the muscle. In the past, a misguided judgment was that the EMG identifying the surfaces ought to be set on the engine unit. In reality, the anode area on the motor point serves as the most noticeably poor area for signal location. In addition, the cathodes ought not to be put at or close to the ligament nor at the edge of the muscle. The muscle filaments become more slender and fewer in number as they approach the ligament of the muscle coming about in a frail EMG flag, demonstrating the truth that anode placement near the ligament is not attainable. On the off chance that the anode is set at the edge of the muscle, the chances of cross talk from other muscles will impressively increment, and those of other muscles will aggravate the resultant flag.

Before proceeding to the signal acquisition stage, it is important to understand the various concerns and factors that affect the EMG signal and signal quality. The amplitude of the EMG signal is 1−10 mV, which makes the signal much weaker. The signal is most prominent in the frequency range 0−500 Hz, 50−150 Hz.

The EMG signal is strongly influenced by sound. The characteristics of electric sound come from a variety of sources. An electromagnetic source may surround the sound, or the sound may be affected by power line interference from wireless transmitters, fluorescent lights, and electrical wires. It is almost impossible to avoid these interventions externally. This particular term is in the frequency range of 50−60 Hz. Motion artifacts can also generate sound. The two main reasons for this noise are the instability of the interface of the electronic skin and the speed of the electronic cables, most of which are in the range of 0−20 Hz. This can be eliminated with the right set of EMG equipment and circuitry. The maximum fidelity of the signal is determined by the EMG signal-to-noise ratio achieved.

The EMG signal is obtained by differential amplification technology. Differential amplifiers must have high input impedance and very low output impedance. Ideally, the differential amplifier has an infinite input impedance and a zero output impedance. Differential amplification is achieved with the help of device amplifiers for high input impedance. A material amplifier amplifies differentially by subtracting voltages V1 and V2. Thus, the common sound signals of V1 and V2 (electrode input) are removed. The tendency of differential amplification of the general rejection signal at both inputs is determined by the common mode rejection ratio (CMRR). 90 dB CMRR is enough to reject the in-phase signal of instrumentation amplifiers, but the latest technology provides 120 dB CMRR even though it is expensive. However, the detection of electronic sounds is not homomorphic, so there are various reasons for not exceeding the CMRR limit.

The monopoly electrode is applied to the skin instead of the reference electrode, which uses only one electron. This method is used because it is simple but is not strongly recommended because it detects all electrical signals around the detection surface. A dipole configuration is used to obtain an EMG signal using two EMG detector planes with reference electrodes. The signals from the two EMG surfaces are connected to the differential amplifier. The two detection surfaces are located at a distance of 1−2 cm from each other. The differential amplifier suppresses the sound signal at both inputs before widening the gap. Exclusive configuration

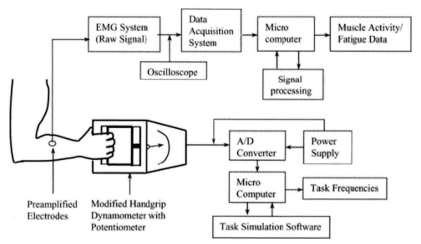

Figure 1.4 Electromyography acquisition system.

restrictions are met through this configuration. This is the most commonly used electrode configuration. In multipole configurations, the reference electrode uses three or more detection planes to receive the EMG signal. This configuration further reduces cross talk and noise anxiety. This configuration gives a highly amplified EMG signal. The signal of three or more EMG detector planes 1–2 cm from each other goes through two or more steps of differential magnification. For example, if three detection planes are used, dual-differential technology is used. This configuration has been used in extensive studies to study the issues of adaptation, conduction velocity, and speed localization of EMG muscle fibers.

Fig. 1.4 shows an EMG acquisition system.

Electrocardiography

Electrocardiography is a quick, simple, painless procedure in which the heart's electrical impulses are amplified and recorded. An electrocardiogram provides information about the part of the heart that triggers each heartbeat (the pacemaker, called the sinoatrial or sinus node), nerve conduction pathways of the heart, and the rate and rhythm of the heart.

Sometimes, the electrocardiogram shows that the heart is enlarged (usually due to high blood pressure) or not receiving enough oxygen due to a blockage in one of the blood vessels supplying it (the coronary arteries).

Fig. 1.5 shows the P–Q–R–S–T wave complex.

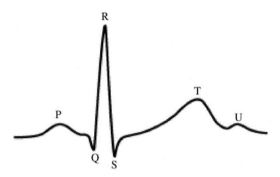

Figure 1.5 P—Q—R—S—T wave complexes.

P wave

The P wave represents depolarization of the left and right atrium and corresponds to atrial contraction. The atria contract a split second after the P wave begins. Because it is so small, atrial repolarization is usually not visible on an electrocardiogram. In most cases, the P wave will be smooth and rounded, no more than 2.5 mm tall, and no more than 0.11 s in duration. It will be positive in leads I, II, aVF, and V1 through V6.

Q—R—S wave complex

The Q—R—S wave (QRS) complex comprises the Q, R, and S waves. These three waves occur in rapid succession. The QRS complex represents the electrical impulse as it spreads through the ventricles and indicates ventricular depolarization. As with the P wave, the QRS complex starts just before ventricular contraction. It is important to recognize that not every QRS complex will contain Q, R, and S waves. The convention is that the Q wave is always negative, with the R wave as the first positive wave of the complex. If the QRS complex includes only an upward (positive) deflection, it is an R wave. The S wave is the first negative deflection after an R wave. Under normal circumstances, the duration of the QRS complex in an adult people will be between 0.06 and 0.10 s. The QRS complex is usually positive in leads I, aVL, V5, V6, II, III, and aVF. The QRS complex is usually negative in leads aVR, V1, and V2. The J-point is the point where the QRS complex and ST segment meet. It can also be thought of as the start of the ST segment. The J-point (also known as junction) is important because it can diagnose ST-segment elevation myocardial infarction. When the J-point is elevated at least 2 mm above baseline, it is consistent with a STEMI.

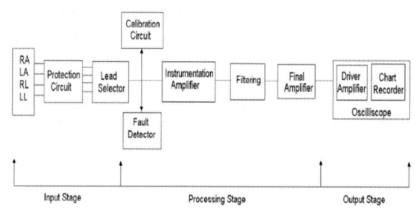

Figure 1.6 Block diagram of electrocardiography machine.

T wave

A T wave follows the QRS complex and indicates ventricular repolarization. Unlike a P wave, a normal T wave is slightly asymmetric; the wave's peak is a little closer to its end than to its beginning. T waves are normally positive in leads I, II, and V2 through V6 and negative in aVR. A T wave will normally follow the same direction as the QRS complex that preceded it (positive or negative/up or down). When a T wave occurs in the opposite direction of the QRS complex, it generally reflects some sort of cardiac pathology. If a small wave occurs between the T and P waves, it could be a U wave. The biological basis for a U wave is unknown.

Fig. 1.6 shows the block diagram of an ECG machine.

Fig. 1.6 represents the block diagram of an ECG system consisting of three stages:

➤ The input stage consists of input devices, such as various electrodes connected to various parts of the human body. The electrodes collect biosignals and then feed them to the input circuit.

➤ The processing stage consists of processing input signals using various circuits and then sending them to output devices.

➤ The output stage consists of an oscilloscope showing the processed ECG signal, which is printed out for the clinician to detect the exact heart defect.

Einthoven's triangle for ECG electrode placement is shown in Fig. 1.7.

Einthoven's triangle is an imaginary formation used in electrocardiography with three limb leads in a triangle formed by the two shoulders and the pubis. The shape forms an inverted equilateral triangle with the heart at the center. It is named after Willem Einthoven, who theorized its existence.

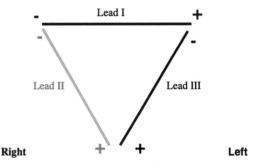

Figure 1.7 Einthoven's triangle for ECG electrode placement.

Einthoven provided the formation of the three limb leads, namely aVR, aVL, and aVF, used in electrocardiography to check for problems in hearts with electrical activity. They function by both shoulders and the pubis. It consists of an inverted equilateral triangle of a heart producing zero potential when the voltages are summed at the middle.

The formation of these three leads must occur through the following measures:

Lead 1

The axis of lead 1 goes from the right shoulder to the left shoulder, with the negative electrode on the right shoulder and the positive on the left. There is a zero-degree angle of orientation:

$$I = LA - RA$$

where LA: left arm
  LL: left leg
  RA: right arm
  RL: right leg

Lead 2

The axis of lead 2 goes from the right arm and reaches the leg on the left side with a negative electrode on the shoulder and a positive electrode on the leg having a 60-degree angle of orientation:

$$II = LL - RA$$

Lead 3

The axis of lead 3 moves from the left part of the shoulder with a negative electrode and consists of a positive electrode, resulting in a 120-degree angle of orientation:

$$III = LL - LA$$

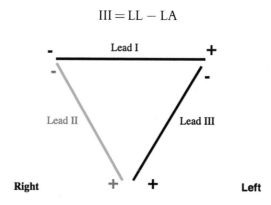

This approach falls in the middle of the field of electrocardiography, but the phenomenon of cardiac vectors is difficult to understand. Hence, this recreates ECG reading conditions.

Heart disorders

Cardiovascular disease is a leading cause of disease in the United States, according to a reliable source, the Centers for Disease Control and Prevention. About 610,000 people die every year in this state.

Heart disease is not discriminatory and is the leading cause of death for many, including whites, Hispanics, and blacks. About half of Americans are at risk of heart disease, and the number is rising. Heart disease can be fatal but can be prevented in most people. We can live a long life with a healthy mind by adopting a healthy lifestyle in the early stages.

Different types of cardiovascular problems are covered in heart disease. Many illnesses and symptoms are in the shadow of heart disease. Types of heart disease are

a. Arrhythmia, an abnormality in the rhythm of the heart
b. Atherosclerosis
c. Cardiomyopathy, a condition that causes the heart muscle to become stiff or weak
d. Congenital heart disease, a heart irregularity that exists at birth
e. Coronary artery disease (CAD), caused by plaque formation in the arteries of the heart; sometimes called ischemic heart disease
f. Heart infections, often caused by bacteria, viruses, or parasites

The term heart disease can refer to a condition of the heart that specifically affects the blood vessels.

Different types of heart disease can cause different types of symptoms.

Arrhythmia: Arrhythmia is an abnormal rhythm of the heart. Depending on the symptoms experienced, the heart rate will gradually increase. Symptoms of arrhythmia include slow events, heartbeat, or racing heartbeat, slow pulse, ignorance mantra, agility, chest pain.

Atherosclerosis: Atherosclerosis reduces the timely supply of blood. In addition to chest pain and shortness of breath, symptoms of atherosclerosis include coldness of the limbs, especially numbness in the limbs, unusual or unexplained pain, and weakness of legs and arms.

Congenital heart disease: Congenital heart disease is a heart problem that develops as follicles increase. Some heart defects cannot be detected. When they have symptoms, others found include blue skin, loop swollen, shortness of breath or dyspnea, fatigue and low energy, and arrhythmia.

CAD is the accumulation of plaque in arteries that remove oxygen-rich blood through the heart and lungs. The symptoms of CAD are chest pain and discomfort, chest tightness, dyspnea, nausea, indigestion, or gassiness.

Cardiomyopathy: Cardiomyopathy is a disease in which the heart muscle grows and becomes stiff, fat, or weak. Symptoms of this condition are malaise, swelling (especially swelling of the ankles and feet), dyspnea, and fast pulse.

Cardiac infections: The term heart infection can be used to describe conditions such as endocarditis and myocarditis. Symptoms of a heart attack are chest pain, chest congestion or cough, heat, cold, and skin rash.

Women often have different signs and symptoms of heart disease than men, especially symptoms related to CAD and other cardiovascular diseases. In fact, a 2003 study looked at the most common symptoms in women who have had a heart attack. The symptoms did not include "classic" heart attack symptoms, such as chest pain or twisting. Instead, the study states that women are more likely to experience anxiety, sleep disorders, abnormalities, or unexplained fatigue. In addition, 80% of the women included in the study experienced these symptoms for at least a month before having a heart attack. Symptoms of heart disease in women can be confused with other symptoms such as those of depression, menopause, and anxiety. The most common symptoms of heart disease in women are agility, paleness, shortness of breath or shallow breathing events, becoming unconscious or unconscious, anxiety, nausea, jaw pain, neck pain, back pain, chest pain, and abdominal pain such as indigestion, gas, or cold sweat.

Heart disease is a collection of diseases and conditions that create cardiovascular problems. Each type of heart disease is caused by a condition in which plaque builds up in a specific artery for a condition that causes atherosclerosis and CAD. Other causes of heart disease are explained below.

The causes of arrhythmia are diabetes, CAD, heart defects including congenital heart disease, medications, supplements, herbal remedies, hypertension, excessive use of alcohol or caffeine, substance use-related disorders, Stress, and anxiety.

Existing heart damage or illness is the cause of congenital heart disease.

This heart disease occurs when the baby grows in the womb. Some heart defects are serious and can be diagnosed and treated at an early stage. It can stay uninterrupted for years to come. The structure of the heart also changes with age. It can cause heart defects that can lead to complications and problems. There are different types of cardiomyopathy. Each type is the result of an individual situation. It is not clear whether this is a very common type of cardiomyopathy and the cause of heart failure. It can be the result of previous heart damage from drugs, infections, heart attacks, etc. It can also be the result of hereditary disorders or uncontrolled blood pressure.

Hypertrophic cardiomyopathy: This type of heart disease builds dense heart muscle. The cause of this type of cardiomyopathy, which tightens the wall of the heart, is often unknown. Possible causes include the accumulation of a type of abnormal protein called scar tissue or amyloidosis.

Bacteria, parasites, and viruses are the most common causes of heart infections. Uncontrolled infections in the body can also damage the heart if not treated properly.

There are many risk factors for heart disease. Some are controllable, and some are not. The CDC states that approximately 4.47% of Americans possess at least one risk factor for heart disease. Common risk factors include high blood pressure, high-cholesterol and low-level high-density lipoprotein, "good" cholesterol, smoking, obesity, lack of exercise. For example, smoking causes a controllable risk. According to the National Institute of Diabetes and Digestive and Kidney Diseases, people who smoke are twice as likely to have heart disease. People with diabetes have an increased risk of hyperglycemia and may also have an increased risk of heart disease. Some factors are angina, sudden heart attack, stroke, and CAD. People with diabetes must manage glucose to reduce their risk of heart disease. The American Heart Association estimates that both high blood pressure and diabetes double the risk of heart disease.

Tests conducted to detect heart disease include the following:

Noninvasive tests

A variety of noninvasive tests can be conducted to diagnose heart disease:

a. An ECG test can monitor the heart's electrical activity and help the doctor see any irregularities.

b. Echocardiography ultrasound can provide a doctor with a better picture of the heart structure.

c. Stress test activity occurs when walking, running, or riding an exercise bike. During the test, the treatment may monitor heart activity in response to changes in physical activity.

d. A carotid ultrasound may be ordered by a doctor to perform a detailed ultrasound of the carotid artery.

e. Halter monitor may be advised by a doctor as a heart rate monitor for up to 24 hours a day. This may provide an extended view of heart activity.

f. Table test: A doctor may order this test for recent fainting or inactivity while standing or sitting. At this point, the people is tied to the table and slowly moves up and down while the heart rate, blood pressure, and oxygen levels are monitored.

g. Computed tomography (CT) scan. This imaging test provides the doctor with a very detailed X-ray image of the heart.

h. X-ray: Like a heart MRI CT scan, cardiac MRI can provide a very detailed picture of the heart and blood vessels.

Invasive tests

If tests are not confirmed without physical, blood, and tests, we may want to have our body examined to determine the abnormal symptoms of treatment. Aggressive tests include cardiac catheterization and coronary angiography. Treatment can be to wrap and insert a catheter into the heart through an artery. The catheter helps examine the involvement of the heart and blood vessels. When this catheter is inserted into the heart, the doctor can perform coronary angiography. In coronary angiography, the pigment is injected into the fine arteries and capillaries surrounding the heart. The colors help create very detailed X-ray images. Electrophysiological research: During this test, the doctor can electronically connect the heart through a catheter. The treatment can send an electrical pulse to how the heart responds when the electrodes are in place.

Electrooculography

Electrooculography is used to record eye movements during electro-nystagmographic testing. It is based on the corneoretinal potential (difference in electrical charge between the cornea and retina), with the long axis of the eye acting as a dipole. Movements of the eye relative to the surface electrodes placed around the eye produce an electrical signal that corresponds to eye position. Recordings of eye movement are accurate to about 0.5 degree, but it is still less sensitive than visual inspection, which can perceive movements of about 0.1 degree.2 Therefore, visual inspection with Frenzel lenses is sometimes still necessary to document nystagmus of low amplitude. Another limitation of electrooculography is that torsional eye movements cannot be monitored.

EOG is an electrical flag produced by the polarization of the eyeball and can be measured on the skin around the eyes. Its magnitude changes relative to the relocation of the eyeball from its resting area. There are two sorts of eye movements: horizontal and vertical eye development. Electrodes are connected to level eye development to detect the cleared out and right eye development. In the interim, the other anodes are joined into vertical eye movement for identifying up-and-down corneal developments. However, numerous exercises have been done with eyes such as blinking, opening, nearing, perusing, seeing, and others. Those activities that require thought or consideration lead to a diminish in blink frequency. Amid perusing is a blinking hindrance that becomes more articulated relative to the reader's intrigue within the fabric. A burst of flickers at that point occurs while the peruser turns the page. The EOG value shifts from 50 to 3500 µV with a recurrence extended to about 100 Hz. Its conduct is essentially straight for gaze angles of 10 degrees. It should be pointed out here that the variables measured within the human body for any biopotential are rarely deterministic. Their sizes shift over time, indeed when all possible factors are controlled.

At this point, the EOG is one strategy that can be utilized as a communication apparatus such as the human—machine interface. In restoration, EOG-based procedures are exceptionally valuable for peoples with serious cerebral paralysis, those born with a congenital brain clutter, or those who have endured severe brain injury. Over a long period, significant incremental gains have been made in advancing assistive technology for individuals with incapacities that have moved conventional systems forward. In addition, computer use in both work and leisure has driven improvements in PC-related handling applications, basically using realistic interfacing. This way, the conventional strategies of control or communication between humans and machines (joystick, mouse, or console) that require a certain control engine on the portion of clients are supplemented with others that permit their use for individuals with severe incapacities. Among these unused strategies, it is necessary to specify voice acknowledgment.

Limitations of electrooculography procedure

The basic presumption of this strategy of recording eye developments is that the development of the electric field within the conducting tissues encompassing the eye is related in a straightforward (as a rule accepted to be straight) way to the developments of the eye itself. Due to the nonuniformity of these tissues and the shapes of the tissues encompassing them, this could be estimated to the organic reality. Be that as it may, for even eye developments inside the extend of 30 degrees, the potential measured is accepted to be direct to the genuine development of the eye within the orbit. The resolution of EOG is considered approximately 1 degree. Because it could be a generally basic procedure, EOG is still commonly utilized clinically for testing eye developments in peoples.

For a settled eye position, the EOG is distant from being steady in size but can be affected by many outside variables. These components include
➢ The commotion produced between the electrodes' contacts and the skin
➢ The metabolic state of the tissues (pO2, pCO2, and temperature)
➢ Visual stimulation
➢ Contraction of facial muscle

In expansion, recorded EOG, especially for vertical eye developments, is very delicate to developments of the eye lids. In outline, many outside variables can complicate the translation of the EOG, so EOG is considered profoundly delicate to artifacts. The impressive artifacts that can be presented through the contact between the terminal contacts and the skin can be minimized by decreasing the resistance between the terminals and the skin.

EOG pigment is used to evaluate the efficacy of epithelium. During dark adaptations, chances of rest are reduced to a minimum after a few minutes ("dark adaptation"). Turning on the light greatly improves your ability to increase comfort ("light peaks") and stops the retina a few minutes after entering the light. The voltage ratio (i.e., the brightest peak divided into dark valleys) is known as the Arden ratio. In reality, the measurement is similar to recording eye movements. The people is repeatedly asked to repeat the eye between two points (center-centered, center-to-right). These conditions are stable, so there are potential changes in rest with changes in recorded capacitance.

The movement of fire changes when an electromagnetic wave comes around the eye. It usually displays a pattern on an electroencephalograph (EEG). Imagination Strategies have been proposed in various ways, including reducing eye movement, preventing eyelids, rejecting data damaged by eyelids, and removing eye patterns from EEG. Recent differentiation of ophthalmology and neurological possibilities by component analysis. There are enough bets on the best way to explain the eye artwork, but there is currently no sensation in this regard. The most popular of these methods for explaining eye movements are conversion/rejection and EOG correction techniques-Nix. Jason instruction presents a secondary feature that has been shown to affect the CNV and P300 components of the ERV, and denying the data is difficult to complete and dazzling to the eye. Link recognition in motion leads to biased patterns. The main limitation associated with EOG correction is that EEG reduces the ocular probability of EEG and neural probability (because the nerve capacity of the EOG channel is reduced and subtracted from the EEG). Many have seen this limitation. Despite the above EOG denial restrictions, not taking EOG properly is a significant priority. However, until recently, the organization of this limitation was explored and assumed to accept EOG denial rather than happiness change, and this study does not support this assumption. In the context of the N1P2 complex. The probability of hearing-related events, accounting for the correct coefficients, was 15%—22% of the EOG removal criteria of the complex, or 4 MV. To implement inequality. In particular, it is distinctly identified as a percentage error (depending on physical parameters such as head and bone size), and this range is considered relatively small, so we call it prose/rejection prose. Several different algorithms have been proposed to improve EOG. What is common in this algorithm is to interpret the work of the eye by subtracting the ratio (b) of one or more EOG channels from all those of EEG. Here, the EOG channel is a record complement that gives a reasonable estimate of eye activity. BSs are estimated for each scale electrode, EOG channel, and subject. The hypothesis is the tendency of Annie's movement at the scale site divided by the length of the same eye movement in the EOG channel, but BS is usually the minimum square regression. The effect of sound on projection is because the eye moves to the right or left during EOG (top or bottom), and the corresponding element muscle strength changes significantly. The reject band filter is used to reject local frequencies (50 Hz). The function of the insulating circuit is to insulate the signal. The line has been applied using force and optical or voltage conversion methods. Bandwidth bandpass filters range from 0.05 to 30 Hz (two active secondary HPF and two active secondary LPF), and the amplifier can increase the amplitude factor by 50 for a weak signal that the filter has passed. Again, the amplified EOG signal can be transmitted directly to the control circuit. Electronic summary is a technique of measuring the possibility of retinal rest. The resulting signal is called an electrooculogram, a device that measures the voltage between two electrodes. Put it on the subject to identify the eyes Protests. Using computers is on the rise today in all fields. Many sophisticated devices, such as touch screens, trackballs, and digitizers that interact with computers, are easy for beginners to professionals and improve quality of life for people with disabilities. Today, there are many support systems for control and autonomous mobile robot guides. This system allow alls users to travel more efficiently and easily. Over the past few years, many applications for developing support systems for humans who have become disabled have rendered the conventional system invalid. These new systems can be seen in things such as video calligraphy systems or infrared eye motion recording, which is based on using a camera to detect the location of the eyes. There are many strategies for speech recognition for initial commands to control equipment or robots; the joystick is the most popular. This technique can control various applications for people with limited upper-body mobility for subtle control of manual actions that are hard to complete. New technology can be adjusted according to the application for different persons, which can address the technology or strategies for each person more efficiently. All these systems allow users to move more efficiently and easily (EOG). A new technology places electrodes on the user's forehead around the eye to record eye movements. This technique is based on the principle of recording polarization probability or corneal retinal probability, and the cornea has the ability to rest.

Radiological recorders
Magnetic resonance imaging

MRI is an imaging technique that uses a magnetic field, radio waves, and a computer system to create various images of the body. The scanner contains a tube surrounded by a large magnetic bar. A movable bed is placed inside the circular portion where the people to be scanned is placed. The magnet creates a strong magnetic field that aligns the protons of hydrogen atoms, which are then exposed to a beam of radio waves. This spins the various protons of the body, and they produce a faint signal that is detected by the receiver portion of the MRI scanner. A computer processes the receiver information, which produces an image. The invention of MRI became a boon to the healthcare sector worldwide. Various scientists and healthcare experts can now diagnose the whole human body with maximum accuracy and detail using this noninvasive technique.

MRI can be used for

- anomalies of the brain and spinal cord
- tumors, cysts, and other anomalies in various parts of the body
- breast cancer screening for women who face a high risk of breast cancer
- injuries or abnormalities of the joints, such as the back and knee
- certain types of heart problems
- diseases of the liver and other abdominal organs
- evaluation of pelvic pain in women, with causes including fibroids and endometriosis
- suspected uterine anomalies in women undergoing evaluation for infertility

MRIs utilize effective magnets that create a solid, attractive field that powers protons within the body to adjust with that field. When a radiofrequency current is at that point beat through the quiet, the protons are invigorated and turn out of equilibrium, straining against the drag of the attractive field. When the radiofrequency field is turned off, the MRI sensors can identify the vitality discharged as the protons realign with the attractive field. The time it takes for the protons to realign with the attractive field, as well as the sum of vitality discharged, changes depending on the environment and chemical nature of the particles. Doctors can tell the contrast between different sorts of tissues based on these attractive properties. To obtain an MRI picture, a people is placed within an expansive magnet and must stay exceptionally still amid the imaging handle so as not to obscure the picture. Differentiate operators (frequently containing the component gadolinium) may be given to a person intravenously sometime before or amid the MRI to extend the speed at which protons realign with the attractive field. The faster the protons realign, the brighter the picture.

MRI scanners are especially well suited to picture the nonbony parts or delicate tissues of the body. They contrast with CT in that they do not utilize the harmful ionizing radiation of X-rays. The brain, spinal cord, nerves, muscles, tendons, and ligaments are seen much more clearly with MRI than with standard X-rays and CT; for this reason, MRI is frequently utilized to picture knee and shoulder injuries. In the brain, MRI can separate between white matter and gray matter and can too be utilized to analyze aneurysms and tumors. Since MRI does not utilize X-rays or other radiation, it is the imaging methodology of choice when visit imaging is required for conclusion or therapy, particularly within the brain. Be that as it may, MRI is more costly than X-ray imaging or CT filtering. One kind of specialized MRI is functional MRI (fMRI.) This is often utilized to watch brain structures and decide which zones "activate" (expend more oxygen) amid different cognitive assignments. It is utilized to advance the understanding of brain organization and offers a modern standard for surveying neurological status and neurosurgical chance.

Although the risk of ionized radiation is currently well managed in most treatment conditions, MRI may be considered a better option than a CR scan. MRI is used in hospitals and clinics to diagnose treatment and stage and track illness without exposure to body radiation. MRI can generate different data than CRT. Risks and discomfort may be associated with MRI scans. Compared with CT scans, MRI scans are usually longer and taller, and the subject usually needs to be in a narrow, confusing tube. In addition, people with certain medical implants or other removable metals inside the body of some people cannot safely do MRI scans. MRI was originally called NMRI (magnetic resonance imaging), but the "atom" was removed to avoid negative associations. If a nucleus is placed in an external magnetic field, radio frequency energy can be absorbed. The result can be detected by inducing a radio frequency (RF) signal into the spin polari- zation radio frequency coil. In clinical and research MRI, hydrogen atoms are often used to produce macroscopic polarized light detected by antennas near the subject of investigation. Hydrogen atoms are naturally abundant in humans and other organisms, especially water and fat. For this reason, most MRI scans scan the locations of water and fats in the body. A group of radio waves stimulates the spin energy transfer of an atom and localizes the polarization of the gradient space of the magnetic field. By changing the pulse sequence parameters, different differences may occur between tissues depending on the relaxation properties of the hydrogen atom. Since the development of the 1970s and 1980s, MRI has proven to be a versatile imaging technology. MRI is most explicitly used in clinical and biomedical research, but it can also create living images. Diffusion MRI and functional MRI enhance the effectiveness of MRI to capture neural pathways and blood flow in the nervous system, respectively, in addition to detailed spatial imaging. The ever-increasing demand for MRI in the healthcare system raises concerns about cost-effectiveness and similarity.

In most therapeutic applications, a proton-generated hydrogen nucleus in the tissue forms a signal, and the signal is processed to form a body image about the density of the nucleus in a particular region. Since protons can differentiate reactions from hydrogen in certain compounds, they are affected by the field from other atoms. To conduct the study, the individuals were placed inside an MRI scanner that created a strong magnetic field around the magnetic field being copied. First, energy from the vibrating magnetic field is temporarily applied to the people at an appropriate resonant frequency. Scans with X and Y gradient coils create the exact magnetic field required for the energy to absorb the desired magnetic field for the people. The excited atomic output emits an RF signal measured by the coil. Gradient coils can process RF signals and reduce position information by observing RF levels and phase changes due to local magnetic field fluctuations. These coils change rapidly during excitation and reaction to perform moving line scans, causing the windings to move slightly due to the magnetic material, resulting in the repeated sound characteristics of MRI scans. An external contrast agent may be allowed to sharpen the image individually.

Diagnostics Used by organs or systems: Peoples were recruited for head and abdominal MR studies. MRI has a wide range of uses for diagnostic treatment, and it is estimated that more than 25,000 scanners are used worldwide. While MRI affects the diagnosis and treatment of many branches, its effects on health can be challenged. MRI is the preferred research study for determining other tumors for the prefunctional stages of rectal and prostate cancer, staging, and determining tissue samples through follow-up and biobanking.

White matter MRI diffusion tanner imaging: MRI is the best diagnostic tool for neurocancer in CT because it provides better imaging of growing fossa, including the brain and cerebellum. Due to the differences between gray and white matter, MRI is ideal for many central nervous system conditions, including demyelinating diseases, dementia, neurological disorders, infections, AD, and epilepsy. Many images are taken in milliseconds, showing how the brain responds to various stimuli, allowing researchers to study functional and structural brain abnormalities in mental illness. Using MRI in guided stereosurgery and radiosurgery, aneurysm dysfunction, and other surgical conditions is known as N-localizer in treating intracranial tumors.

Cardiovascular MR angiography in congenital heart disease: Cardiac MRI complements other imaging techniques such as echocardiography, cardiac CT, and nuclear medicine. It can be used to evaluate the structure and function of the heart. Its applications include assessing myocardial ischemia and survival, cardiomyopathy, myocarditis, iron overload, vascular disease, and congenital heart disease.

Muscular: Applications of the musculoskeletal system include spinal imaging, joint disease assessment, and soft tissue tumors. In addition, MRI technology can be used for clinical imaging of systemic muscle disease.

Liver and gastrointestinal: Hepatobiliary MR is used to identify and identify lesions of the liver, pancreas, and bile ducts. Local or diffuse liver diseases can be assessed using expansion-weighting, antiphase imaging, and dynamic contrast-weighting sequences. Extracellular contrast media are widely used in liver MRI, and new hepatobiliary contrast media offer the opportunity for effective bile imaging. Anatomical imaging of the bile duct is performed using magnetic resonance cholangiopancreatography using the T2-enhanced sequence. Effective imaging of the pancreas occurs after MRI. MR enterography provides an invasive assessment of inflammatory bowel disease and tumors of the small intestine. MR colonography may play a role in identifying large polyps in peoples at risk of colorectal cancer.

Magnetic resonance angiography: Magnetic resonance angiography (MRA) produces arterial images to diagnose stenosis (abnormal stenosis) or aneurysm (expansion of the vessel wall at risk of rupture). MRA is often used to assess cervical and cerebral arteries, chest and abdominal arteries, renal arteries, and legs (known as "run offs"). Images can be created using a variety of techniques, including the administration of paramagnetic contrast media (gadolinium) and "flow-related magnification" (such as the sequence of 2D and 3D flights). The signal in the figure was due to blood clots that had recently entered the aircraft (see also FALS MRI).

A technique of phase accumulation (known as phase-difference angiography) can be used to easily and accurately map the flow velocity. Magnetic Resonance Venography is a similar procedure used in the veins of images. This method results in insufficient stimulation of the tissues, and signals are collected on the surface just above the excitatory surface. Therefore, venous blood recently removed from the stimulus surface is depicted.

The main components of an MRI scanner are the main magnet that polarizes the sample, the bean coil that corrects the symmetry of the original magnetic field, the gradient system used to scan the magnetic field, and the RF system. Samples are excited, and the results detect NMR signals. One or more computers control the whole system. MRI requires one million strong and uniform magnetic fields during the scan. The magnetic field strength of a magnet is measured with a telescope. Most systems run on 1.5T, but commercial systems are available at 0.2–7T. Most diagnostic magnets contain superconducting magnets and require liquid helium to keep them very cool. For peoples with claustrophobia, low magnetic field strength can be achieved with the help of permanent magnets often used in "open" MRI scanners. Field power will also be used in a portable MRI scanner approved by the FDA in 2020. MRIO is usually performed. The prepores (on the order of 10,100 MT) allow ultralow magnetic fields from microtesla to millitesla for adequate signal quality and an instrument using sensitive superconducting quantum interference to supply Larmore's pressure field at 100 μT.

Computed tomography

CT scan uses a combination of X-ray images taken from various angles across the body. It uses computer processing to make cross–sectional images of the various body organs such as bones, blood vessels, and tissues. CT scan is excellent for various internal injuries of the body. It can visualize all the body parts to diagnose the disease and any other injury, and further, doctors can accordingly determine treatment. CT scan uses X-ray machines with computers to create cross-sectional images of the body. These images provide detailed information than X-ray images. The CT scan techniques can check

- head
- shoulders
- spine
- heart
- abdomen
- knee
- chest
  CT scan is carried out for the following:
- diagnose infections, muscle disorders, and bone fractures
- pinpoint the pinpointed location of masses and tumors (including cancer)

- study the blood vessels and other internal structures
- assess the extent of internal injuries and internal bleeding
- guide procedures, such as surgeries and biopsies
- monitor the effectiveness of treatments for certain medical conditions, including cancer and heart disease.

Not at all like an ordinary X-ray—which employs a fixed X-ray tube—a CT scanner employs a motorized X-ray source that pivots around the circular opening of a donut-shaped structure called a gantry. Amid a CT filter, the persistent lies on a bed that gradually moves through the gantry, whereas the X-ray tube pivots around the persistent, shooting contract bars of X-rays through the body. Rather than film, CT scanners utilize uncommon advanced X-ray locators, which are found straightforwardly inverse the X-ray source. As the X-ray is carried out, image is picked up and transmitted to a computer. Each time the x-ray source completes one full revolution, the CT computer employs advanced numerical procedures to develop a 2D picture cut of the quiet. The thickness of the tissue spoken to in each picture cut can shift depending on the CT machine utilized, but more often than not ranges from 1 to 10 mm. The picture is put away when a full cut is completed, and the motorized bed is moved forward incrementally into the gantry. The X-ray filtering handle is at that point rehashed to deliver another picture cut. This handle proceeds until the specified number of cuts is collected. Picture cuts can either be shown independently or stacked together by the computer to create a 3D picture that includes the skeleton, organs, and tissues and any anomalies the doctor is attempting to distinguish. This strategy has numerous focal points that can turn the 3D picture in space or see cuts in progression, making it easier to discover the precise put where an issue may be found.

Computed tomography scan with contrast In a CT scan, thick substances like bones are simple to see. But delicate tissues do not appear as clearly. They may see faint within the picture. To assist them in showing up clearly, an uncommon color called a differentiate fabric is required to square the X-rays to appear white on the filter, highlighting blood vessels, organs, or other structures.

Differentiate materials are more often than not made of iodine or barium sulfate. It might get these drugs in one or more of three ways:

> Injection: The drugs are infused specifically into a vein. Typically done to assist your blood vessels, urinary tract, liver, or gallbladder stand out within the picture.

➤ Orally: Drinking fluid with the differentiate fabric can improve scans of the stomach-related tract, the pathway of nourishment through the body.

➤ Enema: In the case of digestive tracts being filtered, the differentiate fabric can be embedded in the rectum.

After the CT check, one must drink plenty of liquids to assist the kidneys in evacuating the differentiate fabric from the body.

Types of CT scans Spiral tube: CT scan beam and people images in city imaging system spinning tubes, commonly known as spiral cities or helical cities, are imaging techniques that revolve around the central axis of the region where the entire X-ray tube is scanned. These are the main types of scanners on the market due to long-term production and low production and purchase costs. The main limitations of this type are the bulkiness and inertia of the equipment, which limit the rotational speed of the equipment (X-ray tube assembly and detector array on the other side of the circle). Some designs use two X-ray sources and an angle-offset detector array as strategies to improve time resolution.

Electronic beam tomography: Electron beam tomography is characterized by CT, where a deflection coil is used to create an X-ray tube that simply travels between the electron passing cathode and the X-ray tube. This offers a great advantage because this type of spin sweep rate is very fast and reduces the blurring of moving structures such as the heart and arteries. Some designs are compared with the scanner spinning tube type, which is mainly associated with constructing larger X-ray tubes and detector arrays due to higher costs and limited physiology.

Medical uses Since the early 1970s, CT has become an important tool in medical imaging to complement X-rays and medical ultrasonography. Recently, it has been used in preventive medicine and disease screening such as CT colonography for people at high risk of colon cancer and full-speed heart scans for people at high risk of heart disease. Although many facilities provide full-body scans to the general public, this practice goes against the advice and official position of many professional bodies in the region, primarily due to the dose of radiation.

CT scan use has increased dramatically in many countries over the past 2 decades. In the United States, an estimated 72 million scans were performed in 2007, and more than 80 million scans were performed in 2015.

Head: Mathematical tomography of the human brain from the base of the skull to the top. Intravenous administration with contrast medium.

Ordinary Brain Scrolling CT Images. CT scans of the head are commonly used to detect myocardial infarction, tumors, calcification, bleeding, and bone trauma. Low-density (dark) structures may indicate edema and myocardial infarction in the above case, whereas high-density (bright) structures may indicate calcification, bleeding, and bone trauma. This can be seen as the colorlessness of the bone marrow. The cause of the tumor can be identified by inflammation, deformity of the body, or surrounding edema. CT scans of the head are used as CT-guided stereotactic and as an N-localizer in radiosurgery for intracranial tumors, arteriovenous depression, and other surgically treatable conditions.

Neck: Contrast CT is usually the first study of preferences for adult neck masses to play an important role in determining CC thyroid cancer of the thyroid gland. CT scans often accidentally detect thyroid abnormalities and are often the preferred method of screening for thyroid abnormalities.

Lungs: CT scans can be used to detect acute and chronic changes in the parenchyma of the lungs. This is of particular concern here, as ordinary 2D X-rays do not show such errors. Different strategies are used depending on the suspected inconsistency. High spatial frequency reconstruction flakes are used to evaluate chronic interstitial processes such as emphysema and fibrosis. Scans are often performed for both inspiration and expiration. This particular technique is called high-resolution CT, which produces lung samples rather than continuous images. HRCT images of the common chest of axial, coronal, and arc planes, respectively. Bronchial wall thickness (T) and tracheal diameter (D) Thickening of the tracheal wall is found in the lungs and usually (but not always) means inflammation of the bronchi.

In the absence of symptoms, accidental nodules (sometimes called accidents) can raise concerns that they may present a benign or malignant tumor. Perhaps motivated by fear, peoples and physicians may agree to an intensive schedule of CT scans, sometimes every 3 months, beyond the recommended guidelines for observing nodules. At any given time, established guidelines suggest that those with no previous history of cancer and those whose nodules are less than 2 years old are more likely to be free of cancer. For this reason, and no study provides supporting evidence that intensive monitoring provides better results, and because of the risks associated with performing CT scans, peoples are more likely than suggested by established guidelines.

Angiography: Computed tomography angiography is a type of reverse CT used to visualize arteries and veins. Blood travels from arteries to the brain, lungs, kidneys, arms, and legs. An example of this type of test is CT pulmonary angiography used to diagnose pulmonary embolism. Images of pulmonary arteries were obtained using mathematical tomography and iodine-based contrast media.

Heart: A CT scan of the heart is done to gain knowledge of the anatomy of coronary arteries of the heart. Cardiac CT scans are commonly used to detect, diagnose, or track CAD. Recently, CT has played an important role in the rapid development of transcatheter structural cardiac interventions, especially transcatheter repair and heart valve replacement.

Positron emission tomography

PET scan is a technique that can detect the functionality of human tissues and organs. This technique uses a radioactive drug that acts as a tracer to detect its activities. It detects the disease much earlier than any other radiological technique. The radioactive material is injected or swallowed depending on the part of the body to be diagnosed. This material accumulates in those particular areas of the body that contain higher levels of chemical activity corresponding to the diseased area. In a PET scan, these particular areas show bright spots.

PET scans are most commonly used to detect

- cancer
- heart problems
- brain disorders, including problems with the central nervous system

PET images are mostly combined with CT or MRI scans to create better images. A PET scan is an effective way to examine the chemical activity of the human body. PET scan images provide information that is different from other imaging techniques. A PET scan or a combined CT-PET scan enables health care experts to better diagnose illness and assess the body condition. Positron emanation tomography, also called PET imaging or a PET filter, is a sort of atomic medication imaging. Nuclear pharmaceutical imaging employs small amounts of radioactive fabric to analyze, assess, or treat an assortment of infections. These include numerous cancers, heart infections, gastrointestinal, endocrine, and neurological disorders, and other anomalies. Because atomic pharmaceutical exams can pinpoint atomic movement, they have the potential to recognize illness in its earliest stages. They can moreover appear whether a quiet is reacting to treatment. Nuclear pharmaceutical imaging strategies are noninvasive. With the special case of intravenous infusions, they are ordinarily easy. These tests use radioactive materials called radiopharmaceuticals or radiotracers to assist specialists in analyzing and assessing therapeutic conditions.

Radiotracers are atoms connected to, or "labeled" with, a small amount of radioactive fabric that can be identified on the PET filter. Radiotracers gather in tumors or locales of aggravation. They can too tie to particular proteins within the body. The foremost commonly utilized radiotracer is F-18 fluorodeoxyglucose (FDG), an atom comparable to glucose. Cancer cells are more metabolically dynamic and may retain glucose at a higher rate. This higher rate can be seen on PET scans. This permits the specialist to distinguish illnesses that may have been seen sometime recently on other imaging tests. Single-photon emission computed tomography/CT (SPECT/CT) and positron emission tomography/computed tomography (PET/CT) units can perform both exams simultaneously. PET/MRI is a developing imaging innovation. In any case, it is not all-around accessible at this time. A PET scan measures vital body capacities, such as the digestive system. It makes a difference in how well specialists assess organ and tissue functioning. CT imaging employs uncommon X-ray hardware, and in a few cases, a differentiate fabric, to create numerous pictures of the body's interior. A radiologist sees and translates these pictures on a computer screen. CT imaging gives fabulous anatomic information. Combined PET/CT scanners perform nearly all PET filters nowadays. These combined checks offer assistance in pinpointing irregular metabolic action and may give a more precise analysis than the two filters performed independently.

PET is a therapeutic and research tool in the natural and clinical environment. It is used in tumor imaging and clinical oncology to identify metastases and diagnose various types of brain disorders such as dementia. PET is a valuable research tool for learning and strengthening knowledge about the general brain and supporting the development of heart function and drugs. PET is also used in natural research with animals. This helps investigate the same subject repeatedly over a long period, allows the subject to act as their own control, and significantly reduces the number of animals needed for specific research. This method helps reduce the sample size required while improving the statistical quality of the research results. Physical processes lead to physical changes in the body. Since PET can detect expressions of specific proteins and biochemical processes, PET can provide information at the molecular level before any physiological changes are visible. PET scans are performed using radiolabeled molecular probes with different uptake rates depending on the type and function of the tissues involved. Field tracers can give positron emitters relatively large quantities and dimensions among the various physiological structures by injecting them into PET scans. PET imaging is best performed using dedicated PET scanners. It is also possible to capture PET images using conventional dual-head gamma cameras with detectors. The quality of gamma camera PET imaging is low, and it takes a long time to get a scan. However, this method provides a low-cost on-site solution for companies with low demand for PET scans. Another option is to transfer these peoples to another center or rely on mobile scanners. Alternative medical imaging methods include SPECT, mathematical tomography (CT), MRI and fMRI, and ultrasound. Spect is a PET-like imaging technique that uses radioligands to identify molecules in the body. Specter is cheaper than PET and has lower image quality.

PET scans using Tracer 18F-FDG are widely used in clinical oncology. FDG is a glucose analog absorbed by glucose-using cells and phosphorylated by hexokinase, which is significantly enhanced in mitochondrial morphology in the case of rapidly growing defects. A metabolic mesh for radioactive glucose molecules allows the use of PET scans. The concentration of the pictured FDG tracer depends on local glucose and, therefore, reflects the tissue's metabolism. 18F-FDG is used to detect the possibility of cancer spreading to other parts of the body (cancer metastasis). These 18F-FDG-PET scans are most common in standard medicine (equivalent to 90% of current scans) for detecting cancer metastases. One can use the same tracer to determine the type of dementia. Rarely, other radiotracers, usually not labeled with fluorine-1, are used to characterize the tissue density of different types of molecules in the body. The effective radiation dose of the normal dose of FDG used in oncological scans is 7.6 msv. All cells require a hydroxy group replaced by fluorine-1 to produce FDG for the next stage of glucose metabolism, so no further response is observed in FDG. In addition, most tissues (except the liver and kidneys) cannot remove the phosphate added by hexokinase. This means that the phosphorylated sugar cannot leave the cell due to the ionic charge, and the FDG is trapped in the cell that picks it up. As a result of intense radiobalancing of tissues with high glucose uptake, common cancers of the brain, liver, kidneys, and most cancers. Due to the Warburg effect, these tissues have higher glucose levels than most normal tissues, resulting in FDG-PET can be used for cancer diagnosis, staging, and monitoring, especially for Hodgkin's lymphoma, non-Hodgkin's lymphoma, and lung cancer. A 2020 review of research on using PET for Hodgkin lymphoma found that negative results of intermediate PET scans were associated with higher overall and progression-free survival. However, the reliability of the available evidence was very low for medium to moderate and progress-free survival. A few other isotopes and radio trackers have been gradually introduced into oncology for specific purposes. For example, 11C-labeled metomidate (11C-metomidate) has been used to detect tumors of adrenocortical origin. In addition, FDOPA PET/CT (or F-18-dopa PET/CT) has been proven to be a more sensitive alternative to pheochromocytoma detection and detection than MDBG scans.

X-ray

X-ray is one of the traditional ways of imaging. Without doing any incision procedure, one can diagnose inside a human body. There are different purposes for doing the X-ray. The doctor prescribes a mammogram to examine the breast, whereas a barium enema is performed to diagnose the gastrointestinal tract.

X-rays are electromagnetic waves that create images from inside the human body. It shows the images in various shades of black and white as different body tissues n cells absorb different radiation levels. For example, bones look white as the calcium in the bones absorbs the radiations, whereas fat and soft tissues absorb less, giving gray color. Lungs look black as the air absorbs the least radiation. X-rays are mostly used to diagnose broken or fractured bones. A chest X-ray can detect pneumonia.

On November 7, 1899, Wilhelm Roentgen, a professor of German physics, was exposed to X-rays. This was the first paper written on X-rays. Calling Roentgen radiation an "X" indicates that it is an unknown type of radiation. Many of his colleagues (outside of Roentgen's great opposition) advised him to Rntgen Ray, but the name survives. They are still spoken in many languages, including German, Hungarian, Ukrainian, Danish, Polish, Bulgarian, Swedish, Finnish, Estonian, Turkish, Russian, Latvian, Lithuanian, Japanese, Dutch, Georgian, Hebrew, and Norwegian. Roentgen received the first Nobel Prize in Physics for his discovery. There is a reverse explanation for his research, such as the notes in his lab burned after Roentgen's death but possibly reconstructed in his biography. Roentgen is black because there are no scenes. He looked at the cathode ray from the crocus tube. Covered in cardboard, it uses a fluorescent screen painted with barium platinum cyanide to interfere with the tube. He saw a misty green glow about 1 m from the screen. Roentgen noticed that invisible rays from the tube passed through the cardboard and illuminated the screen. He also learned that they could get books and documents at their desk. Roentgen regularly devoted himself to investigating this unknown ray.

Roentgen discovered the therapeutic use of X-rays when he took pictures of his wife's hand on a photographic plate made of X-rays. The image of his wife's hand was the first image of a human body using X-rays. The discovery of X-rays had a great impact. Roentgen biographer Otto

Glasser estimated that in 1896, 49 essays and 1044 articles of the new Kiran were published. This was probably a conservative assumption if you believe that almost every text in the world reported extensively on new discoveries. It was dedicated to 23 articles alone in magazines like *Science*. New discoveries include occult theories such as telepathy and commercial reactions involving new types of rays from miracles.

X-ray photons carry enough energy to ion the atoms and break the molecular bonds. It is a type of ionizing radiation and is therefore harmful to living tissues. Too high a dose of radiation for a very short period can cause radiation damage, but low doses can increase the risk of radiation-induced cancer. In clinical pictures, the risk of this cancer growth is usually influenced by the convenience of the test. The ability of X-rays to kill malignant cells using radiation therapy can be used to treat cancer. It is also used for physical features using the X-ray spectrum.

The length of X-ray decay in water shows the oxygen absorption edge at 540 EV, the strength of the light base depends on flattening at higher photon energy due to Compton scattering, about four orders of magnitude higher for soft X-rays (left half) than for hard X-rays (right half). Hard X-rays can be done without absorbing or spreading excessively thick objects. For this reason, X-rays are widely used to image the interior of visible opaque objects. The most common applications are medical radiography and airport security scanners, but similar technologies are important in industry (such as industrial radiography and industrial CT scans) and research (such as small animal CT). The depth of penetration varies with width throughout the X-ray spectrum. This allows adjustment of the photo power for a photo correct for the application and provides good contrast. X-rays have much shorter wavelengths than visible light, so one can examine very small structures compared with using a regular microscope. This property is used for X-ray microscopy and X-ray crystals to obtain high-resolution images to position crystal molecules.

Uses of X-ray

a. X-ray crystallography, in which patterns formed by X-ray scattering through a densely scattered lattice in the crystal, are recorded and then analyzed to clarify the properties of the lattice.

b. It uses electromagnetic radiation in the soft X-ray band to produce images of very small objects.

c. As a technology in which fluorescent X-rays and X-rays are made and detected in a sample, the output energy of X-rays can be used to identify the composition of the sample.

d. Industrial radiography specifically inspects the industrial parts of welds using X-rays.

e. X-rays of cultural materials support X-rays of paintings to remove previous images during or after painting, and in some cases, the most commonly seen are lead pigments often appearing on white radiographs.

f. The X-ray spectrum is used to analyze the color response of the color in the image; for example, to analyze the color loss of Van Gogh's images.

g. Use of X-rays for inspection and quality control—left chip reveals differences in dye and bonding wire construction.

h. The process of using X-ray machines to create three-dimensional representations of both external and internal components. This is achieved through computer processing of approximate images of objects scanned from multiple directions.

i. Aircraft baggage scanners use X-rays to inspect your bag for safety issues before loading it into the aircraft.

j. Border control uses truck scanners and X-rays to inspect the interior of domestic police trucks.

k. X-ray art and fine art photography, the artistic use of X-rays, such as the work of Stanza Gadik X-ray hair removal. It was a popular method in the 1920s but is now banned by the FDA.

l. X-ray stereophotogrammetry is used to track bone movement based on implants.

m. X-ray photoelectron spectroscopy is a commonly used surface science photoelectric effect-based chemical analysis technique.

n. Radiation injections originate from fission explosions (atomic bombs) that use high-energy X-rays to compress nuclear fuel at fusion ignition points (H bombs).

Although generally considered invisible to the human eye, X-rays can occur in special situations. In an experiment shortly after Roentgen groundbreaking 1895 text, Brand reported after adaptation to the dark, his eyes were brought to the X-ray tube, creating a pale "blue-gray" tinge in his eyes. Upon hearing this, Roentgen reviewed his record book and found that he also saw the effect. When Roentgen noticed the same blue tinge as he placed the X-ray tube on the other side of the wooden door, he noticed that his eyes were gone, but he saw only the effect. He later realized that the effect of the tube was nothing more than a powerful tube that made the flash visible and that subsequent experiments were easy to reproduce. The knowledge that X-rays appear to be unconscious to the naked eye, corresponding to black, is largely forgotten today. This may be due to the desire to reproduce unknown, dangerous, and potentially harmful tests with the help of ionizing radiation. It is not clear whether the proper process of the eye produces visualization. This may be due to conventional detection (stimulation of rhodopsin molecules in the retina), direct stimulation of retinal nerve cells, or secondary detection such as X-rays. Visual light is the second method of deviation of light through conventional retinal detection of the phosphorescence of the eyeball. X-rays are not otherwise visible, but if the X-ray beam is strong enough, one can see the ionization of air molecules. The beamline from ID 11 Wigler of the European Synchrotron Radiation Institute exemplifies such high intensity.

CHAPTER 2

Introduction to biosignals and bioelectrodes

Sudip Paul[1], Angana Saikia[1,2], Vinayak Majhi[1] and Vinay Kumar Pandey[1]

[1]Department of Biomedical Engineering, School of Technology, North-Eastern Hill University, Shillong, Meghalaya, India; [2]Mody University of Science and Technology, Laxmangarh, Rajasthan, India

Contents

Introduction to Biomedical Instrumentation and Its Applications
ISBN 978-0-12-821674-3
https://doi.org/10.1016/B978-0-12-821674-3.00005-X

A bioelectrode is a conductor designed to function as an interface between organic and electronic structures. It both experiences and measures (passively) the electric activity inside the organic shape or stimulates it by inducing external electrical capability. A bioelectrode is usually considered a transducer that converts the ionic cutting edge into modern-day electronic signals or vice versa. The surface electrodes are used as both reusable and disposable electrodes. Reusable electrodes are frequently used to generate electrocardiograms (ECGs) for routine diagnostic purposes. Plate and clamp electrodes are typically used at the limbs, whereas the Welsh electrode is used to measure the chest. Disposable pregelled ECG electrodes are mostly employed for patient monitoring over long time frames inside extensive care or postoperative units.

All rate-switch strategies are ruled through Faraday's regulation (i.e., the quantity of chemical alternate occurring at an electrode—electrolyte interface is at once proportional to the current flowing through that interface) and subsequently are referred to as faradaic processes.

Bioelectrodes function as an interface between organic structures and digital systems. Electrical activity within the biological shape is both sensed and stimulated. The electric structures are both passively sensing or actively stimulating electrical potentials in the biological structure or unit. Bioelectric potentials generated are ionic potentials, and it is important to convert these ionic potentials into digital potentials earlier than they can be measured through conventional techniques. Gadgets that convert ionic potentials into electronic capability are known as electrodes. A transducer that converts the body ionic inside the body into the traditional electronic flowing within the electrode is a bioelectrode.

Bioelectrodes are capable of conducting small cutting-edge throughout the interface between the body and the digital measuring circuit.

The net current that crosses the interface, passing from the electrode to electrolyte, consists of

(1) Electrons moving in a direction opposite to that of current in the electrode

(2) Cations moving in the same direction

(3) Anions moving in a direction opposite to that of current in the electrolyte

Properties of bioelectrodes:

(1) Good conductors

(2) Low impedance

(3) Should not polarize when a current flows through them

(4) Should establish good contact with the body and not cause motion

(5) Should not cause itching, swelling, or discomfort to the patient

(6) Metal should not be toxic

(7) Mechanically rugged

(8) Easy to clean

Electrode—skin/tissue interface

(1) Interface between body and electronic measuring device

(2) Conducts current across the interface

(3) Ions carry current in the body

(4) Electrodes are capable of changing ionic current into an electronic current

Some properties of bioelectrodes are as follows:

➤ Good conductors

➤ Low impedance

➤ No polarization effect during current flow

➤ Better contact with the body surface

➤ Potentials at the metal electrolyte surface must be low

➤ Metals used in the electrode should not be toxic

➤ Mechanically rugged

➤ Chemically inert

➤ Easy to clean

Some materials used for manufacturing bioelectrodes are

➤ Aluminum

➤ Copper

➤ Silver

➤ Gold

➤ Platinum

Biomedical transducers

Electronic devices in the human body measure nonelectrical parameters such as temperature, heartbeat and blood pressure. Transducers are devices that convert biological parameters into electrical signals. The transformation process is transformation. Transducers usually convert energy from one form to another.

Transducers convert the energy of one form or variable into another form of energy or a variable. Transducers must convert physiological changes into electrical signals, thus making the process easier. Several factors are considered when selecting special transducers used to study specific phenomena. These reasons are

a. The required order of precision
b. The size of the amount to be measured
c. For both short-term and long-term monitoring of the patient's body application site
d. The static or dynamic nature of the process under investigation
e. Economic considerations

Any transducer can be used with suitable electronics to provide final power output with convenient amplitude. The relationship between input and output variables is as follows:

a. Linear
b. Logarithmic
c. Square-law

Biomedical transducers have two components:

a. Detection material or detector
b. Transit material

Sensing elements are parts of a transducer that match a physical event or its transformation. Transit material converts generated sensitive elements into electrical outputs. In other words, the transit element acts as a secondary transducer. In therapeutic applications, physiological variables can be converted into several suitable variables that can be easily measured, known as the principle of transit.

Force transducer

Force detection material is used to convert forces into physical variables for the following:

a. Tension using strain gauge
b. Displacement due to removal of member
c. Output voltage using a linear variable differential transformer (LVDT)
d. Photoreceptors by changing the resistance

Pressure transducer

Pressure is a very important parameter used as a transport principle in many treatment devices. The basic principle behind these pressure transducers is that the measured pressure is applied to the flexible diaphragm distorted by the pressure applied to it. This diaphragmatic movement is measured for electrical signals. The deformation is measured by strain gauge or LVDT. In short, three main pressure transducers use the diaphragm.

Capacitance Manometer—here, the diaphragm forms a plate of capacitors where a strain gauge is attached to the diaphragm. A differential transformer is commonly known as an LVDT, where the diaphragm is connected to the core of the differential transformer.

Displacement transducer—displacement transducers can be used in both direct and indirect measurement methods. Direct displacement measurements can determine changes in blood vessel diameter and changes in the size and shape of heart cells. Indirect measurements of the intervention are used to determine the velocity of the fluid through the heart valve. Displacement transducers can be easily converted into pressure transducers by attaching a diaphragm to the moving members of the transducer so that the pressure is applied to the diaphragm. A good application for this is to detect the heart rate indirectly through the movement of the microphone diaphragm.

There are two types of transducers:

a. Active transducers convert one form of energy into another *without* an external energy source; an example is solar cells that convert light energy into electrical energy. Types of active transducers are magnetic induction, piezoelectric, photovoltaic, and thermoelectric.

b. Passive transducers convert one form of energy into another *with* the help of an external energy source using the principle of controlling DC voltage or AC carrier signals. Examples are possible quantity and load cells.

Types of passive transducers:
Resistive method
Inductive
Capacitive

Active transducer

Magnetic induction transducers work with conductors that move within the magnetic field; as the magnetic flux passes through, the conductor changes, creating a voltage proportional to the rate of change in the current.

The opposite magnetic effect also applies. When an electric current flows through a conductor placed in a magnetic position, a mechanical force acts on the F conductor.

Magnetic induction transducer applications include electronic magnetic flux meters, heartbeat microphones, display devices, and biomedical recorder pen motors.

Piezoelectric transducers create charge separation between crystals as they are compressed or tensed, thus creating voltage that creates a piezoelectric effect. Piezoelectric transducers convert displacements or pressures into electrical values. Barium titanium, Rochelle salt, and nickel lithium are piezoelectric transducers. Applications of piezoelectric transducers include piezoelectric transducers as pulse sensors for measuring the human pulse rate.

Photovoltaic transducers emit electrons when light or other wavelengths hit the surface of a metal or semiconductor, which is known as the photoelectric effect. Photoelectric, photoconductive, and photovoltaic devices are types of photoelectric converters. Of these, the photovoltaic power-generating device is an active transducer that produces a voltage proportional to the incidence of radiation to which it is exposed.

Photovoltaic transducer application:

a. Silicon solar cells as pulse sensors for photoelectric plethysmography
b. Light absorption technology to measure the concentration of sodium and potassium ions in samples
c. Thermoelectric converter
d. Transducers working under Seebeck influences
e. The Seebeck effect—two junctions of a thermocouple at two different temperatures generating potential voltage
f. Voltage generated proportional to the temperature difference between the two junctions of a thermocouple

Thermoelectric converters are applied in measurements of the body temperature in remote sensing circuits and biotelemetry circuits.

Passive transducer

Resistive membrane transducer—strain gauges, photo resistors, photodiodes, phototransistors, and thermistors—are resistance-type passive transducers. They have a general principle of operation, and the measuring parameters result from small changes in the resistance of the transducer. Wheatstone Bridge usually measures changes in resistance.

Resistor transducer applications are

a. A strain gauge attached to a finger to measure small changes in blood volume flowing through the finger

b. Arranging veins in the body

c. Light dependent resistors (LDR) or photoreceptor palpation measurement by changing blood volume

Capacitive transducer

Capacitors have two conductor surfaces, and a dielectric medium separates the space between the two surfaces. Capacitive transducers measure displacement changes due to changes in plate operating area, the thickness of the dielectric medium, and the distance between plates.

Capacitive transducer applications include

a. Differential capacitive transducers that measure blood pressure

b. Induction transducers

c. Induction transducers that operate on involuntary changes in coil and number; LVDT as an induction transducer acting as a physical pressure sensor

d. Induction transducer applications that measure tremors in patients with Parkinson's disease

Advantages of electrical transducers include the following:

a. The electrical signal received from the transducer can be easily processed (mainly wide). A layer suitable for output devices can be used as an indicator or recorder.

b. Electrical systems can be controlled with very little power.

c. Power generation can be easily used, transmitted, and processed for measurement.

d. With the advent of integrated circuit technology, the size of electronic systems has become extremely small, and the required operating space has also decreased.

e. Electrical systems do not include moving mechanical parts. Therefore, there is no problem with mechanical wear and no possibility of mechanical failure.

Disadvantages of transducers include the following:

a. Electrical transmitters may be less reliable than mechanical ones because of the maturity and flow of the active components.

b. Signal-processing circuits related to sensor components are relatively expensive.

c. The accuracy and durability of transducers have been expanded using more advanced materials, better technology, and better circuitry.

d. Negative feedback technology improves measurement accuracy and system stability, increasing circuit complexity, space requirements, and of course, costs.

Biosignals and bioelectrodes

Biological systems often involve electrical activity. This activity can be constant DC electric fields, continuous charged particles flow, current or time-changing fields, or currents related to time-dependent biological or biochemical phenomena. Bioelectric phenomena are related to biological structures and distributions of ions or charged molecules in ions. Changes in this distribution are due to specific processes from biochemical reactions or may result from events that alter local physiology. Bioelectric phenomena are associated with almost every organ system in the body. Nevertheless, most of these signals are related to a current event.

Time is not particularly effective in clinical medicine and does not present time-invasive low-level signals. It is easy to measure, and many indications are medically or clinically important. It provides a means of electronic evaluation to help in understanding biological systems. The most well-known of these is the ECG, a signal derived from the heart's electrical activity. This signal is widely used to diagnose cardiac arrhythmias, conduct signals through the heart, and consequential damage from cardiac ischemia and myocardial infarction. Electromyography (EMG) is used to diagnose neuromuscular diseases, and brain waves are important in identifying brain dysfunction and assessing sleep. And these techniques are mainly used in mammalian studies, and bioelectric signals are also generated from plants. These signals usually change steadily or slowly, in contrast to signals that change over time.

The process of conducting electricity in the body consists of ions forming a charge carrier. Bioelectric signals include interactions with these ionic charge carriers and the transfer of ionic currents. It is a conversion function required for cables and electronic devices. It is carried by electrons that contain conductors that come in contact with ionic aqueous solutions in the body. Interaction between electrodes and ions in the body can significantly affect the interaction of electrons in the body. Special consideration must be given to the effectiveness and application of these sensors—the reaction at the interface between the redox of the electronic and the ionic solution. The charge must be transferred between the electronic and the solution—exposure to an ion solution results in local changes in interactions between metals. Ion concentrations in solution near the metal surface do not create charge neutrality but are maintained in this area, causing various electrolytic currents around the metal. Thus, the potential difference is called half-cell potential and is installed in metals and most electrolytes.

The relationship between potential and ion concentration, or more precisely, ion activity. Ionic activity is often considered electronic. The two most common ionic solutions for different activities are ion-selective crescents separated by a membrane to allow one type of ion to pass freely through the membrane. Membranes are based on the relative activity of perusable ions in each solution. This relationship is known as the Nernst equation. If no current flows between the electrode solution and its ions or between the ion-presented membranes, the observation probability must be half-cell or constant probability. However, these possibilities can be changed if the current flows. The difference between the capacitance of zero current and potential in the current flowing state is called overvoltage. And this is the result of a change in the charge distribution of the solution in contact with the electrodes or an ion-selective membrane. This effect is known as polarization and can result in fewer electrodes. Performance is especially under dynamic conditions. There are three basic components to polarization.

Excessive power—ohmic, excessive compression, excessive efficiency. Excessive capacity is the number one concern when measuring bioelectricity. A fully polarized electrode passes an electric current between the electrode and the electrolyte.

Changes to charges distributed to the solution near electrons. Thus, currents do not intersect the electronic–electronic interface. However, the electrical currents allow electrons to flow freely.

A full electronic–electronic interface without changing the charge-distribution electrode-adjacent solution. This kind of electronics can be explained theoretically but cannot be physically generated. However, the electrodes can generate any structure while closely combining features. Electrodes made of precious metals such as platinum are often highly polarized. In charge distribution, unlike bulk electrolytes, electrons are in solution near the surface. Such distributions can create serious limitations in movement and measurement with low-frequency or DC signals. The disappearance of electrodes relative to the electrolytes changes the solution's charge distribution adjacent to the electronic surface. The voltage changes in electrodes are displayed as the measured speed. Therefore, in most biomedical measurements, nonelectrical electrons are preferred for polarizations. A silver—silver chloride electrode has the same properties as a completely unpowered electrode and is practical for many biomedical applications. A silver base structure coated with a layer of ionic compound silver chloride. When chloride is exposed to light, some metal is reduced to silver, so there is a common silver—silver chloride electron. Its surface is metallic silver finely divided into a chloride matrix. It is relatively soluble in aqueous solution and is a stable surface. The minimum polarization connected to this electron reduces the speed artifacts compared with the polarized electrodes as platinum electrodes. In addition, reduced polarization has a small effect.

Electronic barrier frequencies, especially at the lower end. Such silver—silver chloride electrodes can be made starting with a silver base. A layer of silver chloride rises electrically below its surface. Although it is produced by electronics, this method can be used for most biomedical measurements and is not a rigid structure. The silver chloride film can be peeled off repeatedly using the structure of a silver lead surrounded by a water chestnut cylinder made of finely divided silver pressed together with silver chloride powder. In addition to its nonionic behavior, silver—silver chloride electrons show less electricity sound than equilibrium electrodes. This is especially true at low frequencies, so silver—silver chloride is recommended for very-low-voltage measurements of electronic signals, as it consists of low frequencies. More details of silver—silver chloride electrodes and methods for making these devices can be found in the work of Janz and Ives on biomedical instruments.

ECGs and electromyography (EMG) tests conducted at home can measure and screen approximate vital statistics for patient health. Presently, measured ECG or EMG alerts at home can be monitored for long-distance scientific study using the Internet. Biocapable electrodes are vital in monitoring ECGs, electromyograms, and so on.

Table 2.1 shows some biosignals sensed by bioelectrodes and their sources.

Table 2.1 Bioelectric signals sensed by biopotential electrodes.

Bioelectrical signal	Biological source
Electrocardiogram	Heart as seen from the body surface
Cardiac electrogram	Heart as seen from within
Electromyogram	Muscle
Electroencephalogram	Brain
Electrooptigram	Eye dipole field
Electroretinogram	Eye retina
Action potential	Nerve or muscle
Electrogastrogram	Stomach
Galvanic skin reflex	Skin

Electrical characteristics of electrodes

In actual capacitance measurement, at least one current flows through the measuring circuit. A percentage is measured from time to time. Ideally, this current is extremely small, but in practice, it never goes to zero. Thus, a biopotential electrode is required.

Electrons perform the conversion function because current flows through the interface between the body and the electronic device measurement circuit. In the body, it is carried by ions and electrons and by electrodes and their leads. The electrode must act as a transducer to convert the ionic current into an electric current. It entangles the electrodes and interferes with the operation. Theoretically, two types of electrodes are possible. Fully polarized electrodes are a completely unobtrusive classification that means activity occurring in the present time. It passes between it and the electrode. A fully polarized electrode is an electron whose charge electrode does not intersect the electrolyte. The applied interface must be up-to-date. A displacement current is the current, and the electrodes behave like capacitors.

In fully polarized electrodes, electric current flows freely and does not require energy to make conversions. Precious metal electrodes work with completely sealed electronic polarization. These electrical properties create strong capacitances that are not suitable for effective and practical use.

The silver—silver chloride electrode is a practical electron describing its properties. It is a completely nonpolar electrode and can be easily fabricated in the laboratory. Each member of a class electrode—that is, a metal coated with an ionic layer—is a slightly soluble compound of metals with suitable ions. The whole structure is immersed in an electrolyte containing a relatively high concentration of ions.

Electrical properties have been the subject of much research. In many cases, the current—voltage characteristics of the electronic—electronic interface show that there is no wire.

Nonelectrode components require electronic conduction, and the electrode is especially sensitive to the current flowing through it and to the properties of the electron. Relatively high current density can be considered different from low current density. The properties of electrons also depend on the wave. Sinusoidal current measures the circuit operation of the electronics. The properties also depend on the frequency. For sinusoidal inputs, the terminal properties of the electrons are both resistant and responsive.

The skin comprises three main layers surrounding the body, protecting it from the environment and serving as a good interface. The outermost layer or epidermis is the most regenerated electrical interface, with three sublayers. Cells divide and develop into the deepest basal layer; as they form new cells and grow under them, they move outward. As they pass stratum granulosa, they begin to die and continue to lose nuclear material. External velocity is further degraded between the soil and the flattened flat layer of keratin or dead substances on the skin's surface as the process continues. These layers are constantly being formed. It is later in the stratum granulosa and replaces it with new cells. The altered layer of skin on the outer surface contains separated dead matter. The deeper layers of the skin also contain the blood vessels and nerve components of the skin sweat glands, sweat ducts, hair follicles. These layers are similar to other tissues in the body and do not allow the skin to show electrical properties except the sweat glands.

Electrode—skin interface—impedance, noise, and DC voltage

To measure power, and therefore the current, in the body, we need to supply something to the interface between the main unit and the electronic measuring circuit. With biopotential electrodes, biological data can be measured by excluding the transducer function.

A quantitative electrical signal transforms one form of energy into another. The electric current flowing from the body, and consequently in the electron circuit, is called the electrode interface and requires charge transfer. Electrolytes represent a body fluid containing ions (sweat) or an electrolyte solution (gel content) applied electrically into the skin. The metal is placed or added to the electrolyte solution, and the ion—electron exchange is the current through the electrolyte passes through the interface. This means that the electrons are moving in the opposite direction of the electrode current and the cage. It moves in the same direction as the electric current, and the ions move in the opposite direction. Electrode current generates a potential half-cell potential (Ehc).

The resulting charge level is measured relative to the second electron. Explaining the skin requires some understanding of its structure and the skin and tissues beneath it. The skin comprises three main layers: subcutaneous tissue, deep layer, and dermis and epidermis. These layers cover the body to protect it from the environment. The deeper layers of the skin contain blood vessels, nerves, sweat glands, and hair follicles. The upper epidermis comprises dead cells (stratum corneum) on its surface skin, so it has high resistance.

Electrode properties and behavior—skin barriers have already been addressed in other studies and address the use of surface electronics in biopotential recordings. The electrode properties are sensitive to current flow and can be modeled by nonlinear RC circuits whose frequencies and components represent the relevant frequencies.

Biopotential amplifiers—instrumentation amplifiers

In general, symptoms of physical activity are very small and must be amplified by filtering before recording and processing, so a biopotential amplifier is needed to increase the amplitude of the weak potential. The instrumentation amplifier (within amplifier) is widely used as a biopotential amplifier to measure differences caused by biological electrical signals due to its ideal properties to measure signals from low-level output transducers in noisy environments. In-amplifier is a device that amplifies the difference between the two input signals while rejecting a signal common to both input terminals. In fact, common-mode removal (CMR) is the most important feature provided by an in-amp because it is a void property that amplifies the signal with a common signal at both input terminals and the difference between the inputs. Both DC and AC removal modes are important. In compensation for DC voltage produced by communication, for example, due to insufficient CMR of unwanted sources, the interface between the electrodes and the signal properties of the skin and power line causes an error. Fortunately, most modern IC in-amps provide great DC and reject AC normal mode.

Electronic signals can be used to carry various bioelectric signals. The most common use of biopotential electrodes is ECGs for clinical and patient-observation applications. The commercial market provides various ECG electrodes in different shapes, and those discussed in the previous section are commercially available. Electrodes for biotechnology research applications are diverse and specialized for individual studies.

Various types of electrodes have been developed for different biomedical measurements. Some commonly used bioelectrodes for biomedical measurements are described in the sections that follow.

Body surface biopotential electrodes

These electrodes are located at the surface of the body for recording or accumulating biosignals. Through the electrode software, skin integrity is in no way compromised. These electrodes are used for short-term diagnostic recordings, such as medical ECGs, or continual long-term recordings, such as cardiac monitoring. These electrodes are outlined to measure ECG, Electroencephalogram (EEG), and EMG possibilities from the skin's surface; thus, these electrodes are slightly traumatic. Body surface electrodes are outlined and used to measure bioelectric possibilities from the surface of the body. They are accessible in numerous shapes and sizes. The larger anodes are often used to detect ECG possibilities, as these estimations do not depend on the particular localization of the anodes. In any case, smaller electrodes are used to detect EEG and EMG possibilities, as detecting them depends on the area of anode or estimation. Metal plate and suction glass sort terminals are body surface anodes, but they have a common issue in the plausibility of slippage development. These electrodes are sensitive to slippage development and subsequently create inaccurate estimations with movement. Drifting electrodes are used to combat this issue. The purpose of a coasting electrode is to dispose of development artifacts by maintaining a strategic distance from any coordinate contact of the metal cathode with the skin. The contact between the metal electrode and skin is maintained by electrolyte glue or jelly.

Metal plate electrodes

These electrodes include a metal conductor in touch with pores and skin with a thin layer of electrolytic gel between the metal, pores, and skin. The metallic plate electrodes typically used are German silver, gold, and platinum. Some of these electrodes are fabricated from metal foil to make them extra bendy. They are also made in the shape of a suction electrode to make them easier to attach to body surfaces. Metallic plate electrodes are used for diagnostic recordings of biopotentials, such as ECGs, EEGs, etc. Gold-surface metal disk electrodes, which may be conical in form, are used for EEG recordings. The ECG estimation strategy employs either rectangular or circular plate electrodes made of nickel, silver, or German silver materials. It includes a smaller contact zone and does not seal completely and persistently. Electrodes are stuck to the skin with electrolyte glue. Electrode slippage and plate uprooting are the two major drawbacks of this exceptionally delicate electrode, thus driving estimation errors.

Electrodes for chronic patient monitoring

Cardiac video display units measure biosignals, such as ECGs, that call for long-duration monitoring and impose constraints on the electrodes that collect signals. The electrodes must own a strong interface between themselves and the body surface. The mechanical balance of the interface between the electrode and the pores and skin can reduce motion artifacts. Various strategies can reduce interfacial movement among the electrode and the coupling electrolyte or the pores and skin. Recessed electrodes can be used to reduce motion artifacts by recessing the electrode in a cup of electrolytic fluid or gel. The cup is then securely fixed to the skin surface with a double-sided adhesive ring. The motion of the pores and skin with appreciation to the electrode can also affect the electrolyte near the pores and skin—electrolyte interface. However, the electrode—electrolyte interface may be several millimeters away from this region. The fluid motion is not going to have an impact at the recessed electrode—electrolyte interface. The advantages of the recessed electrode may be found out in an easier layout that lends itself to mass production through automation.

Intracavitary and intratissue electrodes

Electrodes may be located inside the frame for biopotential measurements. These electrodes are usually smaller than pores and skin surface electrodes and no longer require unique electrolytic coupling fluid because natural body fluids serve this function. There are many extremely good designs for those internal electrodes. Essentially these electrodes can be categorized as needle electrodes. They penetrate the pores, skin, and tissues to determine where the measurement should be made. These electrodes can be located in a whole natural space or a surgically produced whole tissue area. A catheter tip or probe electrode is located naturally in an area taking up hollow space inside the surface. A steel tip or section on a catheter makes up the electrode. The catheter, or with no lumen hole, the probe, is inserted into the hollow space to make simple contact between the metal electrode and the tissue. A lead cord is run down the lumen of the catheter or the middle of the probe and connects the electrode to outside circuitry. The primary needle electrode consists of a stable needle, generally a stainless metal product with a sharp factor. An insulating material coats the shank of the needle and as much as a millimeter or two of the tip, so the very tip of the needle stays exposed. When this structure is located in tissue such as skeletal muscle, electrical alerts may be obtained from the exposed tip. One can also make needle electrodes by running one or more extra-insulated wires down the lumen of a preferred hypodermic needle.

Microelectrodes

The three microelectrode varieties are etched steel, micropipette, and metal-movie-coated micropipette. The steel microelectrode is a subminiature version of the needle electrode defined in the preceding section. In this case, a strong metal such as tungsten is used and etched electrolytically to provide tip diameters on the order of a few micrometers. The structure is insulated to its tip and can be passed via a cellular membrane to touch the cytosol. The advantage of those electrodes is that they may be both small and sturdy and used for neurophysiologic studies. Their principal disadvantage is their difficult fabrication and high source impedance.

The second and most commonly used microelectrode is the glass micropipette. This shape includes an exceptional glass capillary attracted to a wholly slender factor filled with an electrolytic input. The factor can be as slim as a fragment of a micrometer, and the scale of this electrode greatly depends on the skill of the person drawing the tip. The electrolytic solution inside the lumen serves as the contact between the cellular membrane interior through which the tip has been impaled and a larger traditional electrode placed within the shank of the pipette. High supply impedances and fabrication issues may also afflict those electrodes. A combined shape can be made from these electrodes by depositing a metallic film over the outside floor of a glass micropipette.

Electrodes fabricated using microelectronic technology

Microelectronic current generation may be used to fabricate specific types of electrodes for particular biomedical packages. As an example, dry electrodes with excessive supply resistances or microelectrodes with comparable characteristics can be processed through a microelectronic amplifier for impedance conversion at the electrode itself. A metallic disk with a diameter of 5–10 mm will have a high microelectronic starting impedance for conventional-sized electrodes. Multichannel amplifiers or multiplexers can be used with more than one electrode on an equal probe. Electrodes for contact with individual nerve fibers can be fabricated using micromachined holes in a silicon chip and are simply massive enough to skip a single developing axon. Electrical contacts on the edges of those holes can then pick up electrical pastime from these nerves. These are just a few of the many possibilities that may be realized using microelectronics and three-dimensional micromachining generation to fabricate specialized electrodes.

Table 2.2 shows some applications of biopotential electrodes.

Table 2.2 Applications of biopotential electrodes.

Application	Biopotential	Electrode type
Cardiac monitoring	Electrocardiogram (ECG)	Ag/AgCl with sponge Ag/AgCl with hydrogel
Infant cardiopulmonary monitoring	ECG impedance	Ag/AgCl with sponge Ag/AgCl with hydrogel Thin-film Filled elastomer dry
Sleep encephalography	Electroencephalogram	Gold cups Ag/AgCl cups Active electrodes
Diagnostic muscle activity	Electromyography (EMG)	Needle
Cardiac electrograms	Electrogram	Intracardiac probe
Implanted telemetry of biopotentials	ECG	Stainless steel wire loops
	EMG	EMG platinum disks
Eye movement	Electrooculography	Ag/AgCl with hydrogel

Holter monitor

The Holter monitor is a portable electronic card system invented by Holter and his team. People realized that this device would change the way heart rates were tracked outside the hospital. Historically, electronic cardiography began in 1893 with the study of the single-line galvanometer of Antoben. The Holter monitor device operates as a galvanometer that records electronic cardiographic signals from people performing daily activities, much like continuous portable electrocardiography. A variety of other modalities and advanced gadgets have been developed for this purpose since 1961 and have formed the backbone of rhythm detection and analysis in cardiac electrophysiology.

The Holter monitor is a sort of portable ECG that records the electrical signal of the heart continuously over 24 h. A preferred or "resting" ECG is one of the handiest and quickest checks for assessing the coronary heart. Electrodes (small, plastic patches kept on the pores and skin) are placed at specific points on the chest and stomach. The electrodes are linked to an ECG device by wires. The electrical signal of the coronary heart can then be measured, recorded, and displayed. Natural electrical impulses coordinate the contractions of distinct components of the coronary heart, keeping blood flowing in the manner required. An ECG records these impulses to show how fast the coronary heart is thrashing, the rhythm of the coronary heart beats (consistent or irregular), and the energy and timing of the electric impulses.

Initially, the Holter monitor's briefcase same size of the holter monitor's included an amplifier, a tape recorder, a chest-mounted electrode, a playback unit, and an analysis unit. It has 10 ECG leads, including 6 standard front electrodes and 4 fuselage electrodes. The fuselage lead is meant to prevent signal disruption. However, read differences may occur in the fuselage, and researchers have proposed a mathematical algorithm to overcome such fluctuations.

Holter monitor use has increased significantly, especially when cryptocurrency detects psychotropic atrial fibrillation as the cause of stroke. Antiplatelet drugs are always better in the secondary prevention of stroke due to atrial fibrillation. Therefore, the detection of latent atrial fibrillation to prevent anticoagulants with the help of a Holter monitor can prevent recurrent stroke.

A significant number of patients with symptomatic disease due to dysfunction of the left ventricular contraction, development of arrhythmias, and transient 2 or 3 degree heart block were also examined.

Older devices have used open-reel tape or standard C90 or C120 audio cassettes to record data at speeds of 1.7 or 2 mm/sec. Once the recording was complete, it could be played and analyzed at 60x speed, allowing the 24-hour recording to be analyzed over 24 min. Modern units record EDF files on digital flash memory devices. The data are uploaded to a computer that analyzes the input, calculates the EC complex, calculates summary statistics such as average, minimum, and maximum heart rate, and conducts further engineering research. Field surveys of candidates on record are performed. The size of the recorder depends on the device manufacturer. The average dimensions of today's Halter monitors are about $110 \times 70 \times 30$ mm, but some are $61 \times 46 \times 20$ mm and weigh 99 g, and two AA batteries power most devices. If the battery dies, some containers may be replaced during observation.

Most containers monitor ECGs through just two or three channels (note—different manufacturers use different calculations for leads and lead systems). Today's practice is to reduce the number of reads to ensure patient comfort during recording. As mentioned, Holter surveillance has long used two- and three-channel recordings, but 12-channel Holter systems have recently been released. These systems use the classic Mason—Ricker read system. In turn, this generates signals in the same format during ECG or stress test measurements. These holders may be able to provide information similar to the ECG stress test. It is also suitable for the analysis of patients after myocardial infarction. The recordings of these 12-read monitors show significantly lower resolution than standard 12-read ECGs and provide a confusing ST section presentation, even though some devices are shown to have sample frequencies to 1000 Hz.

Once the ECG signal is recorded (usually 24 or 48 h later), the physician is responsible for analyzing the signal once. Because such long signals can be very time-consuming, the software on each Holter device includes an integrated automated analysis process that automatically determines the various heartbeats, other beats, and more. The success of automated analysis, however, very much depends on quality, and quality depends primarily on the electronic components attached to the patient's body. If attachments are not installed properly, electromagnetic interference can affect the ECG signal and cause substantial noise recording. The deformity becomes more pronounced as the patient moves more quickly. These records are very difficult to process. In addition to electrode mounting and quality, other factors affect signal quality, such as muscle shock, sampling rate, and digital signal resolution (high-quality devices provide high sampling frequencies).

Automated analysis typically provides physicians with information about heart rate, heart rate measurement, heart rate variability, rhythm monitoring, and patient diary (when the patient presses the patient button). Advanced systems also perform spectral analysis, ischemic load assessment, graphs of patient activity, or PQ section analysis. Another requirement is the ability of the pacemaker to detect and analyze emotions. These features are effectuated when a doctor wants to ensure the correct basic pacemaker features.

Another innovation is the inclusion of triangular motion sensors that record patient physical activity. During inspection and software processing, it captures three states of movement: sleeping, standing, or walking. Some modern devices also can record audio patient diary entries that doctors can listen to later. These data help cardiologists better identify events related to patient activity and diaries.

After continuous improvement and progress, Holter monitors are now the size of a small cellphone and have two main types of analytical data. One is the QRS complex, and the other is the R−R interval. Typically, 24−48 L are used but continue to record until one is away from the patient or until electricity is transmitted. The power supply lasts 80−100 h with 10 h of tape-recording activity.

When people report symptoms of heart problems, doctors often use ECGs to help diagnose the problem. A standard ECG shows a snapshot of the overall electrical activity of the heart. If snapshots are not sufficient, a 24-hour Holter observation by a physician may be required. The Holter monitor is attached to the body as with many standard ECGs. The smaller device consists of a thin electrical cord with electrodes attached to the chest space.

A doctor may order 12, 24, or 48 h Holter monitoring depending on symptoms and how often they occur. Some of these symptoms are

a. Fast or slow heartbeat
b. Agile
c. Weakness or instability
d. Dyspnea
e. Chest pain
f. Heart palpitation
g. Different heartbeats
h. Feelings have stopped
i. Dizziness while using pacemaker

Many heart rate problems appear sporadically throughout the day, and if problems occur, a person is less likely to come to the clinic. This indicates the advantage of frequent hall monitoring, as the monitor can record whenever a problem occurs, helping physicians better diagnose the underlying problem.

A 24-hour Halter monitor may look complex, but it is easy to use. The monitor is small enough to fit in the palm of one's hand and usually engages our neck or waist. Many electrical wires known as reeds connect the unit to the electrodes attached to the body. The technician attaches the electrodes to the skin with an adhesive gel and holds them in place. If there is hair in that area, the technician must shave before taking the electronics. The metal in the electrodes selects the activity of the heart as an electrical signal. The electrons then send signals to the monitor, which records them. The electrodes may loosen or fall off and must be reconnected.

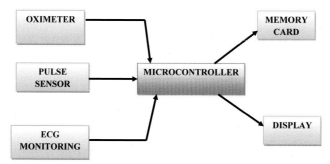

Figure 2.1 Block diagram of a Holter monitoring system.

Holter devices help identify people with heart disease who know about their daily lives. People are usually accustomed to this system, such as staying in a longer hospital. In addition, physicians may order 24-hour Holter monitoring to verify a person's response to an action for heart disease.

Holter screen recording consists of

➤ To assess chest ache
➤ To assess different signs and symptoms that may be heart-related, which include tiredness, shortness of breath, dizziness, or fainting
➤ To pick out abnormal heartbeats or palpitations
➤ To evaluate the chance for heart-associated events
➤ To see how well a pacemaker is running
➤ To determine how nicely treatment for complicated arrhythmias is running

Fig. 2.1 shows the block diagram of a Holter monitoring system.

Operation

The Holter monitor is marginally larger than a deck of playing cards. A few leads, or wires, are attached to the screen. The leads are associated with cathodes put on the skin of our chest with a paste-like gel. The metal cathodes lead our heart's action through the wires and into the Holter screen, where it is recorded. A patient wears a little pocket around the neck that holds the screen itself. It is imperative to keep the screen. The physician will tell you the best way to reattach anodes on the off chance that they become free or tumble off during the testing time frame. We will receive directions that disclose how to deal with the screen and what not to do while wearing it. It is imperative to abstain from washing, showering, and swimming while wearing the screen.

The patient is urged to take an interest in typical exercise during the 24-hour Holter test. The person will be asked to record the exercises on a scratchpad. The physician can then decide whether changes in heart action are associated with practices and developments. Wearing the Holter screen itself includes no dangers. That said, the tape or adhesive that connects the terminals to the skin can cause gentle skin aggravation in certain individuals. A 24-hour Holter monitor test is effortless. Nonetheless, make certain to record any chest pain, quick heartbeat, or other cardiovascular manifestations during the testing time frame.

There is no significant risk associated with wearing a Holter monitor, excluding potential discomfort or skin irritation where the electrodes are placed. However, the Halter monitor should not become wet or damaged. When wearing a Halter monitor, one should not swim or bathe. The wireless Halter monitor sensor must be disconnected and reconnected if a shower is to be taken. Holter monitors are not usually affected by other electrical equipment. However, do not use metal detectors, magnets, microwave ovens, electric blankets, electric razors, and toothbrushes at a time when the Halter can block the monitor's signal from the electrodes. In addition, for the same reason, keep our mobile phone and portable music player at least 6 inches away from the monitor.

The ECG is a medical test used to measure heart rate and rhythm. We also wanted other abnormalities that could affect the normal functioning of the heart. During the ECG, electrodes are placed in the chest to check the heart rhythm. Because the machine is connected for a very short time, one may experience arrhythmias while running that are not visible on an ECG. Arrhythmias and other types of heart symptoms may appear and disappear. Long-term monitoring is required to record these events. Holter monitors give physicians a long-term perspective on how the heart works. The records created by the monitor help physicians determine whether the heart is getting enough oxygen or if the heart's electrical tendency is delayed or accelerated. These irregular tendencies are sometimes called arrhythmias or abnormal cardiac rhythms.

Holter monitor studies are often used when a person has transient symptoms that can be explained by arrhythmia. These symptoms occur frequently and are near-unconscious or unconscious. Rarely, Holter monitor studies can be used to detect episodes of cardiac ischemia that do not cause angina. This is often called "silent ischemia" because it does not cause ischemic symptoms. However, ECG changes due to ischemia in a Holter study are often completely meaningless, and such changes in the

ECG can be confusing. For these reasons, most cardiologists do not order a Holter study for this purpose unless they have treated patients with asymptomatic ischemia, identified stress tests, or a past noncommunal heart attack. Therefore, Holter monitor studies are now often used to diagnose cardiac arrhythmias. If someone is already being treated for heart disease, wearing a monitor can help your doctor determine if the medication is working or if changes are needed. It also helps to understand why we are experiencing other types of symptoms, such as dizziness, fainting, heart palpitations, or heart failure.

The Holter monitor has several small electrode patches that are attached to the skin and the recording device with small cables. The recording device (formerly a small tape recorder, but nowadays, often digital recorders under the card deck) can be worn around the neck or on a belt. Electronic, wire, and recording devices are hidden under clothing. At the end of the test, the electronics and cables are removed, and the recorder is returned to the laboratory for analysis.

Pretest

Arriving at the Holter lab, a technician will place the electrodes (about 50 cm in size) on the chest and attach them to the monitor. Men must shave in small patches to bond the electrodes properly. The technician will describe how to wear the recording device, what to do and what not to do, and how to keep a diary of activities and symptoms. Then patients will be sent on their way.

The test

The usual routine is followed during the test with two exceptions. First, we need to keep the equipment in the hall dry by not taking a shower or having water spilled onto the chest. Second, the patient should keep a diary of all the activities performed and any symptoms experienced while wearing the health monitor. In particular, treatment is most interested in the difficulty of symptoms of lightheadedness, palpitations, synchronization, chest pain, and shortness of breath. The exact time of these symptoms should be compared with the ECG record for the same time.

Posttest

When the test is over, the tools (and diaries) are removed and returned to the Holter lab through the service the lab works with.

Cardiac stress test

A strain test shows how the coronary heart works throughout physical activity. Because exercising makes the coronary heart pumping faster and more difficult, an exercise stress test can monitor problems with blood float inside the coronary heart. A pressure test usually involves walking on a treadmill or driving a stationary motorbike while coronary heart rhythm, blood strain, and respiration are monitored.

A doctor may advocate a strain to observe for symptoms or signs and symptoms of coronary artery disease or an irregular coronary heart rhythm (arrhythmia).

This test is done when one or more of the following are observed:
➤ Acute chest pain in patients excluded for acute coronary syndrome (ACS)
➤ Current ACS handled without coronary angiography or incomplete revascularization
➤ Acknowledged coronary artery disease with worsening signs
➤ Previous coronary revascularization
➤ Valvular coronary heart ailment
➤ Certain cardiac arrhythmias to assess chronotropic competence
➤ Newly identified heart failure or cardiomyopathy

The cardiac stress test (also known as a cardiac diagnostic test, cardio-pulmonary mobility test, or short cardiopulmonary stress test) is a cardiac test that measures the heart's ability to respond to external stress within a controlled clinical environment. A stress response is triggered by exercise or intravenous pharmacological stimulation. The cardiac stress test compares coronary circulation, but the patient loses circulation during most cardiac exercises, reflecting abnormal blood flow to the myocardium (myocardial tissue). The results can be interpreted as a reflection of the general physical condition of the test patient. This test can be used to diagnose coronary artery disease (also known as ischemic heart disease) and evaluate patient diagnosis after myocardial infarction (heart attack). Exercise-inspired stressors practice on a treadmill or a practice bike ergometer row. The level of tension gradually increases with increasing difficulty (decreasing treadmill risk or ergometer resistance) and increasing speed. People who cannot use their legs can practice with bicycle-like cranks bent at their arms. After the stress test is over, the patient is usually advised to gradually reduce the exercise intensity to a few minutes instead of stopping the diminishing activity. The test administrator or attending physician will examine the symptoms and response to blood pressure. Patients can connect to an ECG to measure the heart's response to stress. In this case, the test is commonly referred to as a cardiac stress test, but it is also known by other names such as exercise test, stress test treadmill, exercise tolerance test, stress test, or stress test ECG. Alternatively, the stress test uses echocardiography for ultrasound imaging of the heart (in this case, the test is called an echocardiography stress test or strain resonance) or a gamma camera (called a nuclear stress test).

Echocardiography may involve stress tests. Echocardiography is performed before and after exercise to compare structural differences. Echocardiography at rest is taken before stress. The images obtained are similar to those obtained during full-surface echocardiography, commonly known as transthoracic echocardiogram. Patients express stress in the form of exercise or chemically (usually dobutamine). After achieving the target heart rate, a "stress" echocardiography image was taken. The two echocardiography images are compared to assess abnormal movements of the heart wall. Obstruction is used to detect coronary artery disease.

Nuclear stress test

The most well-known example of nuclear pressure testing is myocardial perfusion imaging. Radiotracers (Tc-99 sestamibi, Myoview, or thallous chloride Tl-201) are usually injected during the test. After a waiting period sufficient to ensure proper distribution of the radiotracer, a scan is taken with a gamma camera to capture blood flow images. Scans obtained before and after exercise are examined to determine the condition of the patient's coronary arteries. Nuclear pressure reduction more accurately identifies local areas of blood flow and indicates the relative number of radioisotopes in the myocardium. The potential for exercise stress and heart damage during the test is patients with ECG abnormalities at rest or problems with rapid movement. Pharmacological stimulants from vasodilators such as dipyridamole or adenosine or positive change agents such as dobutamine can be used. Examiners may include cardiologists, nuclear medicine doctors, nuclear medicine technicians, cardiologists, cardiologists, or nurses. Normal doses of radiation obtained during this process can range from 9.4 to 40.7 ms.

The American Heart Association recommends testing ECG treadmills as a first choice for patients at risk for coronary heart disease, depending on the risk factors for smoking, coronary stenosis, high blood pressure, diabetes, and a family history of high cholesterol. In 2013, the American Heart

Association demonstrated in its Test and Training Practice Standards that high-frequency QRS complex analysis during ECG treadmill testing is an experimental demonstration that helps detect coronary heart disease. Perfusion stress test (using ^{99M}Tc-labeled technetium) is suitable for certain patients, especially those who use abnormal electronic cards, intracoronary ultrasound, or angiography may provide more information about the risk of complications related to cardiac catheterization.

Types of stress testing

Stress test with exercise

A stress test, otherwise called an exercise test or treadmill test, is used by specialists to discover how well a patient's heart functions during physical movement. The stress test can likewise enable a specialist to educate patients about the best kind of physical movements for them. A stress test typically includes walking on a treadmill or riding a fixed bicycle while the patient's breathing, pulse, and heart rhythm are observed. A few patients do not have the option of doing the exercises involved with an activity stress test—for instance, those with joint inflammation. Rather than working out, those patients can take medication to make the heart work more vigorously, as it would during exercise.

A pressure test using a treadmill has the following steps:
➤ The patient is attached to the gadget that screens the heart
➤ A blood pressure cuff is folded over one upper arm
➤ The patient stands on the treadmill
➤ The treadmill begins to move, and the patient strolls gradually
➤ The treadmill speeds up in step-by-step increments
➤ The treadmill may go into a tough, or grade, position
➤ The patient inhales into a cylinder for a moment or two
➤ Subsequent to backing off the speed, the patient gets off the treadmill, waits for a couple of moments, and then rests
➤ The clinical group takes further readings of blood pressure, etc.

Stress test without exercise

A medical attendant completes an activity stress echocardiogram by making an image of the heart during physical effort. The fundamental instrument used is a transducer. Transducers radiate a high-recurrence sound wave that frames an image of the heart once the echoes arrive at the organ. Bats and dolphins explore their environments using a comparable rule. The sonographer (master ultrasound specialist) rubs a gel onto the chest and then moves the transducer within the gel to obtain the clearest image. Some sonographers demand that the patient lies on one side for the test. Others use an intravenous dye to upgrade the clearness of the image. Assuming this is the case, the dye is infused into the circulatory system during the test.

Nuclear stress test

If the initial symptoms continue or deteriorate, the specialist may suggest a nuclear stress test, which presumably provides a more detailed and exact appraisal of the patient's heart. The cycle is equivalent, but a dye is infused into the patient's arm to show the heart and bloodstream on the image. It will likewise show the regions of the heart where blood is not flowing, which may indicate a blockage. This can be discovered by x-beam, single-photon emission computed tomography, or cardiac positron emission tomography (PET).

Two arrangements of the image are taken, each lasting 15—30 min. The primary is taken soon after working out, and the subsequent is taken when the body is very still, either soon thereafter or the following day. This permits a specialist to think about how the heart looks, its regular capacity, and its capacity under pressure. The test takes from 2 to 5 h.

Skin-resistance measurements

Skin resistance is a vital index in affective mental research and could mirror the exchange of emotion by recording its changes. The pore and skin-resistance measurement technique connects the electrode sensor to two hands adjoining the item to document the resistance signal. The behavioral response of the skin, also known as the electrical response of the skin (an older term was "electronic skin reaction"), gives the skin better electrical conduction with external or internal stimuli. Awakening is widely regarded as one of the two main aspects of broad sound and sensory response to activation as a whole. Measuring tension is not the same as measuring emotion, but it is an important issue. Excitement has been shown to be a powerful predictor of attention and memory.

Stimulants sensitive to the movement of the skin are varied and include novel events that are significant or intense. High excitement levels decrease when a person sleeps and during active situations such as anger and mental stress. The excitement level spikes before gradually decreasing in mental stress tasks such as solving math problems (if not particularly difficult). This is because various events can improve skin function (strong sensations, touching events, demanding features, etc.). It is impossible to know why outsiders have made our galaxy shiny unless we participate in a highly controlled experiment.

Most people have no mental control over their skin response in any kind of quick and accurate manner. It may take some time for the reaction to occur and collapse. But we can try exciting and great ideas to see how Galactic Tutor responds. Once dialed to the appropriate baseline, the garage beta should be sensitive enough to respond to widely generated images.

Only two places are widely believed to be easy and reliable for measuring skin flow response, the palms and the soles of the feet. These locations are densely packed with eccrine sweat glands that respond to sensory and other psychological stimuli. In any of these cases, conductivity is measured by placing two electrodes on the side of the skin and passing a small charge between the two points. When things become worse, the skin quickly turns into a somewhat improved electrical conductor. One can then measure and communicate this response.

The conductance response of the skin is measured by the eccrine gland, which covers most parts of the body and is especially dense on the soles of the feet and palms (these differ from the apocrine sweat glands, which are found primarily in the armpits and genitals). The merocrine glands in the palms and the soles of the feet are highly sensitive to important stimuli, and a measurable response has been found before the presence of sweat.

However, if the body is too hot and sweats too much, the overall level of skin flow increases. If necessary, we can reset the baseline. However, skin conductivity is not solely temperature based. In addition, our mobility is higher when our hands are cold (e.g., when nervous) and slower when our hands are warm (e.g., when adequately rested). The garage beta is more like a so-called "mood ring" than just a temperature sensor.

Experimental evidence proves that some individuals have characteristic patterns of electrical baseline activity in contrast. Each has a different baseline, but some have relatively slight differences with comfort and have signs of skin conductors not stimulated by external phenomena or internal thoughts. Individuals in this category are often identified as having puncture wounds. Alternatively, some people's skin has many driving reactions both at rest and in the absence of external stimuli. People with this pattern are known as workers. These two physical predictions have encouraged the first attempt to relate personality styles.

There are a couple of possible reasons why your gal activator may not be glowing. Here is a checklist:

a. Poor communication—make sure the gloves are worn properly. Place it on the left side, cross the index finger through the small hole, insert the thumb through the large hole and attach the Bluetooth Low Energy (BLE) Sense star to the back of the hand. Adjust so that the electrons are flat against the skin. Rotate clockwise until the dial is closed so that the light is as bright as possible.

b. Dry electron—there is a practice of using a gel between the skin and the electrodes in scientific experiments. It turns out that electronic gels do not usually need to be used with gal activators. However, it is helpful to apply a drop of water or saline on the electrodes in many cases to improve communication.

c. Low baseline—of course, some people have very little baseline. I've tried to consider the maximum baseline range on a universal device, but yours may be below the range. Take a deep breath and try to hold it for 3 s. This often enhances the skin conductance response to the level seen on the device. Balloons will maximize our level to "pop."

If we try to address the above suggestions and the cheek activator does not work, there may be a loose wire or other problems. In this case, please feel free to bring it to the Galva Activator Booth or Media Lab students.

The startle reflex is a specific spike in electrical skin response and usually occurs 2—3 s after starting a startle probe or a new stimulus. Shocking probes are usually designed for "amazing" strong, powerful lights or air puffs or drawing sudden attention to a subject. Similar conduction spikes can occur when the subject is "tuned" into a new stimulus. For example, if someone suddenly stands beside us.

Habituations are an event that occurs when many investigations begin at an early stage. The body reacts intensely at first but then continues to react with less intensity for testing. However, after a certain amount of stimulation (which separates from the subject), the body stops responding to the flashing probe, resulting in a characteristic increase in galvanic skin response (GSR). However, when the subject recovers for a while, the startling response is reactivated. The response of the elite has been seen as a very different feature in several sections of schizophrenia.

The Galva activator was invented by Jocelyn Scheirer and Rosalind Picard of the MIT Media Lab. Circuit components were applied by the Media Lab's Dana Kirsch and Blake Brasher. The gloves were designed in collaboration with Philips Electronics. The Philips wearable electronics team was led by Nancy Tilbury (fashion designer) and Johnny Farrington (senior scientist) with the fashion elements of the design process.

Human skin provides some resistance to current and voltage. This resistance depends on the sensitive condition of the body. The skin-resistance tester circuit proposed here measures changes in skin resistance after a change in mood.

The skin's resistance to rest at very high stress levels is more than 2 megohms and less than 500 kilo-ohms. Decreased skin resistance after physiological changes during high pressure is related to blood flow and permeability. It improves the electrical conductivity of the skin. This circuit helps to monitor the skin's response to relaxation techniques. It is extremely sensitive, reacts to moments of sudden stress, and responds to deep AH circuits.

Skin-resistance meter

In 1849, du Bois-Reymond of Germany first saw that human skin was electrically active. He immersed the subject's limbs in a zinc sulfate solution and found that an electric current flowed through the limbs when muscles contracted and relaxed. Therefore, he believed that muscular events were responsible for his electrodermal activity (EDA) observations. Thirty-nine years later, in 1878 in Switzerland, Hermann and Luchsinger showed a relationship between EDA and sweat glands. Hermann later showed that the electrical effect on the palate was the strongest, indicating that sweating was an important factor.

Vigouroux (France, 1879) treated patients with emotional disorders and was the first researcher to associate mental activity with EDA. In 1888, a French neurologist proved that sensory stimuli could alter skin resistance and inhibit drug activity. In Russia in 1889, Ivan Tarkhanisvil developed a meter to observe changes in the electrical capacity of the skin in the absence of external stimuli and to observe changes in real time. Scientific research on EDA began in the early 1900s. One of the first references to EDA instruments was in the study of sound analysis by C.G. Jung, published in 1906. Jung and his colleagues used a meter to verify the sensitivity of patients listed by word during word association. The controversial Austrian psychologist Wilhelm Reich studied the concept of EDA in 1933 and 1935 at the Institute of Psychology at the University of Oslo. We have confirmed the existence of discharges. By 1972, more than 1500 articles on the electrical activity of the skin had been published in professional journals, and today EDA is considered the most common method for investigating human psychological phenomena. As of 2013, EDA still monitors the growth of clinical applications.

A GSR sensor measures sweat gland activities associated with emotional arousal. To measure GSR, we take benefit of the electrical residences of the pores and skin. In particular, pores and skin resistance vary with sweat gland activity, i.e., the greater the sweat gland activity and extra perspiration, the less the pore and skin resistance. The maximum common measure of a GSR sign is not resistance but conductance. Conductance is the alternative of resistance and is measured in siemens (conductance = 1/resistance). The conductance makes the signal interpretation less difficult because the greater the sweat gland activity, the better the skin conductance. The most commonplace method to measure a GSR sign for emotional research functions is primarily based on a consistent voltage gadget. The GSR sensor applies a consistent voltage-typically 0.5 V to the two electrodes that might be in contact with the pores and skin.

Skin sensitivity is not consciously controlled. Rather, it is replaced by autonomous activity that drives human behavior and cognitive and sensory states at the subconscious level. Therefore, skin conduction provides direct insights into the psychological control of the autonomic nervous system. Human organs, such as the fingers, palms, and third parts of the feet, exhibit various bioelectric phenomena. They are an EDA meter, which shows electrical changes in conductivity between two points over time. Two errors go all the way through the body over the surface. In active measurements, small amounts flow through the body. In some studies, instead, current reactions to human skin include recent carcasses.

Although electrical changes cannot distinguish any particular emotion, there is a link between sensory arousal and sympathetic nerve activity. These autonomic and sympathetic changes alter sweat and blood flow, affecting GSR and galvanic skin potential (GSP). The number of sweat glands varies from body to body and is highest in hand and foot areas (up to 200 cm per sweat gland 2). A properly tuned device can record and display subtle changes.

Changes in the joints between the electrical resistance and electrical potential of the skin produce the skin's electrical activity. Galvanic skin resistance is an ancient term for the electrical resistance recorded between two electrodes when very weak currents flow rapidly between the two electrodes. The electrodes are usually placed about 1 inch apart, and the recorded resistance depends on the sensitivity of the subject. GSP refers to the voltage measured between two electrodes without external currents. Measure by connecting electronically with a voltage amplifier. This voltage also varies depending on the sensitive state of the subject.

EDA is a general measure of the activity of the autonomic nervous system and has been used for a long time in the history of psychology. It is used to assess the nervous state of the body. EDA is also studied to determine pain in premature infants. EDA monitoring is often combined with recordings of heart rate, respiratory rate, and blood pressure. This is because they are all autonomous dependent variables. EDA measurements are a component of modern polygraph devices and are often used as false detectors. EDA reflects by gradually altering the sympathetic activity of the tonic and gradually increasing the sympathetic activity. Tonic activity can be expressed in electronic station level units and step-by-step changes in electrodermal response (EDR) units. Gradual change (EDR) is a short-term change in EDA that manifests itself in response to a specific stimulus. EDR may appear spontaneously without external stimuli. This type of EDR is called "redundant EDR." Skin conductance response (SCR) gradually becomes effective when investigating polygamous transverse processes. Tonic change is based on gradual parameters. Spontaneous variants of unnecessary EDR can be used to assess tensile EDA. In particular, it uses the frequency of "redundant EDR" as an indicator of EDA over a period of 30–60 s. The tonic EDA is considered useful for general stimulation and exciting testing.

EDA is a general measure of the activity of the autonomic nervous system and has long been used in psychological research. The EDA is a sensitive psychological measure of autonomous change. EDA is also being studied as a way to determine the pain of premature infants.

EDA monitoring is often combined with recordings of heart rate, respiratory rate, and blood pressure. This is because they are all autonomous dependent variables. EDA measurements are a component of modern polygraph devices and are often used as false detectors. The ammeter is a custom EDA measuring device used by the Church of Scientology as part of its "monitoring" and "safety check" exercises.

Human skin offers some protection from current and voltage. This obstruction changes with the passionate condition of the body. The skin-resistance meter measures changes in our skin obstruction following changes in our psychological state.

In a relaxed state, the obstruction offered by the skin is as high as 2 megohms or more, which lessens to 500 kilo-ohms or less when the emotional stress is excessively high. The decrease in skin resistance is identified with expanded bloodstream and porousness followed by the physiological changes during high pressure. This builds the electrical conductivity of the skin. This circuit is valuable to screen the skin's reaction to unwinding methods. It is delicate and shows reaction during abrupt moments of stress. Indeed, even a profound sigh will give a reaction in the circuit.

Galvanic skin response

When a weak electric flow is consistently applied between two cathodes set about an inch apart on the palm, the recorded electrical obstruction between them, alluded to as the GSR, shifts in agreement with the enthusiastic condition of the subject. The physiological premise of the GSR is a change in autonomic tone, to a great extent thoughtful, happening in the skin and subcutaneous tissue in light of a change in the emotional condition of the subject. Changes in fringe autonomic tone change perspiration and the cutaneous bloodstream, which thus change GSR. As for substantial tangible upgrades (e.g., torment, pressure, contact), changes in feeling evoke changes in peripheral autonomic tone and subsequently the GSR. A basic model is the vasodilation of cutaneous veins of the face (becoming flushed) and expanded perspiration that regularly happens in the enthusiastic condition of shame.

The time course of the GSR signal is the consequence of two added substance measures—a tonic base level driver, which varies gradually (seconds to minutes), and a quicker differing phasic segment (fluctuating in practically no time). Changes in phasic action can be recognized in their constant information stream, as these blasts have a lofty grade to a particular pinnacle and a moderate decrease comparative to the baseline level. Scientists center around the dormancy and amplitudes of the phasic overflows concerning upgrade beginning when exploring GSR signal changes because of tangible improvements (pictures, recordings, sounds). When there are critical changes in GSR movement in light of an upgrade, it is alluded to as an event-related SCR. These reactions, also called GSR peaks, can give data about emotional arousal to stimuli. Different peaks in GSR action not identified with the introduction of an improvement are alluded to as non-stimulus-bolted SCRs. Using the skin conductance values or the quantity of GSR peaks, it is conceivable to add quantitative information to investigations of passionate excitement. It is easier to reveal new discoveries and reach new conclusions about human conduct with more information within reach.

Patient monitoring systems

The continuous patient parameters measured include coronary heart price and rhythm, respiration fee, blood strain, blood oxygen saturation, and many others. Instantaneous selection-making is important for effective patient care; electronic monitors are used regularly to gather and show physiological statistics. Increasingly, such statistics are collected using noninvasive sensors from much healthier patients with a health center's medical-surgical gadgets.

The inclusion of telecommunication devices in health care alleviates the problem experienced using medical examiners to monitor patients concurrently. It permits them to study sufferers without being physically present at their bedsides, be it inside the health facility or in their homes. Several sensors are used in devices to reveal an affected person's vitals ranging from heart rate, temperature, ECG, respiration, noninvasive blood stress, oxygen saturation, and so on. The deployment of wireless fitness tracking removes the geographical boundaries to receiving specialist care. Currently, wireless fitness video display units transmit vital physiological signs to clinical employees and simplify the size and increase the tracking performance of sufferers. Continuous monitoring of the affected person's health is vital during treatment. Consequently, the Wi-Fi health tracking device performs a major function in delivering high-quality care even for patients in rural regions.

Patient monitors measure, record, appropriate, and show mixes of biometric qualities, for example, pulse, SpO^2, circulatory strain, temperature. High-ability, multiwork screens are normally used in emergency clinics and centers to guarantee a significant level of value-tolerant consideration. Versatile patient screens are intended to be minimal and power proficient. This permits them to be used in far-off regions or by paramedics to help determination in the field as well as empowering observation and communicating information to medical care suppliers in different areas.

Microchips offer an expansive arrangement of exceptionally coordinated microcontrollers, designs innovation, programming libraries, and availability arrangements. This wide contribution additionally empowers advancement in the plan of independent biometric gadgets, such as circulatory strain screens, beat oximeters, ECG/EKG, and others. Computer chips convey arrangements that help in quiet observation capacities such as low-noise analog signal conditioning, touch-sensing technology,

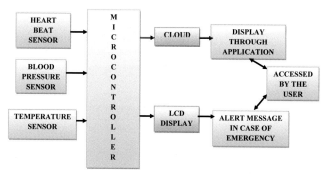

Figure 2.2 Block diagram of a patient monitoring system.

LCD control, wired and wireless connectivity, motor control, and high-speed memory.

Fig. 2.2 shows the block diagram of a smart patient health monitoring system.

Five classifications of patients need physiological observation:

➤ Patients with insecure physiological administrative frameworks; for instance, a patient whose respiratory framework is smothered by medical sedation

➤ Patients with a presumed dangerous condition; for instance, a patient with respiratory failures

➤ Patients at great danger of building to a dangerous condition; for instance, patients following an open-heart medical procedure or an untimely baby whose heart and lungs are not completely formed

➤ Patients in a basic physiological state; for instance, patients with various injuries or septic shock

➤ Mother and child during labor and delivery processes

Classes of patient monitoring systems

There are two types of patient monitoring systems:

➤ Single-parameter monitoring system—this system measures the blood pressure in a human body, electrocardiography testing, oxygen saturation in the blood, etc.

➤ Multiparameter patient monitoring system—this system is used in several critical physiological signs of patients. It shows various vital parameters of the human body such as electrocardiograph, respiration rate, and blood pressure. This system plays an important role in the field of healthcare devices.

Current parameters used in patient monitoring systems are

➤ ECG 3/5/10 leads
➤ Respiration
➤ Invasive blood pressure
➤ Noninvasive blood pressure
➤ Pulse oximeter

Some future advanced trends in patient monitoring systems are

➤ Blood gas analyzer
➤ Drug dosage calculator
➤ Drug management system
➤ Real-time patient location system
➤ Wearable patient monitoring system
➤ Telemetry/telemedicine

Remote patient monitoring

Numerous individuals comprehend that distant patient observation permits medical care experts to obtain vitals and other significant information from patients being checked from the comfort of their own homes. These patient screens help doctors reduce medical clinic readmissions and maintain strength for patients who experience the ill effects of a ceaseless sickness by maintaining steady correspondence between patient and doctor.

Distant patient observation accomplishes this compelling correspondence through a particular arrangement of capacities:

➤ Collection—once the client enacts the patient screen, data are gathered, put away on inside frameworks, and organized for transmission to the patient's medical care doctor.
➤ Transmission—the information gathered is then sent to the medical care office through the web, telephone, text, or another type of correspondence.
➤ Assessment—internal algorithmic programming and medical care experts will look at the information and demonstrate any territories of concern. Data relative to these worries is then sent back to the patient or the patient's medical care group.
➤ Notification—if the doctor finds an intermittence in the information that requires prompt consideration, cautions will be sent to assigned crisis responders.
➤ Activity—a clinical crisis group will help the patient clarify what has occurred and share approaches to forestall a comparable future episode.

The innovation engaged with this correspondence has definitely improved recently after quite a long while. Ongoing distant patient screens are presently ready to send exact information all the more productively, take clinical readings through less obtrusive techniques, and dissect continuous information to distinguish anomalous readings quicker and all the more precisely.

Remote patient monitoring (RPM) and home monitoring are a set of remote patient monitoring technologies and practices that enable healthcare providers to remotely detect real-time changes in patient health data and use them in treatment plans. It is an integral part of the broader telecommunications industry and the e-health domain. It is an integral part of the broader telecommunications industry and the e-health domain. RPM focuses on patients with limited access to treatment facilities for chronic diseases, postoperative patients, elderly patients, and patients in rural areas. Recent studies have shown that remote observation is possible if applied to the target range and can lead to substantial changes in some areas:

a. Reading reduction of 38%
b. Less than 25% emergency room inspections
c. Increase patient satisfaction by 25%
d. Care costs reduced by 17%
e. Drug compliance improved by 13%

Given the potential benefits, it is not surprising that about 88% of healthcare providers have already invested in or are considering RPM technology. These technologies must create an integrated system that covers the entire RPM cycle to reach their goals. It focuses on patients with limited access to healthcare facilities, postoperative patients, elderly patients, and rural areas. Recent studies have shown that remote observation is possible if applied to the target range.

The remote patient monitoring system can be considered a kind of Internet of Things system. Commonly, there are four main components:

a. Personal medical device with Bluetooth module
b. Patient mobile applications
c. Cloud storage
d. Hospital software

Modern remote patient monitoring devices receive a great deal of health information from heart rate to blood pressure without the supervision of a healthcare provider. They can be embedded under the skin or in user-friendly wearables such as the Apple Watch or Fitbit. The new FDA policy applies to the following types of RPM devices:

a. Electronic thermometers
b. Electrocardiograph
c. Electroencephalograph

d. Heart monitors

e. Apnea monitors

f. Important power meters

g. Oximeters

h. Audiometers

i. Blood pressure monitors

j. Breathing frequency monitors

k. Electronic stethoscopes

All wearers must somehow send patient data to their healthcare provider. Healthcare providers often provide dedicated mobile applications. Many technologies can transmit data from RPM devices to smartphones, but BLE communication is the most common.

Patient-targeted mobile applications collect data from wearable and smart sensors and transmit them to clinical staff. This integral part of the RPM system meets the following criteria:

a. BLE support for data exchange with devices

b. Compliance with IEC 62304 safety standards that specify development requirements for medical software and medical device software

c. HIPAA compliance to safeguard protected health information

d. Protection against connectivity issues, caching processes, etc.

e. Integration of healthcare provider systems through secure APIs built to FHIR industry standards ensures the distinction between different healthcare software products.

The application provides a useful view of the presented data, educational information, treatment surveys, medication prescription reminders, and physician—patient interoperability tools (video, audio, or chat) to attract patients.

Cloud storage

Cloud archives (databases) typically retrieve raw patient-generated data from mobile applications. However, some systems provide devices with a direct connection to the cloud. That is, all the captive information is collected in the cloud. In this case, the patient does not need to download the dedicated mobile application.

Hospital applications adjacent to the hospital

Like inpatient apps, hospital applications must comply with HIPAA standards and interdisciplinary medical rules. In addition, the RPM web solution should be integrated with the hospital's EMR system to exchange information through FHC-based APIs and avoid creating data silos.

Decision support module—vital signs data obtained from the repository are compared with doctor-prescribed thresholds.

Report module—the system stores all measurements and manual entries daily in the appropriate report and sends them to the doctor.

Notification module—when the decision support module detects the warning signal, the notification module is triggered, a warning message is generated, and the doctor is notified of the abnormality via SMS, email, or in-app notifications.

Analysis module—the module uses BI tools and data visualization techniques to demonstrate patterns and trends and provide real-time measurements, which helps physicians anticipate dangerous events and make informed treatment decisions.

Home monitoring systems for seniors

By 2060, the US Census Bureau predicts that those aged 65 or older will represent a quarter of the country's population. As a result, demand for technologies that support home care will continue to grow. Several solutions designed for the elderly have already arrived to improve quality of life, reduce caregiver stress, and reduce treatment costs. The ultimate view system includes visual sensors, cloud servers, and mobile applications for Android and iOS. The sensor runs an intensive learning algorithm that analyzes the activities of the elderly. If a person leaves a "safe zone" or an emergency arises, the smart device sends a notification to the family and the care provider. The sensors also include face recognition, daily activity statistics, and risk assessment. Grand care systems provide remote monitoring solutions for home healthcare providers, the elderly community, service providers with disabilities, and families. The core of the system is customizable touch screen technology with video, drug management, rule-based alerts, and other services. Alternatively, sensors can be added to alert a caregiver if something goes wrong.

Cardiac monitoring systems

Heart disease is a leading cause of death worldwide. Not surprisingly, many organizations prefer distant cardiac monitoring over other approaches. Birdie DX integrates a thin patch that captures ECG data with the patient management portal. The patch is placed just above the chest and records specific low-amplitude signals to create the specific P-waves needed to diagnose arrhythmias accurately. Biotricity's Bioflux is an FDA-approved cardiac telemetry device with a built-in algorithm for ECG data analysis. Data are sent to the secure cloud immediately, allowing healthcare providers to access data through their existing systems. In addition, hospitals may use branded Bioflux software, which is an online portal for physicians and a mobile application that alerts physicians to dangerous events and allows them to view reports.

Chronic and post-acute-care management systems

These measures help patients with heart disease, chronic pain, chronic vascular pulmonary disease (COPD), cancer, or COVD-19 spend more time at home without going to the hospital, even those who have had surgery. Biofourmis helps physicians remotely track patients' vital symptoms through the clinic's mobile app or web dashboard. The brand's armband biosensor, Avrion, captures blood oxygen, blood pressure, body temperature, and many more physiological parameters 24 h a day, 7 days a week. Patients report their symptoms to the care group via a mobile app.

The system applies machine learning algorithms to combine data from sensors with the patient's medical history to create unique real-time profiles. This profile is used by the conscious physician to predict unintended consequences and suspicious events. Monitoring is usually continued for 3—6 months after discharge. CAREMINDr helps clinics of all sizes launch RPM programs. It provides an easy-to-use mobile app for monitoring and managing risky situations such as diabetes, high blood pressure, asthma, COPD, and COVID-19. Patients use their smartphones to access data from their home medical devices (such as blood glucose meters, thermometers, and blood pressure monitors). The application supports billing and enforces CPT codes for long-term care and remote patient monitoring.

Data health is a scalable mobile platform that is easily integrated with multiple wearable and treatment devices. Important signals from sensors detect and predict failures by mixing patient EHR data. The system provides a personalized mobile application with a cloud-based interface for patients and care teams. The hospital department can also deploy its own patient mobile application. The system provides other services such as text messaging and video calling.

Huma (formerly Medop) is a modular solution for identifying symptoms and can be integrated with more than 400 medical devices to assemble the required matrix for specific cases. Patients must download mobile applications to capture vitals, manage their conditions, and communicate with healthcare professionals. This application has been integrated with hospital platforms for patient management. Huma also provides a specific application for monitoring the symptoms of COVID-19.

Guidelines to painless remote patient monitoring implementation are the following:

1. Whether choosing an off-the-shelf full-stack solution, building components from different vendors on your system, or developing customized software to suit your needs, it is easy to switch to remote care. These changes include multiple therapies for physician and patient education, changes in clinical workflow, and integration of all hardware and software into the interface. You can reduce effort and expense with this global process.

2. Start with the pilot—not all RPM programs show success, even during epidemics. Gradually testing the system with some participants to avoid costly mistakes proves that the initiative yields better clinical results.

3. Do not push for perfection—focus on the actual needs of those who use your system. Good product design is enough for them. One may retain some improvements or additional features for later use.

4. Create a video tutorial or run a training webinar—before launching RPM, you must ensure that your patients and therapists are properly educated about how to use the system and understand the data. Invest in detailed video tutorials and manuals. You can also provide a training webinar for your staff.

5. Think ahead about access to technical support—it can be provided by any system vendor, hospital IT department, or third-party IT specialist. However, it is important to provide as much continuous technical support as possible. Otherwise, the RPM will not work. If technical aspects become a problem for users, patients will stop sending data, and doctors will stop the "prescription" of remote observation.

Find a reliable technology partner—for healthcare providers who lack technical knowledge, the safest option is to establish long-term partnerships with IT consultancies with healthcare software expertise.

CHAPTER 3

Transducers and amplifiers

Sudip Paul[1], Angana Saikia[1,2], Vinayak Majhi[1] and Vinay Kumar Pandey[1]

[1]Department of Biomedical Engineering, School of Technology, North-Eastern Hill University, Shillong, Meghalaya, India; [2]Mody University of Science and Technology, Laxmangarh, Rajasthan, India

Contents

Introduction to Biomedical Instrumentation and Its Applications
ISBN 978-0-12-821674-3
https://doi.org/10.1016/B978-0-12-821674-3.00008-5

Sensors

In its broadest definition, a sensor is a gadget, module, machine, or subsystem that distinguishes occasions or changes in current circumstances and sends the data to other hardware, usually a PC processor. A sensor is constantly used with other gadgets.

Sensors are used for ordinary items—for example, contact with delicate lift catches (material sensor), and lights that dim or brighten by contact with the base, as well as endless other uses, the vast majority of which are rarely thought about. With advances in micromachinery and simple-to-use microcontroller stages, sensor employment has been extended beyond the customary fields of temperature, pressure factors, and stream estimation—for instance, into MARG (magnetic, angular rate, and gravity) sensors. Besides simple sensors, potentiometers and power-detecting resistors are also still broadly used. Applications encompass assembling and apparatus, planes and aviation, vehicles, medication, advanced mechanics, and numerous other parts of our everyday life. A wide range of sensors estimate material substances and properties. A couple of models incorporate optical sensors for estimating refractive lists, vibrational sensors for estimating liquid thickness, and an electro-compound sensor for observing the pH of liquids.

A sensor is a system that can identify a specific event or change in the environment and send all of the collected information to an electronic device. A sensor can be a device, module, machine, or subsystem. The sensor can measure the physical and chemical properties of substances. Sensors can be of various types: optical, vibrational, electrochemical, etc.

Basic criteria for good sensor design

 (i) A sensor should be highly sensitive in measuring physical or chemical properties.
(ii) It should be insensitive to unmeasured properties—for example, a temperature sensor's values should not be influenced by humidity.

(iii) And finally, it should not influence the measured properties—that is, a temperature sensor should not produce any heat, because otherwise, the sensor itself could affect the measured object's temperature.

In general, sensors are designed to transform a value with the sensing parameter linearly, although sensor sensitivity is defined as the ratio between the output signal and the measuring parameter. For example, for a temperature sensor that varies voltage in response to changes in temperature, the sensitivity will be the ratio of the output voltage to the temperature:

$$\text{Sensitivity} = \frac{\text{Output Parameter}}{\text{Sensing Parameter}}$$

This sensitivity is the slope of the sensor's transfer function equation. Sometimes we also add or subtract an offset to adjust the initial value of a sensor. If a temperature sensor senses a value from -20 to $50°C$ and varies the output from 0 to 5 V, the sensor must be set so the initial value, $-20°C$, corresponds to 0 V.

A sensor's affectability demonstrates how much the sensor's yield changes when the information amount being estimated changes. For example, if the mercury in a thermometer moves 1 cm when the temperature changes by $1°C$, the affectability is 1 cm/$°C$ (essentially the incline dy/dx assuming a linear function). A few sensors can likewise influence what they measure; for example, a room temperature thermometer embedded in a hot cup of fluid cools the fluid while the fluid warms the thermometer. Sensors are generally intended to find small effects.

Sensor specifications

(i) **Accuracy**. In an absolute sense, how well the sensor measures the environment. How good is the data when compared with a recognized standard? For instance, a temperature sensor of up to $0.001°C$ is expected to agree with a temperature standard within $0.001°C$, such as a triple-point-of-water cell or a temperature measured by a platinum resistance thermometer (PRT) standardized to the recognized calibration standards or with the same accuracy properly calibrated by the second sensor. You may compare the results with other readings.

(ii) **Range**. This refers to the maximum and minimum value range within which a sensor works well. Often sensors work well outside this range but require special or additional calibration—such as a salinity sensor positioned in salinity in ppt at an outfall below where the practical salinity unit (PSU) scale is defined (2 PSUs). Typically, however, if you try to operate a sensor outside its range, it will not work (it will instead provide a constant output at maximum, change the sensitivity significantly, or provide uncertain results) or may be damaged, e.g., 130 m pressure sensor positioned at 200 m depth.

(iii) Resolution. This is the ability of the sensor to see small differences in readings. For example, a temperature sensor can have a resolution of 0.00001°C but only be accurate to 0.001°C. This is why you can believe the size of relatively small changes in temperature that are smaller than the sensor's accuracy. The resolution is often controlled by quantization in digitizing a signal—e.g., 1 bit is equal to 0.0005°C. This is not a function of the sensor but a sampling process.

(iv) Repeatability. This is the sensor's ability to repeat prior measurements when it returns to the same environment. This is often related to the accuracy, but it is also true that a sensor may be inaccurate yet repetitive in making observations.

(v) Drift. This is the low-frequency change in a sensor over time. It is often associated with the electronic aging of components or reference standards in sensors. Drift typically decreases with the age of the sensor as the parts mature. A smoothly flowing sensor can be corrected for drift, such as seabird temperature sensors that flow at around 1 m°C/yr (and easily change over many years), thus allowing for more accurate readings. Drift is also caused by biofuels that cannot be properly repaired, but we often try.

(vi) Hysteresis. A linear up-and-down input for the sensor results in output that supersedes input, e.g., you obtain one curve at increasing pressure and another at decreasing pressure. Many pressure sensors have this problem, and it can be ignored for some healthier people. This is often seen in conductivity, temperature, and depth (CTD) measurement when the pressure on the deck after recovery is different from reading before reading. This is not a problem with the sensor's response time, but some sensors have inherent properties that are undesirable. This can also be a temperature-sensitivity problem in CTD.

(vii) Stability. This is another method of fast flow. That is, you always obtain the same output from a given input. Drift and short- and long-term stability are ways of expressing sensor noise as a function of frequency. Sometimes this is expressed as a guaranteed accuracy over a certain period. Drift is often a problem with pressure sensors at high pressures. All sensors flow over time—hence, the standardization of PRT in triple-point-of-water and gallium melt cells.

(viii) **Response time**. This is a simple approximation of the frequency response of a sensor assuming an exponential behavior. We will discuss it in more detail below.

(ix) **Self-heating**. To measure the temperature the resistance in the thermistor, we must put a current through it. Flow through a resistor causes heat dissipation in the thermistor, causing it to heat up or self-heat. This is particularly important in temperature measurement. If the velocity of water changes, the amount of preventive cooling will change, and temperature sensing will change as a function of the velocity—anemometer effect.

(x) **Settling time**. Time to reach the static output of the sensor once it is turned on. Therefore, if you are conserving power by turning the sensor off between measurements, you will need to turn on the power and wait for the sensor to reach stable output.

Variance properties of sensors

In general, sensors are not identical to others in their category. Like the characteristic graph of a similar kind, temperature sensors will always vary from one another. Depending on the variation type, they can be categorized by different factors, and all these factors are considered the error factors of the sensors.

(i) **Sensitivity error**: Sensitivity error is the most common error for any kind of sensor. This error is measured by the deviation of the actual sensitivity slope versus the ideal sensitivity curve.

(ii) **Noise error**. Noise error is defined by the random variance in sensor output over time.

(iii) **Drift error**. Sensor output that slowly changes independent of the measured parameter is called drift error. In general, it is caused by the materialistic deterioration of the sensors. This error is a slowly processed error that may take months or years to develop.

(iv) **Hysteresis error**. The output error is called hysteresis error when a sensor value is affected by the previous input value. In other terms, we can say that hysteresis error is the deviation between two output values taking in a specific input value once during increasing order and another in decreasing order. This error mainly occurs from the use of magnetic or elastic materials in sensors.

Table 3.1 Differences between sensors and transducers.

Properties	Sensor	Transducer
Definition	The sensor can sense physical changes in the surrounding environment, convert them, and provide readings in the same format.	The transducer senses the signal in one form of energy and converts the signal into another form, usually from electrical to nonelectrical and vice versa.
Component	Sensor itself	Sensor and signal conditioning
Function	It senses physical changes and corresponding electrical signals	Can convert energy from one form into another
Applications	Patient monitoring, liquid dispensing in drinking machines, infrared toilet flushing	Steering systems on vehicles, engine controls, HVAC monitoring, ramp or bridge lifting systems
Examples	Motion sensor, pressure switch, proximity sensor	Microphone, pressure transducer, linear transducer, potentiometer

(v) Quantization error. This kind of error results from analog-to-digital signal conversion. During the conversion from analog to digital, some data loss occurs during bitwise conversion. Initially, the error is varied on the resolution bit of the conversion. The higher bit resolution has a higher sample rate for the quantification of an analog value. For example, a 10-bit converter for 0−5 V can fragment the voltage into 1024 samples, so each sample will have 0.00488 V. This sample value is the multiplicative quantization step through the path of 0−5 V. The process will be like 0 V, 0.00488 V, 0.01464 V, 0.0152 V,… 5 V. Among these, any adjacent values will be ignored or mapped to the nearest higher or lower value. For example, 0.01469 will be mapped to 0.01464, the closest of the two low and high values, 0.01464 and 0.0152.

(vi) Max−min range error. Each sensor has its own measuring limit. This range is called the dynamic range of the sensor. If a sensor crosses this scale for either the maximum or the minimum value, the sensor's output may not vary with the input value or may provide a garbage value. This type of error is called max-min range error.

(vii) Dynamic error. Dynamic error results from rapid changes in the output signal within the dynamic sensing range over time.

(viii) Aliasing error. This type of error also occurs during digital signal conversion or signal transmission where the signal may not be distinguishable. This may occur due to the influence of the sampling frequency. It may also result from signal distortion during the reconstruction of the converted signal. Sometimes noise frequency may be matched to a multiplication of the sampling frequency and create errors periodically to the output signal.

(ix) Offset error. If the output signal has a constant value error for any input value in the dynamic sensing range, the value is called offset error. It is also called y-intercept error.

(x) Nonlinearity deviation. This error is measured by the sensitivity difference between ideal and actual.

All of these types of errors may occur in any sensor. No sensor can be ideally made. To minimize these errors, several procedures may be incorporated depending on the error types. In general, calibration is the most common procedure to make the sensor work efficiently within the desired range. In addition, various signal-processing tools and equipment can be used to minimize errors.

Transducers

A transducer is a device that converts one form of energy into another. Usually, the transducer receives a signal in one form of energy and converts it into another. This device senses physical measurement and then converts it to another—for example, nonelectrical to electrical. Transducers are usually used for automation, control systems, and measurement purposes where electrical signals are converted into many physical quantities like energy, force, torque, light, motion, and position.

Transducers are primarily of two types:

(i) Mechanical transducers—these transducers convert physical quantities into mechanical quantities. For example, a thermocouple that produces a temperature-dependent voltage and does not require external power.

(ii) Electrical transducers—these transducers convert physical quantities into electrical quantities. For example, a linear variable differential transformer, or linear variable displacement transducer (LVDT), converts the position from a mechanical reference into an electrical signal (Table 3.1).

Classification of sensors

Depending on physical laws or convenient specific properties, sensors can be classified into different categories: active and passive, contact and noncontact, and absolute and relative. But sensors are also classified based on their electronic applications. Here we discuss some sensors commonly used in electronic devices.

Active sensor. A sensor that requires external power to operate, e.g., carbon microphones, thermistors, strain gauge, and capacitive and inductive sensors. The active sensor is also called a parametric sensor (a function of the output parameter—like resistance).

Passive sensor. It generates its own electric signal and does not require a power source, e.g., thermocouple, magnetic microphone, piezoelectric sensor, photodiode. It is also called a self-generating sensor.

Contact sensor. A sensor that requires physical contact with the stimulus, e.g., strain gauges and temperature sensors

Noncontact sensor. This requires no physical contact—e.g., most optical and magnetic sensors, infrared thermometer, etc.

Absolute sensor. A sensor that reacts to an excitation on an absolute scale, such as thermistors and strain gauges

Relative sensor. The stimulus is considered sensory relative to a fixed or variable reference. The thermocouple measures the temperature difference; pressure is often measured relative to air pressure.

Temperature sensors

A temperature sensor is a sensor that can measure the temperature of the surrounding environment. This sensor helps us detect temperature's ups and downs by providing analog or digital output. Temperature sensors are mostly divided into two types:

(i) Contact-based temperature sensors—these sensors require physical contact with the object to detect the temperature.

(ii) Noncontact temperature sensors—these sensors do not require any physical contact. They can measure temperature by the convection or radiation method.

In addition to the two primary categories, temperature sensors are further divided into four subcategories.

Thermostat

A thermostat is an electromechanical temperature sensor. In a thermostat, two thermally different metals are stuck together back to back. They can be two different metals like nickel, copper, tungsten, etc., and the metallic strips are bonded together to form a bimetallic strip. In cold conditions, both strips become closed, and current flow starts, but in hot conditions, the metals expand more, which prevents current flow. Thermostats use different types of sensors to measure temperature. In one form, a thermostat in the form of a mechanical thermostat, a coil, conducts direct electrical contacts that control the heating or cooling source. Electronic thermostats, instead, use a thermistor or other semiconductor sensor for amplification and processing to control heating or cooling equipment. The thermostat is an example of a "bang-bang controller" because the heating or cooling equipment output is not proportional to the difference between the actual temperature and the temperature setpoint. Instead, the heating or cooling device operates at full capacity until the set temperature is reached, then turns off. Therefore, increasing the difference between the thermostat setting and the desired temperature does not change the time to achieve the desired temperature. The rate at which the temperature of the target system can change is determined by both the capacity of the heating or cooling equipment to accumulate heat from a target system or to add or remove the capacity of the target system, respectively.

Thermostats may include hysteresis when the device is near the setpoint to prevent extremely fast cycling. Instead of immediately changing from "to" at a set temperature to "off" and vice versa, a thermostat with hysteresis will not switch until the temperature has changed slightly past the set temperature point. For example, a fridge set at $2°C$ cannot turn on the cooling compressor until the temperature of its food compartment reaches $3°C$, and until the temperature is lowered to $1°C$. It will continue until then. This greatly reduces the risk of equipment wear from frequent switching, although it introduces target system temperature oscillations of a certain magnitude.

For improved comfort of occupants of hot or air-conditioned locations, biometal sensor thermostats may include an "anticipatory" system to slightly heat the temperature sensor while the heating device is running or to slightly heat the sensor when no cooling system is operating. When properly adjusted, it reduces any excessive hysteresis in the system and reduces the magnitude of temperature variations. Electronic thermostats have an electronic equivalent.

Thermistor

The second type of temperature sensor is the thermistor, usually made of polymer or ceramic materials like oxides of nickel, manganese, or cobalt coated in glass. Most thermistors have a negative temperature coefficient of resistance (NTC), and very few have a positive temperature coefficient of resistance (PTC). For NTC, the resistance value goes *down* when the temperature increases, and for PTC, the resistance value goes up with the increase in temperature. To obtain a measurable output voltage, we must pass a current through the thermistor, so we can say this is a passive resistive device.

With NTC thermistors, the resistance decreases as the temperature increases and thermal movements from the valve band collide, usually due to an increase in electron conductivity. NTC is commonly used as a temperature sensor or an inrush current limiter with a circuit in series. With PTC thermistors, the resistance increases because the increase in temperature is usually due to thermal lattice movement, especially impurities and imperfections. PTC thermistors are usually installed in series within a circuit and are used in the form of resolvable fuses to protect against overcurrent conditions.

Thermistors are usually produced with powdered metal oxides. With massive improvements in formulas and techniques over the last 20 years, NTC thermistors can now achieve accuracy over a wide temperature range such as ±0.1 or 0.2°C to 0 to 70°C with excellent long-term stability. NTC thermistor elements come in many styles, such as axially leaded glass-encapsulated (DO-35, DO-34, and DO-41 diodes), glass-coated chips, epoxy-coated with bare or insulated lead wire, and surface-mounted, as well as rods and disks. The typical operating temperature range is C 55 to +150°C, although some glass-body thermistors have a maximum operating temperature of +300°C.

Thermistors differ from resistance temperature detectors (RTDs)—the material used in a thermistor is typically a ceramic or polymer, whereas RTDs use pure metals. The temperature response is also different; RTDs are useful over wide temperature ranges, whereas thermistors usually achieve greater accuracy within a limited range, typically from -50 to 130°C.

As a first-order approximation, assuming that the relationship between resistance and temperature is linear,

$$\Delta R = k\Delta T$$

where ΔR is change in resistance, and k is the first-order temperature coefficient of resistance.

Thermistors can be classified into two types based on the sign of k. If k is positive, the resistance increases with increasing temperature, and the device is called a positive temperature coefficient (PTC) thermistor, or posistor. If k is negative, the resistance decreases with increasing temperature, and the device is called a negative temperature coefficient (NTC) thermistor. Resistors that are not thermistors are designed to have a k as close to 0 as possible so that their resistance remains nearly constant over a wide temperature range.

Resistive temperature detector

RTD is another kind of temperature sensor. Every metal changes its resistance with temperature change, and RTDs feature this characteristic. RTD sensors are made with high-purity conducting metals such as platinum, copper, or nickel. An RTD usually consists of a film, but sometimes a wire wrapped around a ceramic or glass core for greater accuracy. Platinum RTD is the best but is expensive. Nickel and copper RTDs are cheaper but are not as accurate as platinum RTDs. Platinum RTDs are also called PRTs. This PRT has a standard resistance value of 100 Ω at 0°C. A constant current flowing through an RTD can obtain an output voltage because RTDs are passive resistive devices like thermistors. The material has the exact resistance/temperature relationship used to indicate temperature. Because RTD elements are fragile, they are often placed under protective screening. RTD, which has high accuracy and repetition, is gradually replacing thermocouples in industrial applications below 600°C. Unlike a thermocouple, however, an RTD requires a current sensor to produce a passive sensor and output voltage. Conditioning-circuitry thermocouples use the same high-performance signal for RTDs with a TD of less than 0.385%/°C; however, the RTD voltage drop is much larger than the thermocouple output voltage. A system designer can select larger RTDs with higher outputs, but larger RTDs exhibit slower response times. Moreover, although the cost of RTDs is higher than that of thermocouples, they use copper leads, and the termination of junctions does not affect the thermoelectric effects on their accuracy. And finally, because their resistance is a function of absolute temperature, RTDs do not require cold-junction compensation. Caution should be exercised when using current excitation as the cause of current heat through RTD. This self-heating changes the temperature of the RTD and appears as a measurement error. Therefore, careful attention should be paid to the design of the signal conditioner circuitry so that the self-heating is kept below 0.5°C. Self-manufacturers specify different RTD values and self-heating errors for sizes in stationary and moving air. The minimum current for the required system resolution must be used to reduce self-heating errors, and the largest RTD value chosen must provide an acceptable response time. Measurements of temperature by mechanical shock or corrosion when corrosion liquids, moisture, or gases like sweat, blood, and air are exhausted, or heat-sensitive resistance wires protect RTDs from atmospheric damage. Or a protective case can be used to cover the film and is usually made of ceramic, glass, synthetic resin, brass, or stainless steel.

Semiconductor-based temperature sensor

Semiconductor temperature sensors come in IC form, i.e., IC, popularly known as IC temperature sensors. These electronic devices are made in the same fashion as electronic semiconductor devices like current micropro-cessors. More than 1000 devices can be made with thin silicon wafers. Different manufacturers have come up with a whole new range of semi-conductor temperature sensors. However, the most popular include the AD590 and LM35. Semiconductor-based temperature sensors are embedded with IC's. These sensors can be of two types—local temperature sensors and remote digital temperature sensors. Local temperature sensors may use either analog or digital output. For remote digital temperature sensors, the transistor is located away from the sensor chip, and it uses the physical properties of the transistor. Semiconductor thermometers have a major feature—they provide logically linear output. They are available in medium to small sizes and are not capable of measuring high temperatures. Their range of temperatures is usually limited to -40 to $120°C$. If they are properly calibrated, they provide fairly accurate temperature readings. They offer very small interrelationships. Semiconductor temperature sensors are not appropriately designed for thermal communication with external surfaces. These temperature sensors provide an easy interface with other electronic devices such as amplifiers, controllers, digital signal processors, microcontrollers, etc. This type of temperature sensor is considered ideal for applications with etc. embedded where they are installed within the device. Other temperature sensors differ from such as thermocouples and RTD.

Pressure sensor

A pressure sensor is a device that measures actual applied pressure to the sensor and converts it into an output signal. These sensors are made with a pressure-sensitive element that measures the accurate pressure that is applied to the sensor. Pressure is an expression of the forces required to prevent the expansion of a fluid and is usually expressed as energy as a unit region.

Pressure sensors usually act as transducers. It produces a signal as an action of applied pressure. Pressure sensors are used to control and monitor thousands of applications. Pressure sensors can be used indirectly to measure other variables such as liquid/gas flow, velocity, water level, and altitude. Pressure sensors can alternatively be called pressure transducers, pressure transmitters, pressure transmitters, pressure indicators, piezometers, manometers, and many more. There is also a series pressure sensor designed to measure in dynamic mode to capture very fast changes in pressure. An example application of this type of sensor is the measurement of the combustion pressure of an engine cylinder or gas turbine. These sensors are usually made from piezoelectric materials such as quartz. Some pressure sensors act as pressure switches to turn certain pressures on or off. For example, a water pump can be controlled by a pressure switch that can be activated when water comes from the system to reduce the pressure in the reservoir. According to the working principle, pressure sensors can also be divided into different types.

- **Absolute pressure sensor**: Most people are usually accustomed to dealing with the different kind of pressure compared with the general atmospheric pressure around us. For example, "absolute" pressure sensors and absolute pressure sensors that measure relative pressure to absolute emptiness can be somewhat confusing. In addition, it is extremely difficult to measure and confirm the absolute pressure sensor because it is impossible to achieve zero absolute pressure (absolute emptiness). Perfect pressure is often easier to understand if you have a clear idea of common differential and gauge pressures. Differential pressure is the pressure difference measured between two pressure sources. It is usually expressed in pounds per square inch difference. When the source has atmospheric pressure, it is called a gauge or relative pressure and is usually expressed in pounds per square inch of gauge (mind). Thus, the pressure difference is a special case of pressure-only differential pressure, but it is always associated with local atmospheric pressure. In the case of the same relationship, the absolute pressure can also be considered the split pressure where the measured pressure is compared with the absolute emptiness. Complete pressure sensors are commonly used to measure changes in barometric pressure or altimeters. These applications cannot simply refer to ambient atmospheric pressure, so they must specify the pressures described above. Absolute pressure is called the pressure measured against absolute emptiness. For example, a psi of 10 pounds per square inch is 10 psi more than absolute emptiness. This is about 4.7 psi higher than the standard atmospheric pressure of 14.7 psi at sea level. 0 psi is the pressure of a complete vacuum used as a unit of pressure measurement per pound per inch (psi). It is clear that this unit can be converted to other common pressure units such as mmHg, KPA, bar.

- **Gauge pressure sensor**: Measure the relative pressure to the atmospheric pressure. When it shows zero, that means it is the same as the atmospheric pressure. The exact pressure or vacuum produced by some applications is less important. Instead, I just want to understand how much pressure or vacuum changes compared with atmospheric pressure. Atmospheric pressure also changes around the world due to changes in our altitude and climate. For example, consider a vacuum pump used during or after surgery. These are used to remove liquids, gases, and even tissues. Typically, what is needed to prevent injury is a small, finely controlled vacuum. You will need to make a decision about local atmospheric pressure. The atmospheric pressure is higher in sea-level hospitals than in mountain hospitals. The gauge pressure sensor measures the port pressure relative to the local atmospheric pressure. This can be compared to the use of a multi-meter DC measurement limit. In this case, the display shows the voltage across the positive probe against the negative probe. Gauge pressure sensors are usually packed with ports where pipes can be connected (see figure on the right) and open vents in the atmosphere. The pipe may be connected to where the system is measured. When installing sensors in the application, it is important that the vent is exposed to the atmosphere. This may require holes in the printed circuit board or product housing.

- **Differential pressure sensors**: Calculate the difference between two pressure. This can also be used to measure pressure drop, flow rate, etc. A differential pressure sensor measures the pressure difference between two points in the system. This is usually because this difference can be used to measure the flow of liquid or gas in a pipe or duct. Alternatively, it can only be used to detect obstruction or seizure valves. If the pressure at the front of the valve is greater than the pressure at the front of the valve, there must be something that prevents the medium from moving between the two measurement points. Typically, two measured pressures are applied to the opposite side of a single diaphragm. Isolation of the diaphragm is related to the stable condition, determining the pressure difference is positive. Alternatively, two complete pressure sensors can be used to measure the differential pressure if desired, and the industrial control system can calculate the differential pressure. This can happen in situations where two different types of sensors are needed to measure the medium, such as the liquid and the gas or the measuring atmosphere. Most gauge pressures are technically differential pressure sensors, but true differential pressure sensors, which measure the difference between medium and atmospheric pressure, are used to detect differences between two separate physical fields. For example, when an object moves from one side to the other, differential pressure is used to check the release or loss of pressure. In the field of treatment, differential pressure sensors are used to treat deep vein thrombosis, infusion pumps, respiratory organs, and respiratory detectors.

- **Sealed pressure sensors**: Like gauge pressure sensors, these also measure relative pressure, but to a fixed pressure, not atmospheric pressure. Ceiling pressure is less common than the previous three, but it still exists in the world of pressure. The ceiling pressure is not necessarily a vacuum but uses a given reference point. This makes it possible to measure pressure in changing places in response to changes in the atmosphere. There is no need to vent the sensor because the reference point is predefined. The sealed pressure is called the atmospheric pressure stuck behind the diaphragm; it is also the absolute pressure. It does not use a vacuum, however, as it is a given reference point, so the pressure does not need to be pushed back to flow or out. Seal gauge pressure can take on both positive and negative values. If the value is positive, it is called overpressure. In that case, the measured pressure will be higher than the standard atmospheric pressure, and the absolute pressure will be equal to the negative atmospheric pressure.

$$P_o = P_{abs} - P_{atm}$$

If the measuring gauge pressure is negative, it is called underpass or partial emptiness. The measured pressure is then less than the standard atmospheric pressure and is calculated by subtracting the absolute pressure from the atmospheric pressure, $P_u = P_{atm} - P_{abs}$.

It is not necessary to use the minus sign indicating partial emptiness. If the vacuum cleaner operates at an absolute pressure of 0.8 bar, it can also be said to operate at the recognition of 0.2 bar. Typically, the vacuum inside the cavity measures fewer impurities due to temperature changes, so a full pressure sensor is preferable to a sealed gear.

Different types of technologies are used in pressure sensors to obtain better results:

- **Strain gauge-based pressure sensors**: These use a pressure-sensitive element that can either be a diaphragm or for metal foil gauges measuring bodies. Strain gauge-based pressure transducers are suitable for use in a wide range of industries. The extraordinary versatility and flexibility of the design allow you to manage the most diverse and challenging applications. Pressure sensors and transducers are available in smaller or more accurate models with 0.05% accuracy. It can measure not only differential pressure but also extremely low or high pressure. It is often the most expensive and effective solution to the challenges that require durable, reliable, and accurate pressure-measuring devices. Strain gauge-based pressure transducers convert the pressure into a measurable electrical signal. Their study is based on piezoactive effects. In other words, the ability to change the resistance value of the strain gauge according to the physical deformation of the material due to stress. Through the wiring of strain gauges in the Wheatstone bridge configuration, small modifications applied to these resistors can be used to generate an accurate electrical signal in proportion to the applied pressure. The same principle is applied to force cells and load cells. A pressure transducer is created when the Wheatstone bridge is attached to a specially designed metal element. And after proper compensation and proper testing for temperature changes, it is ready to succeed in the case.

- **Capacitive pressure sensors**: Capacitive pressure sensors measure pressure by detecting changes in electrical power caused by the movement of the diaphragm. The capacitors have two separate parallel conductor plates with a slight gap. When you change a variable, the capacitance changes accordingly. The easiest interval to control. This can be done by making a diaphragm of one or both plates that is flexible with changes in pressure. Typically, one electron is a pressure-sensitive diaphragm, and the other electron is fixed. An example of a capacitive pressure sensor is shown on the right. An easy way to measure changes in capacitance is to make it part of a tuning circuit consisting of a capacitance sensor and an indicator. It can change either the frequency of the pendulum or the alternating current (AC) coupling of the resonant circuit. When you change a variable, the capacitance changes accordingly. The easiest interval to control. This can be done by making a diaphragm of one or both plates that is flexible with changes in pressure. Typically, one electron is a pressure-sensitive diaphragm, and the other electron is stationary. An example of a capacitive pressure sensor is shown on the right. An easy way to measure changes in capacitance is to make it part of a tuning circuit consisting of a capacitance sensor and an indicator. It can change either the frequency of the pendulum or the AC coupling of the resonant circuit. It is possible to sensor capacitor plates that are highly sensitive to temperature changes by selecting capacitor plate components with low thermal expansion properties. This structure requires less hysteresis to ensure measurement accuracy and repetition. Because the diaphragm itself is a detection component, there is no problem with the additional components attached to the diaphragm. Therefore, capacitive sensors can be operated at higher temperatures than other types of sensors. The change in capacitance can be measured by connecting a sensor to a pendulum of a frequency-dependent circuit or a control line tank circuit. In both cases, the capacitance depends on the resonant frequency pressure of the circuit. The pendulum requires additional electronic components and a power supply. Resonant LC circuits can be used as passive sensors without their own power supply. The allowance of the material between the plates can vary with pressure or temperature and can also cause errors. Because the relative permeability of air and most other gases increases with pressure, the change in capacitance with pressure increases somewhat. Absolute pressure sensors with a gap between the plates work ideally in this case.

- **Piezoresistive pressure sensors**: Piezoresistive strain gauges are one of the most common types of pressure sensors. They measure pressure using changes in the electrical resistance of a substance. These sensors are suitable for a variety of applications due to their simplicity and visibility. These can be used to measure perfect pressure, gauge pressure, relative pressure, and differential pressure in high- and low-pressure applications. This article describes the different types of piezoelectric pressure sensors available, how they work, and their related features. The basic principle of piezoelectric pressure sensors is to use strain gauges made of conductive material. When the strain gauge is pulled, its electrical resistance changes. Strain gauges can be attached to the diaphragm that recognizes changes in resistance as well as distortion of the sensor component. Changes in resistance are converted to output signals. There are three different effects that contribute to changes in conductor resistance. The resistance of a conductor is proportional to its length, so as it expands, the resistance increases. As the conductor expands, its cross-sectional area decreases, and its resistance increases. The resistance of some elements increases as they become stronger. These final piezoelectric effects vary from material to substance. Strain gauge components can be made of metal or semiconductor components. Changes in the resistance of metal strain gauges are mainly due to changes in material length and shape as cross-sectional regions. For some metals, such as platinum alloys, the piezoactive effect can more than double the sensitivity. The piezoelectric effect is dominant for semiconductor materials and is usually ordered in width rather than size contribution. Piezoresistive strain gauge measurements are made using the Wheatstone bridge circuit. An excitation voltage must be supplied to the bridge. If there is no tension and all the resistors on the bridge are balanced, the output will be 0 V. As the pressure changes, the resistance across the bridge changes and produces an output voltage or current uniform. You can improve performance by using two or four sensitive components on the bridge and applying equal and opposite pressure to each welded component. This amplifies the output signal and reduces the effect of temperature on the sensor components. In metal detectors, the pressure on the diaphragm pulls the wire and changes its resistance. The sensor material can be fixed directly to the surface with an adhesive or deposited directly into the diaphragm by separating the conductor. The latter method eliminates the potential problem of poor adhesion to high temperatures and simplifies the production of small devices. Metal wire sensors can also be made by wrapping a wire between displaced posts due to pressure changes. This structure works even at high temperatures because no 18 is required to connect the wires to the level. Semiconductor materials for semiconductor detectors, mostly silicon, can also be used to make strain gauge pressure sensors. The properties of the detector, especially the size of the piezoelectric effect, can be adjusted by doping. In other words, a carefully controlled amount of impurities (dopants) is added to the semiconductor. Lightly doped silicon results in higher resistance and higher gauge factor. However, it improves the heat sensitivity of both the resistance and the gauge factor. Semiconductor sensors similar to metal wire sensors can be made by placing silicon wire gauge material in the diaphragm. These can be manufactured directly on a silicon surface using the same manufacturing methods used to manufacture electronic semiconductor devices. It makes it possible to make very small sensors cheaply with precisely controlled features like sensitivity, linearity, and temperature response. The gauge factor of a typical metallic strain gauge sensor is about 2–4. If the ideal maximum strain is a few ppm, it means an output change of about 1 mV for each volt. Silicon-based sensors are usually doped to deliver a gauge factor of about 100−200. It provides a good compromise between sensitivity and thermal properties. The output from the silicon sensor is about 10 mV/V. Piezoresistive strain gauge pressure sensors have the advantage of being more powerful. Their performance and calibration are constant over time. The disadvantage of these sensors is that they receive more power than other types of pressure sensors. This may mean that they are not suitable for battery-powered or portable systems.

- **Resonant pressure sensors**: Due to the applied pressure, the detection method uses resonance frequency changes to measure changes in pressure and gas concentration. This technique can be used in conjunction with ball collectors, as listed above. Alternatively, the resonance technique reveals the resonance element of the medium itself, and the resonance frequency can be adopted depending on the density of the medium. The sensor is made of vibrating wire, vibrating cylinder, quartz, and silicon MEMS. This technique is considered to provide very stable measurements over time.
- **Piezoelectric**: Piezoelectricity is the charge produced by materials when mechanical pressure is applied. Piezoelectric pressure sensors capture this effect by measuring the voltage of the piezoelectric component produced by the applied pressure. These are extremely strong and are used in a wide range of industrial applications. It can be measured as voltage proportional to the pressure. There is also an inverse piezoelectric effect that changes shape when a voltage is applied to the material. The given static power provides the sensor a uniform charge. However, due to incomplete insulation, internal sensor resistors, and connected electronics, it will disappear over time. As a result, piezoelectric sensors are not suitable for measuring static pressure. The output signal gradually drops to zero, even in the presence of constant pressure. However, they are sensitive to dynamic pressure changes at different frequencies and pressures. This dynamic sensitivity means it is great for measuring even small changes in pressure in a very high-pressure environment. Unlike piezoelectric resistors and capacitive transducers, piezoelectric sensor components do not require an external voltage or current source. These generate output signals directly from the applied pressure. Proportional to the output pressure from the piezoelectric material. A charge amplifier is required for this detection to convert the signal to voltage. Piezoelectric pressure sensors include an internal charge amplifier that provides a voltage output that simplifies the electrical interface. It requires sensor power. The built-in amplifier simplifies the use of the

sensor. For example, you can use a long signal cable to connect to the sensor. The amplifier can also connect signal conditioner circuitry to filter the output, control the temperature and compensate for the different sensitivities of the sensitive material. However, the presence of electronic components is limited to the operating temperature 120 component C and below. In a high-temperature environment, you can use a charge mode sensor. It provides the output charge as a direct output signal. Therefore, an external charge amplifier is needed to convert its outputs to voltage. Care must be taken when designing and implementing external electronic devices. The high-impedance output of the sensor means that the circuit is sensitive to weak connections, wiring, electromagnetic interference, and radio frequency (RF) interference. The low-frequency response of the sensor is determined by the discharge time of the amplifier. The piezoelectric effect requires a material with a specific nonuniformity of crystal structure that contains natural crystals such as quartz and tourmaline. In addition, specially prepared ceramics can be made their piezoelectric with proper polarization. These ceramics are more sensitive than natural crystals. It can produce useful output with 0.1% distortion. Due to the rigidity of piezoelectric components, the number of steps required to obtain a usable output signal is negligible. This makes the sensor extremely strong and tolerant to extreme stress conditions. This means that they respond quickly to changes in pressure. Pressure sensors can be affected by external forces such as acceleration and noise in piezoelectric components. Microsensors can be made using thin films. Zinc oxide was one of the first materials used. Made of materials such as lead zirconate titanate (PZT), it is mainly replaced by ceramic due to its large piezoelectric effect. The output is wide and linear with about 1% accuracy, typically 0.7 kPa to 70 MPa (0.1–10,000 psi). Ceramic sensors may become less sensitive over time. However, it is usually quite small. Usually less than 1% per year. Exposure to high pressure and high temperature in the beginning may also cause a slight decrease in sensitivity. The effect can be avoided by turning the sensor to the maximum expected pressure and temperature before applying the sensor. The frequency response of the piezoelectric sensor decreases to a lower frequency because it cannot hold the generated charge. At higher frequencies, the piezoelectric component has a peak that corresponds to the resonant frequency. The visibility, high frequency, and fast response times of piezoelectric pressure sensors mean they can be used in a wide range of industrial and aerospace applications exposed to high temperatures. These are often used to measure kinetic pressure, such as turbulence, explosion, and engine ignition. These require a quick response, roughness, and a wide range of activities.

- **Optical**: An optical sensor measures the physical change in an optical fiber to detect strain due to applied pressure. Optical pressure sensors detect pressure changes through their effect on light. In the simplest case, it could be a mechanical system that blocks light as the pressure increases. For more advanced sensors, the phase difference measurement allows very accurate measurements of small pressure changes. Increasing pressure with intensity-based optical pressure sensors gradually shuts off the light source. The sensor then measures the change in the received light. For example, in the general method shown below, the pressure is applied to the aperture, blocking more light from the mounted opaque van LED. The decrease in light intensity is detected by the photodiode, and the pressure is measured directly. Such a common optical pressure sensor requires a reference photodiode that is not blocked by the blade. This allows the sensor to compensate for changes in the light source, such as aging of the light source and fluctuations in the supply voltage. These mechanical systems are relatively large. Very small versions can have a reflection film and two optical fibers. One is the light source, and the other is the reflected light. The pressure bends the membrane and returns the amount of light to the detector. Other fiber optic sensors use interferometry to measure the change in optical path length and light phase due to changes in pressure. The rest of this section will focus on these. Fiber optic pressure sensors can be classified as either external, where the sensor is outside, or inside the fiber, where the fiber itself changes with pressure. Very sensitive optical measurements can be made using interferometry. That is, it measures the change in episodes between two separate paths of light. Distance changes can be detected depending on the part of the wavelength of light. There are two common types of pressure sensors that use interferometers. These are the Fabry–Pérot interferometer (FPI) and the fiber Bragg grating (FBG). FPI is an external sensor that uses interference between multiple rays reflected backward between the two surfaces of the cavity. As the difference between them increases, the interference changes the amount of light received at a certain wavelength. It is one of the best optical sensor technologies. It is simple, accurate, and easy to compress for different sizes and pressure ranges. FBG is an internal sensor, and the fiber has a regular reflective structure that expands the fiber and is affected by contraction. It changes the wavelength of reflected light. The Fabry-Perrot cavity has two parallel reflective surfaces that may form at the tip of an optical fiber. A semireflective surface is attached to the fiber (M1), and a reflective film is formed at the opposite end of the cavity (M2). This membrane forms a pressure-driven diaphragm. Distinguishing between mirrors creates a difference in the path of each ray (E1 and E2), resulting in a relative shift between the mirrors. The resulting interference increases or decreases the light of a certain wavelength. Many reflections and many interfering rays allow very high-resolution measurements. Bragg gratings can be made into a fiber using various periodic changes in the fiber removal index. It reflects or transmits light wavelengths depending on the wavelength-to-break ratio. As a result, the spectrum of reflected light varies depending on the fiber, and the interval increases. The fiber can be attached to the diaphragm that expands the fiber when pressure is applied. Compressing the fiber also changes the effect of grating, creating two peaks in the spectrum. Output from both types of sensors can be measured in two ways. When using a monochromatic or narrow light source, the amplitude of the output signal will vary because the length of the cavity or the gap between the gratings indicates the amount of reflection of the wavelength. Wideband light sources like the white band can also be used. In this case, the frequency of constructive or destructive interventions depends on the pressure. This can be measured with a spectral analyzer. These structures, especially fabric-perforated cavities, are also suitable for silicon manufacturing technology and can be used to manufacture small optical sensors as microelectromechanical system (MEMS) devices. You can create waveguides on a micrometer scale like mechanical components, such as optical fibers, cantilevers, and membranes. Due to their small size, these sensors can respond very quickly to pressure changes. Light-emitting diodes, solid-state lasers, photodiode detectors, and electronics can all be integrated into a single device.

Displacement sensors

Displacement sensors (displacement gauges) are used to measure the movement of an object and the occurrence of a reference position. The displacement sensor can be used to measure amplitude to determine the height, thickness, and width of an object in addition to the range of motion. It is important to select the most suitable equipment according to the application, required accuracy, and use environment. It calculates the amount of displacement by a variety of elements and converts it into measurable distance units. According to the working principle, displacement sensors can be of various types.

Optical displacement sensor

(i) **Position-sensing device (PSD)**: From a particular source, light is emitted that is condensed by the lens and directed to the object. A one-dimensional position-sensing device (PSD) receives the reflected light from the object, which is condensed by another lens. When the object changes its position, PSD changes as well. By calculating the changes in PSD, the displacement of the object can be measured. If the two outputs are A and B, calculate A/(A + B) and use appropriate values for the span coefficient "k" and the offset "C" as shown below:

$$\text{Displacement} = \frac{A}{(A+B)} \, k.C$$

(ii) **Complementary metal-oxide semiconductor (CMOS)**: A sensor that applies the CMOS method is more efficient than a PSD because CMOS is not affected by the color or texture of the object. Reflected light from the object received by the sensor as individual pixels in the CMOS.

Linear proximity sensor

When AC passes through a coil, magnetic flux occurs in the coil. If this magnetic flux flows through a metal body, then an eddy current is generated. This eddy current creates a magnetic field, and the magnetic field has a tendency to oppose changes in the current. If the gap between the object and the sensor decreases, eddy currents increase, and obviously amplitude of the oscillation circuit decrease.

Ultrasonic displacement sensor

Displacement sensors (displacement gauges) are used to measure the amount of movement that occurs between an object and a reference position. The displacement sensor can be used to measure amplitude to determine the height, thickness, and width of an object in addition to the range of motion. It is important to select the most suitable equipment according to the application, required accuracy, and use environment. Ultrasonic sensors emit low-frequency sound pulses at regular intervals. They propagate in the air at the speed of sound. When they hit an object, they are echoed by the sensor as an echo signal. The echo signal calculates the distance to the target based on the time between sending the signal and receiving the resonance. Ultrasonic sensors are good for suppressing background interference because the distance to an object is determined by measuring the time of flight rather than the intensity of the sound. It can detect almost any element that reflects sound regardless of color. Even transparent materials and thin foils are fine for ultrasonic sensors. Microsonic ultrasonic sensors are suitable for target distances of 20 mm to 10 m and measure flight time so that measurements can be detected with point-to-point accuracy. Some sensors can resolve the signal with an accuracy of 0.025 mm. Ultrasonic sensors can be seen through the air filled with dust and ink. Thin deposits in the sensor membrane do not reduce its effectiveness. With a blind area of only 20 mm and very narrow beam propagation, the sensors enable a completely new application today: microheater plates and test tube wells can be used in the packaging industry to measure the level and easily detect dew. Even thin strands can be reliably detected. A fixed source transmits an ultrasonic wave toward a target object, and the receiver receives the reflected wave from it. An ultrasonic displacement sensor provides the measurement of shifting of the object by calculating the required time and time and the speed of sound.

Contact displacement sensor using a linear variable differential transformer

LVDT is a device that marks mechanical linear motion as displacement and converts it into an electrical signal. Based on this method, the contact displacement sensor converts the signal into electrical signals and changes the target size. The LVDT type communication displacement sensor has a core in the middle and a coil around the core. A contact is attached to the main tip to form a spindle. The spindle is pressed against the target by a mechanism of spring. As the shape of the target changes, this spring can be used to slide up and down. Spring LVDT Type Intro. The movement of the root indicates a change in the target size. As a result, displacement can be measured based on the detection of changes in the output signal level. Measurement of the displacement through direct contact of a particular object with the sensor is done by contact displacement sensors. These kinds of sensors are more accurate and perfect than contactless sensors.

Position sensors

These sensors measure the accurate position or displacement of an object in terms of linear travel, rotational angle, or three-dimensional space. Position sensors are also used in automated production units of industry and in traffic control systems. According to the working principle, position sensors can be different types, as follows.

Potentiometric position sensor

These sensors use resistance in resistive tracks with a wiper attached to the particular object. When the wiper changes its position along the resistance track for movement of the object, the value is changed. This way, the measured resistance indicates the object's position.

Inductive position sensor

Inductive position sensors detect the position of an object by changes in the characteristics of a magnetic field that is induced in the coils of the sensor. One type is called an LVDT. In an LVDT position sensor, three separate coils are wound on a hollow tube. One is the primary coil, and the other two are secondary coils. They are wired electrically in series, but the phase relationship of the secondary coils is 180 degrees out of phase with respect to the primary coils. A ferromagnetic core or armature is placed inside the hollow tube, and the armature is connected to the object whose position is being measured. An excitation voltage signal is applied to the primary coil, which induces an EMF in the secondary coils of the LVDT. By measuring the voltage difference between the two secondary coils, the relative position of the armature (and the object to which it is attached) can be determined.

Capacitive position sensor

As the name suggests, this sensor works on changing the capacitive value. Capacitors comprise two plates separated by the dielectric material. This capacitive value changes in two ways, first by altering the dielectric constant of the capacitor and second by altering the overlapping area of the capacitor plates. Capacitive positioning sensors are designed by keeping these two principles in mind. During displacement, the overlapping area of the capacitor changes; by counting those changes, the position can be measured.

Magnetostrictive position sensor

Ferromagnetic materials such as iron, nickel, and cobalt exhibit magneto-striction, which means the material will change its size or shape when it is put in a magnetic field. A current pulse is sent down the waveguide, and a magnetic field is created in the wire that interacts with the axial magnetic field of the permanent magnet (position magnet). The result of the field interaction is a twisting. This twisting causes a strain in the wire that generates a sonic pulse. By this sonic wave, we can measure the position.

Hall effect position sensor

The Hall effect states that when a thin flat electrical conductor has a current flowing through it and is placed in a magnetic field, the magnetic field impacts the charge carriers, forcing them to accumulate on one side of the conductor relative to the other to balance the interference of the magnetic field. In a Hall effect position sensor, As the object moves, the position of the magnet changes relative to the Hall element in the sensor. This movement of position then changes the strength of the magnetic field that is applied to the Hall element that in turn becomes reflected as a change to the measured Hall voltage.

Motion sensors

The active electronic motion detector consists of an optical, microwave, or acoustic sensor and transmitter. However, because the passive has only one sensor, only the signature is passed by the emission or reflection from the moving object. Changes in light, microwave, or sound field near the device are explained electronically on the basis of one of several technologies. Most low-cost speed detectors can detect speeds at a distance of about 15 feet (4.6 m). Specialized systems are more expensive but have increased sensitivity or a longer range. These sensations of radio waves occur at frequencies that penetrate most walls and barriers and are detected in many places, allowing tomography motion detection systems across very large areas. Motion detectors are widely used in commercial applications. A common application is to enable automatic door openers in corporate or public buildings. Motion sensors are widely used to enable indoor lights such as store lights, lobbies, and stairs instead of real occupancy sensors. This type of smart lighting system saves energy by storing light only during the timer, after which a person can leave the area. Motion sensors can be placed inside the theft alarm sensor used to alert homeowners or security services when detecting the movement of potential intruders. These identifiers can trigger security cameras to record potential intrusions. This is a sensor used in many devices to detect and measure movement in the vicinity. According to the working theory, there are various kinds of motion sensors.

Passive infrared motion sensor

A passive infrared (PIR) motion sensor consists of a thin pyroelectric film material that emits electricity and responds to IR radiation. When the influx of electricity occurs from the motion of any living object's emitted infrared variations, the sensor activates the high voltage. The PIR sensor can detect changes in the amount of infrared light that varies depending on the temperature and surface properties of the object in front of the sensor. When an object, such as a person, passes in front of a background like a wall, the current temperature in the sensor field rises from room temperature to body temperature and returns again. Sensors detect changes in output voltage as a result of infrared radiation trigger similar temperatures, but objects with different surface features may also have different infrared emission patterns, so moving an object against the background can trigger the detector.

PIR comes in a variety of configurations for many applications. The most common models have several fresnel lenses or mirror sections, an effective range of about 10 m (30 feet), and a field of view of fewer than 180 degrees. Widefield models with 360 degrees are available and are usually designed for roof mounting. Some large PIRs built with single-segment mirrors can convert infrared power more than 30 m (100 ft) from PR. There is a PIR designed with reverse orientation mirrors, which can be selected individually to allow wide coverage (110-degree width) or very narrow "screen" coverage or to "size" a section.

Ultrasonic sensor

Ultrasonic transducers emit ultrasonic waves that can hear sound at a higher frequency than the human ear and receive images of nearby objects, like Doppler radar, which will show the speed of heterodyne detection in acceptable areas. The detected Doppler shift is also a low audible frequency because the ultrasonic wavelength of about 1 cm is equal to the wavelength used by the microwave motion detector. A potential drawback of ultrasonic sensors is that they may be sensitive to movement in areas where coverage is undesirable, for example, due to the reflection of sound waves around angles. Such extended coverage may be desirable for optical control, where the goal is to detect occupancy within this region. However, for automatic door openers, for example, sensor selection is a good choice for traffic on the way to the door. Ultrasonic (US) sensors can be of two types. Active US sensors generate ultrasonic wave pulses, and when that pulse reflects from a certain object, the motion of that object can be measured. A passive US sensor pays attention to particular sounds such as glass breaking.

Microwave sensor

Electromagnetic radiation from microwave motion sensors emits a wave that is reflected to the receiver. The receiver analyzes the rising waves. If the object is moving in the room, these waves are going to change. Microwave detectors can detect instantaneous fluctuations. Ideally, the recipient should receive the same type of wave repeatedly. Sensitivity can be higher or lower depending on how the microwave motion sensor will be operated. They can detect changes in a completely empty room or need to be adjusted to move frequently to avoid false-positives. More advanced microwave sensors can detect whether someone is away from the sensor or running randomly. These detectors help to detect and differentiate between normal movement and intrusive movement. This feature of these sensors makes them more reliable. The use of microwave sensors is completely safe. These can be used both inside and outside the property and can be placed in relatively large areas. It can also be configured to detect a variety of activities, such as ignoring certain areas of the home where pets and children may be active.

Humidity sensor

Humidity sensors measure the humidity of air at a certain temperature. It works by detecting the electrical current in the air. The ratio of the water vapor mass and the volume of the air is called Absolute Humidity (AH). If m is the mass of the water vapor, and V is the volume of air and water vapor mixture, then AH is given by

$$AH = m/V$$

AH does not account for temperature, but it changes with temperature and pressure. The ratio of the actual water vapor pressure in the air and the maximum water vapor pressure in the air at a certain temperature is called relative humidity.

Humidity sensors are primarily of three types:
- Capacitive: a capacitive humidity sensor works through a thin strip of metal oxide between two electrodes. It measures relative humidity. The electrical capacity of a metal to oxidize varies with the relative humidity of the atmosphere.
- Resistive: resistive humidity sensor use the ions that are present in the salt to detect the electrical impedance of atoms. According to the change in humidity, the resistance of the electrode also changes.

• Thermal conductivity: Thermal conductivity humidity sensors measure the AH of air. Two thermal sensors with NTC are used here. One is put in a sealed nitrogen chamber, and the other is put in an open environment. The difference between these two readings is the AH.

Moisture sensor

Moisture sensors measure the number of water molecules in any material. But usually, this is used to measure the shortage or abundance of water content in the soil. In this process, capacitance is used to determine the dielectric permittivity of the medium. In soil, dielectric permittivity is a function of the water content. A voltage proportional is created to the dielectric permittivity. The sensor averages the water content over the entire length of the sensor. The relationship between the calculated properties and soil moisture must be adjusted and can vary depending on environmental factors such as temperature, soil type, and other electrical conductivity. These sensors are commonly used to determine volumetric water volume, and another group of sensors calculates a new property called groundwater potential. Typically, these sensors are designated as potential sensors for groundwater, including gypsum blocks and tension gauges. This sensor uses capacitance to measure the dielectric permeability of the soil. The function of this sensor can be performed by using this sensor on the earth, and the amount of water in the soil can be expressed as a percentage.

Amplifiers

An amplifier is an electronic device or circuit that increases the amplitude of voltage or current of a signal by increasing its power gain. It is a two–port electronic circuit. An amplifier uses the power supply to the input terminal and provides increased amplitude of the signal to its output.

The gain of the amplifier is the difference in amplified signals between the input and output terminals. There are three different kinds of amplifier gain that can be measured: voltage gain (Av), current gain (Ai), and power gain (Ap), depending on the quantity being measured. Examples of these different types of gains are provided below.

Voltage amplifier gain:

$$\text{Voltage Gain (Av)} = \frac{\text{Output Voltage}}{\text{Input Voltage}} = \frac{\text{Vout}}{\text{Vin}}$$

Current amplifier gain:

$$\text{Current Gain (Ai)} = \frac{\text{Output Current}}{\text{Input Current}} = \frac{\text{Iout}}{\text{Iin}}$$

Power amplifier gain:

$$\text{Power Gain (Ap)} = \text{Av} \times \text{Ai}$$

The power gain (Ap) or power level of the amplifier can also be expressed in decibels (dB). The bel (B) is a logarithmic unit (base 10) of measurement that has no units. Decibels is 1/10th of a bel. To calculate the gain of the amplifier in decibels or dB, the following expressions are used:

Voltage Gain in dB: av = 20 logAv
Current Gain in dB: ai = 20 logAi
Power Gain in dB: ap = 10 logAp

Biological amplifier

A biological amplifier, or bioamplifier, is an electrophysiological device used for amplifying the electrophysiological signal while keeping its signal integrity. In any living body, several types of electrical activities may occur. The biosignal has a very low potential, and thus to understand this signal, we must amplify it by keeping integrity similar to that of the signal itself. To achieve this goal, we use a bioamplifier. The most commonly used bioamplifier is an op-amp.

Biological amplifier basic characteristics

  (i) Input impedance value of the biological amplifier should be high because a higher impedance value reduces the loss of the signal. Depending on the application, the range of values should be between 2 and 10 MΩ.
  (ii) For security purposes of a patient from electrical shocks bioamplifier should be made up with isolation and protection circuits.
 (iii) Bioelectric amplifiers send signals in millivolt or microvolt ranges. So to study the signal, the voltage gain value of the amplifier should be higher than 100 dB.
 (iv) A constant gain should be present throughout the whole bandwidth range to obtain an accurate measurement.
  (v) A bioamplifier should have a small output impedance.
 (vi) There should not be any signal drift and noise in a biological amplifier.

(vii) If the common-mode rejection ratio (CMRR) value of the amplifier is more than 80 dB, then the interference from the common-mode signal will be decreased.

(viii) Gain Calibration is very much necessary for a biological amplifier.

Operational amplifier

Operational amplifiers (op-amps) are widely used for signal conditioning, filtering, or performing mathematical operations such as addition, subtraction, consolidation, and inequality because they have a linear device with almost all the features required for standard DC amplification. Op-amps, or abbreviated op-amps, are voltage amplification devices designed for use with external output components such as resistors and capacitors between output and input terminals. These reaction elements

Table 3.2 Comparison of ideal and practical operational amplifiers.

Ideal operational amplifier	Practical operational amplifier
An ideal op-amp has	In a practical op-amp
Infinite voltage gain so that it can amplify input signals of any amplitude	Voltage gain is not infinite, but typically 10^5 to 10^8, so it is not able to amplify input signals smaller than 100 µV.
Infinite input resistance, so that almost any signal source can drive it and there is no loading of the preceding stage	Input resistance is typically 10^6–10^{12} Ω (for FET input op-amps such as uAF771), so it still draws some current, and not all sources can drive it.
Zero output resistance, so that output can drive an infinite number of other devices	The output resistance is typically 75 Ω for standard op-amps, so it has a limit to deliver current to output devices.
Zero output voltage when the input voltage is zero	It is not able to give zero at output when input is zero due to mismatching of input transistors.
Infinite bandwidth, so that any frequency signal can be amplified without attenuation	Op-amp has its own gain-bandwidth product, so input frequency should not exceed that particular frequency range at the desired gain.
Infinite common-mode rejection ratio, so that the output common-mode noise voltage is zero	CMRR is typically 90 dB, so it still provides output voltage even if both input terminals are shorted.
Infinite slew rate, so that output voltage changes occur simultaneously with input voltage changes	The slew rate is typically 0.5–90 V/µS (for improved op-amp such as LM318), so output cannot be changed simultaneously with input, and there is some delay.

determine the activity or behavior of the amplifier's result and are named "op-amps" depending on the different reaction configurations of resistance, capacitive, or both. The amplifier can perform a variety of operations.

An op-amp is a three-terminal device with two high-impedance inputs. Inputs are called inverted inputs and are denoted by a negative or "minus" sign (−). Other inputs are called noninverting inputs and are marked with a positive or "plus" sign (+).

The third terminal represents the output port of the op-amp that can carry both voltage, current, or sync. In linear op-amps, the output signal is an amplified factor, which is multiplied by the value of the input signal (a) and is based on the nature of these input and output signals, which have four different behaviors. Classification. Amplifier gain.

Voltage—voltage "input" and voltage "output"

Current—current "input" and current "output"

Transconductance—voltage "input" and current "output"

Transresistance—current "input" and volt "output"

Most circuits working with OPAMPS are voltage amplifiers, so the tutorials in this section are limited to voltage amplifiers (Vin and Vout).

The output voltage signal from an op-amp is the difference between the signals applied to two separate inputs. In other words, the output signal of the amp is the difference between the two input signals. This is because the input stage of the op-amp is a differential amplifier.

The circuit below shows a common form of a differential amplifier with two inputs marked V1 and V2. Two identical transistors, TR1 and TR2, are connected to the emitter at the same operating point and return to the common rail through the reactor rear.

The circuit is powered by a dual power supply + VC-based, which ensures continuous supply. The voltage displayed at the output, the voltage of the amplifier, is the difference between the two input signals because the two base inputs are out of phase with each other. Thus, as the forward bias of transistor TR1 increases, the forward bias of transistor TR2 decreases and vice versa. Second, if the two transistors match perfectly, the current through the common emitter resistor will remain constant. Like the input signal, the output signal is balanced, and the collector voltage is in the opposite direction (reverse phase) or in the same direction (step), so the output voltage signal is found between the two collectors. A perfectly balanced circuit is a zero difference between the voltages of two collectors. This is known as a common mode of operation because of the common-mode gain of the amplifier when the input is zero. However, the actual amplifier always has fluctuations, and the ratio of change in output voltage is abbreviated as the change in normal mode input voltage to CMRR.

The op-amp itself has a high open-loop DC gain, and negative feedback can be applied to create an op-amp circuit with very accurate gain characteristics depending only on the feedback used. Note that the term "open-loop" means that the reaction path or loop is open because no reaction material is used around the amplifier.

Op-amp only responds to voltage differences between two input terminals, commonly known as differential input voltages. Then, if the capacitance of the same voltage is applied to both terminals, the resulting output will be zero. The advantage of op-amps, commonly known as open-loop differential gain, is denoted by the symbol (A_{OL}).

In an op-amp, multiple amplifiers are integrated in a very complex way. There are many transistors, FETs, and resistors are present in the op-amp circuit. Op-amp has two inputs, one output, and two voltage supply terminals. If two signals are applied, one at the inverting and another at the noninverting terminal, an ideal op-amp will amplify the difference between the two applied input signals. The difference between two input signals is called the differential input voltage. The equation provides the output of an op-amp.

$$\text{Vout} = A_{OL} (V_1 - V_2)$$

Vout = voltage at the output terminal; A_{OL} = open-loop gain that constant for a specific amplifier.

Op-amps should have features like infinite voltage gain, infinite input resistance, zero output resistance, infinite bandwidth, and zero noise.

Op-amps should have features like infinite voltage gain, infinite input resistance, zero output resistance, infinite bandwidth, zero noise, etc. The characteristics of an ideal operational amplifier and practical operational amplifier are provided below (Table 3.2).

Inverting operational amplifier

As you know, op-amps have very high open-loop gain (A_{OL}), which can be 10^6 or higher. However, this very high gain is not really effective because it uses an amplifier as an input signal to control both instability and difficulty; producing an output voltage of a few microvolts (μV) or more is enough. The open-loop DC gain of the op-amp is so high that you can lose this high gain by connecting both the appropriate resistors of the amplifier behind the output terminal to the inverting input terminal and controlling the overall gain. This is called negative feedback and creates an effect resulting in a very stable op–amp–based system.

Negative feedback is part of the feeding back process of the output signal to the input, but to make the feedback negative, you must use the outside to return to the OP–AMPs negative or "inverting input" pin with a feedback resistance Rf. This feedback connection between the output and the inverting input terminals forces the differential input voltage to zero.

This effect creates a closed-loop circuit in the amplifier so that the gain of the amplifier is called the closed-loop gain. The close-loop inverting amplifier then uses the negative feedback to accurately control the overall gain of the amplifier, but the gain of the amplifier decreases. As a result of this negative feedback, the input terminal receives a signal that is different from the actual input voltage. This is because the addition of input voltage and the negative response voltage provides a label or summary of the condition. Thus, the input resistor Rin should be used to separate the actual input signal from the inverting input.

Because it does not use a positive noninverting signal input, it is connected to a common ground or zero voltage terminal, but due to the effect of this closed-loop feedback circuit, the value of inverting inputs is equal. The inverting input has the same potential as the grounded reference input and does not form the virtual ground connecting point. In other words, the op-amp is a "differential amplifier".

In the inverting amplifier circuit, the app amp is connected to the feedback to create a closed-loop activity. When using op-amps, two very important rules should be kept in mind when disabling an amplifier. These are "no current flows through the input terminals" and "V1 is always equal to V2." However, in a real op-amp circuit, both of these rules are slightly violated. This is because the junction of the input signal and the response signal (X) is similarly possible as a positive (+) input at 0 V or ground, so the junction is "virtual." With this virtual ground node, the value of the input resistance of the amplifier input is equal, so the closed-loop gain of the inverting amplifier can be set according to the ratio of the two external resistors. There are two very important rules to keep in mind about an amplifier or op-amp, and that is it.

There is no current flow at the input terminal.

Differential input voltage zero $V1 = V2 = 0$ (virtual earth).

One can then use these two rules to obtain an equation to calculate the closed-loop gain of an electronic amplifier using the first principles.

$$i = \frac{Vin - Vout}{Rin + Rf}$$

$$i = \frac{Vin - V2}{Rin} = \frac{V2 - Vout}{Rf}$$

$$i = \frac{Vin}{Rin} - \frac{V2}{Rin} = \frac{V2}{Rf} - \frac{Vout}{Rf}$$

$$\frac{Vin}{Rin} = V2 \left[\frac{1}{Rin} - \frac{1}{Rf} \right] - \frac{Vout}{Rf}$$

$$i = \frac{Vin - 0}{Rin} = \frac{0 - Vout}{Rf}$$

$$\frac{Rf}{Rin} = \frac{0 - Vout}{Vin - 0}$$

$$\frac{Vout}{Vin} = -\frac{Rf}{Rin}$$

$$\text{Close Loop Gain}(Av) = \frac{Vout}{Vin} = -\frac{Rf}{Rin}$$

$$\text{Output Voltage (Vout)} = -\frac{Rf}{Rin} \times Vin$$

The negative signal of the equation is 180 out of phase and thus represents the corresponding output signal at the input. This response is because the value is negative. The formula for output voltage also implies that the circuit must be linear with amplifier gain voltage = Vin × Gain. This property is very useful for converting small sensor signals to very large voltages.

Noninverting operational amplifier

In this configuration, the input voltage signal (Vin) is applied directly to the noninverting (+) input terminal. This means that the output gain of the amplifier will be "positive" at the reverse price of the "inverter amplifier" circuit, as seen in the latest tutorial, where the value of output gain is negative. As a result, the output signal is "in phase" with the input signal. The response control of the noninverting op-amp is obtained by regenerating a small portion of the output voltage signal of the inverting (−) input terminal through the Rf − R2 voltage divider network. This closed-loop configuration creates a noninverting amplifier circuit with very good stability and very high input impedance, and positive values do not flow to the input terminals (ideal state), resulting in low output impedance.

As you know, in the case of a standard op-amp, no current flows through the input terminal of the amplifier; V1 is always equal to V2. This is because the connection between the input signal and the response signal (V1) has the same potential; this can be called a virtual shot. In other words, the junction is an addition to the virtual ground; with this virtual ground, the resistors R and R2 form a common voltage divider network with a noninverting amplifier.

Then, using the equation to calculate the output voltage of the voltage divider network, the cloud-loop voltage gain (AV) of the noninverting amplifier can be calculated as follows:

$$V1 = \frac{R2}{R2 + Rf} \times Vout$$

As we know $V1 = Vin$

$$\text{Voltage gain } (Av) = \frac{Vout}{Vin} = \frac{R2 + Rf}{R2}$$

$$Av = \frac{Vout}{Vin} = 1 + \frac{Rf}{R2}$$

From the above equation, the overall closed-loop gain of the non-inverting amplifier is always high but never less than 1. It is positive in nature and is determined by the ratio of values of the equations Rf and R2.

If the value of resistor Rf is zero, then the gain of the amplifier equals 1 unit. When the resistor R2 is zero, the gain reaches infinity, but in reality, the amp is limited to the open-loop differential gain (AO) of the amplifier.

If the response resistor Rf is equal to zero and the resistor R2 is equal to infinity, then the result is a clear gain of 1 unity of the circuit. The output voltage returns to the inverting input terminal as a negative feedback configuration. This configuration creates a special kind of noninverting amplifier circuit called a voltage follower, known as a "Unity Gain Buffer."

Because the input signal is directly connected to the noninverting input of the amplifier, the output signal is not reversed, and the output voltage is equal to the input voltage. Therefore, Vout = vin. This makes the voltage following circuit ideal as a constant voltage source or voltage regulator due to its input-to-output isolation properties.

The Unit gain voltage maintains the input signal voltage at the following configuration output terminals, which has the advantage that it can be used only when impedance matching or circuit isolation is more important than voltage or current amplification. In addition, the input impedance of the voltage follower circuits is very high, equal to the input resistance of the op-amp, which is gain ($R2 \times Ao$), and usually greater than 1 MΩ. Since the output impedance of the op-amp is very low, it is assumed that the ideal op-amp load will not be affected by change.

$$Av = \frac{Vout}{Vin} = 1 + \frac{Rf}{R2}$$

$$Av = \frac{Vout}{Vin} = 1 + \frac{0}{\infty}$$

$$Av = \frac{Vout}{Vin} = 1$$

$$Vout = Vin$$

In this noninverting circuit configuration, the input impedance R2 increases infinitely, and the feedback impedance decreases to zero. The output is directly connected to the negative electronic signal, so the response is 100%, equal to the Vin, and has a definite gain of 1 unit.

An operational amplifier as voltage adder

We know that a single input voltage (Vin) was applied to the inverting input terminals in the inverting amplifier. If you add more input resistors to the inputs, each input will be equal to the value (Rin) of the original input resistor; the end result is an additive amplifier.

In this adder amplifier circuit, the output voltage Vout will be proportional to the input voltage V1, V2, V3, and so on. We can then modify the basic equation of the inverting amplifier to obtain these new equations:

$$If = I1 + I2 + I3 = \frac{V1}{Rin} + \frac{V2}{Rin} + \frac{V3}{Rin}$$

$$Vout = - \left[\frac{Rf}{Rin}V1 + \frac{Rf}{Rin}V2 + \frac{Rf}{Rin}V3 \right]$$

$$Vout = - \frac{Rf}{Rin}[V1 + V2 + V3]$$

You now have an op-amp circuit that amplifies the individual input voltages and produces an output voltage signal proportional to the algebraic sum of the three separate input voltages V1, V2, and V3. If individual inputs are required as their own resistor Rin, then Rin can be added as an input impedance to simply connect the input. This is because the input signals are effectively separated from each other by the Virtual Ground node at the inverting input of the op-amp. If all resistors are equal and Rf is equal to Rin, then direct voltage addition is also obtained. Note that if the connecting point is connected to the inverting input of the op-amp, the circuit will generate many negative input voltages. Similarly, if the connecting point is connected to the noninverting input of the op-amp, a positive amount of input voltage will be generated. The equation will be like-

$$\text{Vout} = -\,\text{Rf}\left[\frac{V1}{R1} + \frac{V2}{R2} + \frac{V3}{R3}\right]$$

Only two or more voltage signals may need to be connected without amplifying a connecting circuit. By placing all the resistors in the circuit at the same value, a voltage gain of the op-amp is obtained, and the output voltage is equal to the direct sum of all the input voltages.

$$\text{If }\ \text{Rf} = \text{R1} = \text{R2} = \text{R3}\ \text{ then}$$
$$\text{Vout} = -\,[V1 + V2 + V3]$$

Noninverting voltage adder

We can also use the op-amp's noninverting input to generate an adder of noninverting input. As we have seen above, if the inverting time amplifier produces a negative amount of input voltage, the sum of the noninverting of the amplifier configuration will produce a positive amount of input voltage.

As the name implies, noninverting adder amplifiers are based on the configuration of noninverting op-amp circuits where input (AC or DC) is applied to noninverting terminals as required. Negative feedback and gain are achieved by feeding a portion of the output signal (Vout) to the inverting terminal.

What are the advantages of a noninverting configuration over an inverting amplifier configuration, in addition to the most obvious fact that the output voltage of the inverter is in phase with the output and input? The output voltage is determined by the weighted sum of all the inputs and the ratio of resistance itself; this is the biggest advantage of noninverters. This is because there is no virtual ground position on the input terminal, so its input constraint is much higher than the value of an amplifier configuration. Changes in the closed-loop voltage gain of the op-amp do not affect the total input of the circuit. However, a further calculation is not recommended when choosing a heavy gain for each individual input at the total junction, especially if we have three or more inputs with different weight factors. However, the math involved is very low if all inputs have the same resistance. If the closed-loop gain of a noninverting op–amp is equal to the number of inputs, then the output voltage of the amp is equal to the sum of all the input voltages. This is the case when the number of inputs of a noninverting amplifier is equal to the gain: where the op–amp has two inputs, op–amp gain will be 2, and for three inputs, op–amp gain is 3. This is because the current flowing through each input resistor is a function of the voltage across all its inputs. If all input resistors are the same (R1 = R2), the conventional current cannot flow to the high-impedance noninverting inputs of the amp and is discarded because the output voltage is the sum of those inputs.

$$iR1 + iR2 = 0 \text{ as applying KCL}$$

$$\frac{V1 - V+}{R1} + \frac{V2 - V+}{R2} = 0$$

$$\left[\frac{V1}{R1} - \frac{V+}{R1}\right] + \left[\frac{V2}{R2} - \frac{V+}{R2}\right] = 0$$

If R1 = R2   considered as R

$$\left[\frac{V1}{R} - \frac{V+}{R}\right] + \left[\frac{V2}{R} - \frac{V+}{R}\right] = 0$$

$$\frac{2V+}{R} = \frac{V1 + V2}{R}$$

$$V+ = \left[\frac{V1 + V2}{2}\right] \times \frac{R}{R}$$

$$V+ = \frac{V1 + V2}{2}$$

As we know that the voltage gain of a noninverting op–amp is

$$Av = \frac{Vout}{Vin} = \frac{Vout}{V+} = 1 + \frac{RA}{RB}$$

$$Vout = \left[1 + \frac{RA}{RB}\right] \times V+$$

$$Vout = \left[1 + \frac{RA}{RB}\right] \times \frac{V1 + V2}{2}$$

If RA = RB, then Av will be 2, and the output will be the sum of input voltages.

$$\text{Vout} = [1+1] \times \frac{\text{V1} + \text{V2}}{2}$$
$$\text{Vout} = \text{V1} + \text{V2}$$

Therefore, for a three-input noninverting voltage adder amplifier configuration, if the closed-loop voltage gain is set to 3, the Vout is equal to the three input voltages V1, V2, and V3. Similarly, for four-input and five-input heat for closed-loops, the voltage gains will be 4 and 5. In addition, note that if the amplifier of the adder RA is equal to zero and RB is infinite, then in the absence of voltage, the output voltage Vout will be the average of all input voltages; that is, Vout = (V1 + V2)/2.

Differential amplifier

So far, we have only used the single input of an op-amp to connect to an amplifier and used the "inverting" or "noninverting" input terminal to amplify a single-grounded input signal. However, standard op-amps have two inputs, inverting and noninverting, so we can connect the signal from both inputs at the same time; this type of op-amp circuit is called a differential amplifier. In fact, as we saw in the previous section, all op-amps are "differential amplifiers" because of their input configurations. However, if we connect one voltage signal to one input terminal and another voltage signal to another input terminal, the result is proportional to the "difference" between the two input voltage signals V1 and V2. The differential amplifier then amplifies the difference between the two voltages by subtracting; this is an op-amp circuit, as opposed to a voltage adder amplifier that sums the input voltages.

The voltage transfer function of the differential amplifier is

$$\text{I1} = \frac{\text{V1} - \text{Va}}{\text{R1}}$$
$$\text{I2} = \frac{\text{V2} - \text{Vb}}{\text{R2}}$$
$$\text{If} = \frac{\text{Va} - \text{Vout}}{\text{R3}}$$

As we know that Va = Vb is the same voltage due to the virtual shot between two input terminals:

$$\text{Vb} = \text{V2} \left[\frac{\text{R4}}{\text{R2} + \text{R4}} \right]$$

If $V2 = 0$, then op-amp will act as an inverting amplifier, so consider as the first case, then

$$Vout1 = -V1\left[\frac{R3}{R1}\right]$$

If $V1 = 0$, then op-amp will act as a noninverting amplifier, considered the second case

$$Vout2 = Vb\left[1 + \frac{R3}{R1}\right]$$

$$Vout2 = V2\left[\frac{R4}{R2 + R4}\right]\left[\frac{R1 + R3}{R1}\right] \text{ as } Vb = V2\left[\frac{R4}{R2 + R4}\right]$$

Now according to the superposition theorem, both the input voltage V1 and V2 will be responsible for output, so

$$Vout = Vout1 + Vout2$$

So the transfer function will be

$$Vout = -V1\left[\frac{R3}{R1}\right] + V2\left[\frac{R4}{R2 + R4}\right]\left[\frac{R1 + R3}{R1}\right]$$

If we consider that the value of $R1 = R2$ and $R3 = R4$, then the transfer function will look like this:

$$Vout = -V1\left[\frac{R3}{R1}\right] + V2\left[\frac{R3}{R1 + R3}\right]\left[\frac{R1 + R3}{R1}\right]$$

$$Vout = -V1\left[\frac{R3}{R1}\right] + V2\left[\frac{R3}{R1}\right]$$

$$Vout = \frac{R3}{R1}(V2 - V1)$$

If the resistance R1 becomes equal to R2, the amplifier will have unit gain, and then the transfer function will be

$$Vout = (V2 - V1)$$

In addition, note that the sum of the output voltage will be negative if the input V1 is greater than the input V2, and the sum of the output voltages will be positive if the input V1 is greater than V2.

Instrumentational amplifier

An instrumentation amplifier (in-amplifier) is a very high-gain differential amplifier with having very input impedance with a single output. Instrumentation amplifiers are primarily used to amplify much smaller differential signals from strain gauges, thermocouples, piezoelectric devices. Unlike standard OP-AMPS, where the closed-loop gain can be obtained from the connected feedback resistance between the output and input terminals, there is an internal feedback resistance attached to the positive or negative input terminal of an instrumentational amplifier. The input signal is effectively applied to two differential inputs V1 and V2, separately. The CMRR of the instrumentational amplifier is very good, and the zero output for V1 = V2 when it exceeds 100 dB.

The two noninverting amplifiers form a differential input phase that acts as a 1 + 2R2/R1 gain for differential input signals and a buffer amplifier to achieve uniformity for common-mode input signals. Because amplifiers A1 and A2 are closed-loop negative feedback amplifiers, the input voltage Va is equal to V1 can be expected. Similarly, the voltage of Vb is equal to the value of V2. Because the op-amp does not carry current to the input terminal (virtual ground), the same op-amp must be connected to the output with three resistance networks R2, R1, and R2. This means that the voltage at the top of R1 is equal to V1, and the voltage at the bottom of R1 is equal to V2. This resistive input creates a voltage difference between V1 and V2, a voltage drop equal to the differential input voltage against R1. This is because the voltage VA and VB at the same junction of each amplifier is equal to their positively applied voltage. However, when a common-mode voltage is applied to the amplifier input, the voltages across R1 are equal, and there is no current through this resistor. Amplifiers A1 and A2 act as a buffer for unity because no current flows through R1 (and therefore no current through R2 registers). The input through the output of A1 and A2 are at three resistor ends. This is because the differential gain of the network circuit can be changed by changing the value of R1. The voltage output from the differential op-amp A3, which acts as a subtractor, is only amplified by the difference between its two inputs (V2−V1) and the gain of A3. The gain of A3 would be 1 if R3 = R4.

The transfer function of the network is as follows:

$$Vout = (V2 - V1)\left[1 + \frac{2R2}{R1}\right]\left(\frac{R4}{R3}\right)$$

Tuned amplifier

Tuned Amplifiers amplifies a specific signal with a particular frequency. Selection of a specific signal is the main feature of this amplifier, and this is done by tuned circuit and resonant frequency.

The tuned amplifier is used to tune the amplifier. Tuning means liking. Such a process is called selection, when a specific frequency must be selected when rejecting another frequency from a set of available frequencies. This selection is made using a circuit known as a tuning circuit. When an amplifier circuit is replaced with a load circuit, such an amplifier can be called a tuned amplifier circuit. It is nothing but an LC circuit; it is also known as a tuned circuit and a resonant circuit or tank circuit. Select a frequency. The tuning circuit can amplify the signal in a narrow frequency band focusing on the resonance frequency. When the index response corresponds to the capacitor response of the tuning circuit at the equilibrium frequency, this type of frequency is called the resonant frequency. It is represented by fr.

The advantages of tuned amplifiers are
- Reactive components such as L and C reduce power loss and make the tuned amplifier more efficient.
- Providing high impedance to resonant frequencies increases the selection and amplification of the desired frequency.
- A small collector power supply conducts VC due to the low resistance of the parallel tuning circuit.

It is important to note that these advantages do not apply in the presence of highly resistant collector understandings.

A tuned circuit uses the property of resonance to select a particular range of narrow bandwidth frequency signals. The center frequency of the selected frequency band is considered the resonant frequency of the tuned circuit. The resonance frequency of a tuned circuit is given as

$$fr = \frac{1}{2\pi\sqrt{LC}}$$

Chopper amplifier

In the case of electronics, the amplitude of the signal must be managed whether it is large or small. Consider the case of a scene where you must amplify a signal in the millivolt range where the signal that correspondence changes very quickly over time. These signals cannot be bypassed by coupling capacitors and cannot be amplified using an RC coupled amplifier. Even the method of direct coupling is not suitable for this situation.

Therefore, to overcome this complication, the technique was used to cut the signal into a few small pieces through a chopper amplifier.

A chopper amplifier is an electrical circuit that applies a modulation scheme to reduce low-frequency noise signals and DC operations. Modern chopper amplifiers use different conditions to achieve this, but the basic principle is that the chopper amplifier works in conjunction with the original amplifier. These amplifiers use modules and switches for cutting purposes. The resulting signal is sent to the filter, shortened, and transmitted to the chopper and the main amplifier as an input. Chopper amplifiers are mainly classified into mechanical, nonmechanical, and differential chopper amplifiers.

Mechanical chopper amplifiers are the primary solution to provide conditions such as minimum drift and offset conditions using a mechanical relay switch. Compared with mechanical chopper amplifiers, nonmechanics employ photoconductors or photodiodes for modulation and demodulation. If the photodiode does not come in contact with light, it will not flow with any current. Then, when the photodiode comes in contact with light, a current flows, and the resistance decreases. This scenario matches the switching function.

In the case of biomedical, the two barriers that many people encounter when recording biopotentials are drift and noise. Noise occurs while the patient is moving. On the other hand, drift is a baseline change caused by various thermal effects. For DC amplifiers, if the input is DC 0 V, it will be caused by a sudden peak or shift. Therefore, in this state, the chopper amplifier works best by solving the drift problem of the DC amplifier.

The DC signal is fed to the first block of the circuit and is replaced as the AC signal. Then, in the next block, the truncated AC signal is amplified,

Table 3.3 Comparison of positive- and negative-feedback amplifiers.

Positive feedback	Negative feedback
Gain of the amplifier increases	Gain of the amplifier decreases
Decreases the gain stability	Increases the gain stability
Increase the noise and distortion	Decrease the noise and distortion
Bandwidth decreases	Bandwidth increases
Mostly used in oscillator	Mostly used in amplifier

and finally, the AC signal in the demodulator section is converted to the DC signal.

Chopper amplifiers have a wide range of uses, some of which are

- These devices are used as battery chargers
- Used in battery-controlled electric vehicles
- It is mounted on the railway track
- DC voltage can be used for amplification purposes
- Brushless DC torque motors use chopper amplifiers as drivers and activator devices use them as stepper motors
- DC also controls the speed of the motor
- It acts as a variable frequency driver and a class D type electronic amplifier
- Converted condenser filter
- Used in applications that require a minimum voltage limit

Feedback amplifier

A feedback amplifier is an amplifier having a feedback way that exists between output to input. Like all other amplifiers, feedback amplifiers increase the strength of a signal. But in this amplifier, the ratio of the feedback signal and the input signal works as a feedback factor that measures the sum of feedback. There are two parts to a feedback amplifier: one is an amplifier, and the other is a feedback circuit.

From the above figure, the gain of the amplifier is represented as A. The gain of the amplifier is the ratio of output voltage Vo to the input voltage Vi. the feedback network extracts a voltage $Vf = \beta Vo$ from the output Vo of the amplifier.

This voltage is added for positive feedback and subtracted for negative feedback, from the signal voltage Vs. Now,

$$Vi = Vs + Vf = Vs + \beta Vo$$

$$Vi = Vs - Vf = Vs - \beta Vo$$

The quantity $\beta = Vf/Vo$ is called a feedback ratio or feedback fraction.

Let us consider the case of negative feedback. The output Vo must be equal to the input voltage $(Vs - \beta Vo)$ multiplied by the gain A of the amplifier.

Hence,

$$(Vs - \beta Vo)A = Vo$$

Or

$$AVs = A\beta Vo = Vo$$

Or

$$AVs = Vo(1 + A\beta)$$

Therefore,

$$Vo/Vs = A/(1 + A\beta)$$

Let Af be the overall gain (gain with the feedback) of the amplifier. This is defined as the ratio of output voltage Vo to the applied signal voltage Vs, i.e.,

$$Af = (\text{Output voltage})/(\text{Input signal voltage}) - Vo/Vs$$

So from the above two equations, we can determine that the equation of gain of the feedback amplifier with negative feedback is given by

$$Af = A/(1 + A\beta)$$

The equation of gain of the feedback amplifier, with positive feedback, is given by

$$Af = A/(1 - A\beta)$$

These are the standard equations to calculate the gain of feedback amplifiers.

In a feedback amplifier, feedback can be of two types, positive and negative feedback.

Positive feedback: In positive feedback, the feedback energy is in phase with the input signal and support it. Gain is increased in case of a positive signal.

Negative feedback: In negative feedback, the feedback energy is out of phase with the input signal and oppose it. Gain is reduced in case of negative signal (Table 3.3).

Rectifiers

A rectifier is a device that converts a bidirectional AC into a unidirectional direct current (DC). Rectifiers are widely used in electronic devices because the most electronic device is operated by DC current and rectifier helps to convert from AC to DC.

Two primary voltages are widely used these days, AC and DC. These voltage types can be converted from one type to another using special circuitry designed for that particular conversion. This transformation happens everywhere. The main supply nature obtained from the power grid varies, and household appliances typically require a small DC voltage. This process of converting alternate current to direct current is named. Further processing before converting AC to DC includes filtering, DC−DC conversion, etc. The most common part of electronic power is the bridge modifier. DC power is modified to supply power to various electronic components from AC mains available in many electronic circuits. Common bridge rectifiers are used in a variety of electronic AC-based power devices. Another way to look at rectifier circuits is that they convert current instead of voltage. This makes it more intuitive because you are accustomed to using a current to define the properties of your components. In short, the rectifier takes a current that contains both negative and positive elements and modifies it so that only the positive elements of the current remain.

Rectifiers are mainly classified into single-phase and three-phase rectifiers. Further, these are classified into three types: uncontrolled, half-controlled, and fully controlled rectifiers. Now we will briefly describe some of these rectifiers.

Uncontrolled rectifier

Which rectifier cannot control its output voltage that is called an uncontrolled rectifier. The rectifier works with a switch. There are different types of switches, some of which are widely controlled and some of which are out of control. A diode is a one-way device that allows only one side to flow. The operation of the diode is uncontrollable because it operates as long as it is biased. This type of rectifier is called an uncontrolled rectifier because if you configure a diode in a rectifier, the rectifier is not completely under the control of the operator. The power cannot be changed according to the load requirements. Therefore, this type of rectifier is usually used with uninterrupted or static power. Uncontrolled rectifiers use only diodes and provide only a constant output voltage based on the AC input. A diode is a unidirectional device that allows current flow in only one direction. This diode rectifier is an uncontrolled rectifier. Uncontrolled rectifiers can be further divided into half-wave and Full-wave rectifiers.

Controlled rectifier

A controlled rectifier can control the power fed to the load. It is used to convert AC supply into unidirectional DC supply in an inverter. Controlled rectification is the process of converting AC to direct current (DC) based on the required voltage and current demand. To design the controlled feature, The silicon-controlled rectifier (SCR) is basically a thyristor, a four-layer (P—N—P—N) semiconductor device that contains three PN junctions in series called a Shockley diode. If a voltage higher than the trigger voltage is not applied to that voltage, the shackle diode is closed with too much resistance. When the voltage exceeds the trigger value, the resistance goes to a very low value, and the device is turned on. The transistors of the components help them to continue and continue. The structure is like a pair of interconnected bipolar transistors, one with PNP and the other with NPN, so no current flows through any base-emitter junction, and the other remains constant until it is turned on. The transistor cannot be turned on. When sufficient voltage is applied and one transistor breaks, it starts moving, allowing the base transistor to pass through the other and saturating both transistors to maintain the state.

The shackle is achieved by a slight addition advancing from the diode to the SCR. In fact, it is only a third wire connection to an existing NPR structure. If the SCR gate remains floating, the SCR behaves like a startling diode. This may exceed the critical rate of voltage increase between the anode and cathode due to breakover voltage or shoddy diode. One or both of the internal transistors are reduced until the dropout current shake is in cutoff mode like a diode. However, the gate terminals are directly connected to the base of the lower transistor and can be used as an alternative to SCR latching. By applying a small voltage between the gate and the cathode, the lower transistor is consequently forced by the base current, the upper transistor conducts, and the current is supplied to the base of the lower transistor, thus activating. Conversion. Ho. Depending on the gate voltage. The gate required to start the latch–up must be much lower than the current flowing from the cathode to the anode with the SCR, so the SCR amplitude is measured. Although SCR is a DC unidirectional device, most SCR applications are for AC power control. Multiple SCRs can be used in one or more adaptations to conduct current through both half-cycles of an AC wave if a bidirectional circuit current is required. The main reason for using SCR in ACR power control applications is the thyristor's unique response to alternative changes. Being a one-way (one-way) device, it has at most reached half-wave power at half-cycle loads of AC. In the AC half-cycle, the pole of the supply voltage is positive, and the downside is negative. However, to demonstrate the basic concept of time-proportional control, this simple circuit is better than full-wave power control, which requires two SCRs. If there is no trigger at the gate and the AC supply voltage is below the breakout voltage rating of the SCR, the SCR will never turn on. The SCR gate includes a gate cathode resistor to block the reverse flow through the standard rectifying diode and to connect the SCR gate to the anode. This puts SCR on every positive start. It will be triggered immediately after the half-cycle.

Half controlled rectifier

This is a hybrid rectifier where we use diode and thyristor together for converting AC current to DC. Using a half-controlled rectifier single quadrant of AC signal can be controlled, whereas in a fully controlled rectifier, both quadrants can be controlled. The single-phase half-wave-controlled rectifier circuit consists of an SCR/thyristor, an AC voltage source, and a load. Understanding can be perfectly resistant, persuasive, or a combination of resistance and emotion. Consider a resistive load for simplicity. The requirement to turn on the SCR is that the SCR must have front-biased and gate signals installed. In other words, SCR is introduced only when it is biased and dismissed or gated. The SCRT is only turned on when the current through the SCR decreases and a reverse voltage is applied longer than the SCR is turned off.

Half-wave rectifier

The half-wave rectifier is a rectifier that allows one-half of the AC voltage signal and blocks the other half, and converts AC voltage to DC voltage, so we obtain the pulsating DC voltage as an output signal. A single diode works for a single-phase supply, but in the case of a three-phase supply, three diodes are needed. Only half of the input waveforms reach the output, so three diodes are needed. By adding more filters in the half-wave rectifier circuit, the rippling of the AC frequency from the output can be reduced.

Full-wave rectifier

A full-wave rectifier converts both positive and negative half-cycles of the AC into DC (direct current). Output voltage supply is double for a full-wave rectifier compared with the half-wave rectifier. Full-wave rectifier further divided into Bridge Rectifier and Center-tap rectifier.

Like half-wave circuits, full-wave rectifier circuits generate output voltages or currents that contain pure DC or specific DC components. Some primary advantages of full-wave rectifiers over the corresponding half-wave rectifiers are that some average (DC) output voltages are higher than half-wave and produce a full-wave rectifier with a waveform that produces a smooth output waveform.

Center-tap rectifier

This type of rectifier circuit uses a transformer with secondary winding tapped at the center point. Two diodes are connected in the circuit so that each one uses one half-cycle of the input AC voltage. For rectification, one diode uses the AC voltage showing the upper half of the secondary winding, while the other diode uses the lower half of the secondary winding. Full-wave rectifying circuits now use two diodes, one for each half of the cycle. A multi-rotating transformer is used in which the secondary windings are evenly split in two with a common center tap connection (C). As a result of this configuration, each diode operates when its anode terminal is positive with the center point C of the transformer. It produces an output during both half-cycles and is 100% efficient twice with a half-wave corrector. The input AC given to the full-wave rectifier is too high. The step-down transformer in the rectifier circuit converts high-voltage AC to low-voltage AC. The anode of the center tap diode is connected to the secondary winding of the transformer and is connected to the load resistor. During the positive half-cycle, diode D1 is biased to the front when connected to the top of the secondary air, and diode D2 is biased to the rear when connected to the bottom of the secondary air. Thus, diode D1 acts as a short circuit, and D2 does not act as an open circuit. During the negative half-cycle, when the upper half of the secondary circuit is negative, and the lower half of the circuit is positive, diode D1 is reverse biased, and diode D2 is front-biased. Thus, full-wave rectifiers supply DC voltage in both positive and negative cycles.

The average value of DC output voltage will be as follows:

$$V_{dc} = V_{dc}R_L = \frac{2}{\pi} I_{max} R_L$$

The current RMS value can be calculated using the following formula:

$$I_{rms} = \frac{I_{max}}{\sqrt{2}}$$

The rectification efficiency of a full-wave rectifier is twice that of a half-wave rectifier. The efficiency of a half-wave rectifier is 40.6%, whereas that of a full-wave rectifier is 81.2%. Full-wave rectifiers have a low ripple factor and require a common filter. For full-wave modifiers, the ripple coefficient is 0.482, but for half-wave modifiers, it is about 1.21. The output voltage and power obtained with a full-wave rectifier is higher than that of a half-wave rectifier.

Bridge rectifier

This type of single-stage rectifier uses four separate diodes or SRCs connected in a closed-loop "bridge" configuration to produce the desired output. The main advantage of this bridge circuit is that it does not require a special center taped transformer. This reduces the size and cost. A single secondary winding diode is connected to the load on one side of the bridge network and on the other.

Four diodes labeled D1 through D4 are placed in a "series pair," and only two diodes carry current during each half-cycle. During the positive half-cycle of the power supply, the diodes D1 and D2 in forward biased series conduct, and the diodes D3 and D4 are reverse biased to carry the load.

During the negative half-cycle of the power supply, D3 and D4 are forward biased and conduct the series, but D1 and D2 switch in the reverse bias, and "off" the current flowing through the load is in the same direction as before.

Because the current flowing through the load is one-way, the voltage across the load is also predictive, similar to the previous two diodes full-wave rectifiers. Thus, the average DC voltage across the load is 0.637 V.

In the previous section, we saw that a single-phase half-wave rectifier produces an output wave every half-cycle. It was not practical to use this type of circuit for a stable DC power supply. However, the full-wave bridge rectifier provides a larger DC value (0.637 Vmax) and lower ripple superposition, while the output waveform is twice the frequency of the input power. When improving the average DC output of the rectifier capacitor, you can use a smoothing capacitor to filter the output waveform to reduce the AC variation in the rectifier output. A smoothing or reservoir capacitor parallel to the load at the output of a full-wave bridge rectifier further increases the average DC output level because the capacitor acts like a storage device.

Power supply

A power supply is an electrical device that supplies electricity to those components that use electric power. A power supply is different from a power source. The main function of a power supply is to receive the current from a source and convert it to accurate voltage, frequency, or format to that component that is called power load. The power supply can be of different types.

DC power

DC power supply flows electric charge in one direction, so it supplies energy with fixed polarity. This power supply can obtain power from an AC or DC source. When we need a large power supply, this DC can be used for processes like the smelting of aluminum and other electrochemical processes. The battery is a very common example of a DC power supply.

AC power

AC power supply flows electric charge periodically in a reverse direction. AC power supply provides variable current and frequency to a load. Commonly we use AC power sources for electrical testing in aviation, lighting, laboratory testing, military, and factory production, and our normal electricity supply is AC supply. From an AC power source, we can generate AC supply from 45 to 500 Hz.

Linear-regulated power

The linear-controlled power supply is named because it uses linear non-switching technology to control the voltage output from the power supply. The term linear power supply refers to the fact that the power supply is controlled to supply the correct voltage to the output. The voltage is detected and fed back to this signal. It is usually compared with a reference voltage in any shape of a differential amplifier, and the resultant signal is used to keep the output at the required voltage. In some cases, the voltage can be detected at the output terminal, and in other cases, it can be obtained by the direct load. Remote sensing is used when there is a possibility of ohmic loss between the power supply and the load. When different capacities are required, different linear power supplies have different circuits and have different circuit blocks, but they always include basic blocks and some optional additions.

Since many controlled power supplies receive power from the AC main input, linear power supplies are step-down, or in some cases, have step-down transformers. It separates the power supply from the main input for protection. Transformers are usually relatively large electrical components, especially when used in high-power linearly controlled power supplies. Transformers can add significant weight to the power supply and can be very expensive, especially for high power. Depending on the rectification method used, the transformer may be single secondary or center taped. Additional windings may be present if additional voltage is required.

Inputs from AC power sources are optional and must be converted to DC format. Different types of rectifier circuits are available. The simplest form of rectifier that can be used to supply power is a single diode that provides half-wave correction. This method is not commonly used because the output is more difficult to smooth enough. Full-wave correction typically uses both parts of the cycle. It provides a waveform that can be smoothed more easily. There are two main approaches to providing a half-wave correction. One is to use a transformer with a center tap and two diodes. The second is to use a power transformer and a bridge rectifier with four diodes.

After the AC signal is restored, the DC must be slowed to eliminate different voltage levels. A large reservoir capacitor is used for this. The smooth material of the circuit uses a large capacitor. As this charge, the incoming wave rises to the top. As the voltage of the modified wave moves away, the capacitor begins to charge as soon as the voltage drops below the capacitor, and until then, the next rising wave holds the voltage until it is formed from the rectifier. Smoothing is not entirely accurate, and there are always some residues, but it can withstand large changes in voltage.

Nowadays, most power supplies provide a stable output. It is very simple with modern electronics, does not cost too much to include a linear voltage controller, and provides a constant voltage output regardless of the load at the specified limit. Many electronic devices, such as sensitive electronic devices used in biomedicine, require a controlled power supply because they require a properly regulated power supply.

There are two main types of linear power supplies.

Shunt Regulator: Shunt regulators are not used too much as the main component of linear voltage regulators. To provide such linear power, variable components are placed above the load. The source resistors are placed in series with the input, and the shunt controller is modified to maintain a constant voltage across the load.

Series Regulator: This is the most commonly used format for linear voltage regulators. The name implies that a series of components are placed in a circuit and has a variety of resistors by control electronics to create the correct output voltage for the current.

Switched-mode power

Switched-mode power supply (SMPS) is becoming more common and, in most cases, replaces the traditional theoretical linear AC–DC power supply due to reduced power consumption, reduced heat dissipation, and reduced size and weight. Switched-mode power supplies are now available in most PCs, power amplifiers, televisions, DC motor drives, etc., and switched-mode power supplies are becoming more mature, so almost anything that requires a highly efficient power supply is using SMPS.

By definition, an SMPS is a type of power supply that uses semi-conductor switching technology instead of the standard linear method to supply the required output voltage. A basic switching converter consists of a power switching phase and a control circuit. The power switching stage circuit performs a power conversion from input voltage Vin to output voltage Vout with output filtering.

The main advantage of switched-mode power supplies is that they are standard linear regulators used by switch transistors or power MOSFET. This reduces the cost of electricity. This means that when the switching transistor is fully turned on and the current flows, the voltage drop is minimized when the transistor is completely turned off. Therefore, the transistor acts as a standard switch.

The result is an input voltage negate that uses one or more switched-mode power supplies, phase down, phase up, and three basic switched-mode circuits, that is, buck, boost, and buck–boost topologies in contrast to linear regulators that only provide phasedown voltage regulation. It shows how transistor switches, inductors, and smoothing capacitors are connected to the basic circuit. The AC input supply signal is transmitted directly to the rectifier and filter circuit combination without the use of a 50 Hz transformer. There are many variations on this output, and it is necessary to increase the capacitance value of the capacitor to handle the input variation. These irregular DC are given to the central switching part of the SMPS.

High-speed switching devices such as power transistors and MOSFETs are employed in this section, which can be turned on and off according to the variety, and this output is provided to the primary side of the transformer in this section. The transformer used here is very small and lightweight, in contrast to the device used to supply 60 Hz. These are so efficient that they have a high energy conversion rate. The output signal from the switching section is corrected and filtered again to obtain the required DC voltage. It is a controlled output voltage that is fed to the control circuit; it is the feedback circuit. The final output is obtained after considering the response signal. The output sensor senses the signal and connects it to the control unit. The signal is separated from the second section, so sudden spikes do not affect the circuit. The reference voltage is provided to the error amplifier as the input of the signal, which compares the signal to the required signal level. The final voltage level is maintained by controlling the chopping frequency. This error is controlled by comparing the inputs provided to the amplifier. The output of the amplifier helps you decide whether to cut or increase or decrease the frequency. The PWM oscillator produces a standard PWM wave constant frequency.

Programmable power

DC power supplies provide a controlled DC output for reinforcing supply components, modules, or equipment. A good DC power supply must provide stable and accurate voltage and current with minimum noise for all types of loads such as resistance, low-impedance, high-impedance, steady-state, and variable. Its limitations in being reached are defined in its specifications. The power supply has two main settings: output voltage and current range. How we configure these with the load determines how the power supply works. Most DC power supplies have two modes of operation. In constant voltage (CV) mode, the power supply controls the output voltage based on the user's settings. In constant current (CC) mode, the power supply controls the current. Whether the power supply is in CV mode or in CC mode depends on both user settings and load resistance.

CV mode is the normal operating state of the power supply that controls the voltage. The output voltage is constant and is determined by the user's voltage settings. The output current is determined by the impedance of the load.

CC mode is considered a safe mode, but it can be used in other ways. In CC mode, the output current remains constant and is determined by setting the user's current limit. The voltage is determined by the impedance of the load. If the power supply is in CV mode and it exceeds its current user limit setting, the power supply will automatically switch to CC mode. If the power load falls below the current limit setting, the power supply may return to CV mode. The most important parameters for the application are the maximum voltage, the maximum current, and the maximum supply that can generate the power supply. We must make sure that the power supply can power the required voltage and current level.

DC-powered users turn on the potentiometer to set the output voltage or current. Today, microprocessors receive inputs from users. A digital-to-analog converter (DAC) takes a digital setting and converts it to analog values. It is used as a reference for analog controllers. The resolution and accuracy values of the setting depend on the quality of this conversion and adjustment process. There is a resolution and accuracy detail associated with each voltage and current setting (sometimes known as a limit or program value). The resolution of these settings determines the minimum amplitude where the output can be adjusted, and the accuracy represents the degree to which the values of the output match the international standard. In addition to the output settings, there are measurement or readback specifications that differ from the output specification. Most DC power supplies have built-in measuring circuits for measuring both voltage and current units. These circuits measure the voltage and current supplied by the power output. The measurements produced by the circuit are often referred to as readback values because the circuit reads the voltage and current in the power supply. Most professional power supplies include circuits that use analog-to-digital converters, and the specifications of these internal devices are similar to those of a digital multi-meter. The power supply displays readings on the front panel and can transmit them to a remote interface if it is equipped. Determining accuracy determines how close the regulated parameters are to the theoretical values defined by international standards. Uncertainty in power output is mainly due to DAC error conditions, including quantization errors. Accuracy is tested by measuring tuned variables with a traceable and accurate measurement system connected to the power supply output. The accuracy of the settings is as follows: (% of setup + offset). For example, consider a power supply with a voltage setting accuracy specification of $\pm$ (0.03% + 3 mV). Once ready to supply 5 V, the output value is uncertain (5 V) (0.0003 + 3 mV), or 4.5 mV. The set resolution is the smallest change in voltage or current setting that the power supply can select. This parameter is sometimes referred to as programming resolution during interface on the bus.

Uninterruptible power

An uninterruptible power supply (UPS) is an electrical device that provides emergency power to the load in case of any input or major failure. UPS is different from auxiliary or emergency power systems or standby generators that provide short-term protection from input power outages by providing power stored in batteries and supercapacitors. Most uninterruptible power supplies have a relatively short battery life, but a standby power supply is sufficient to properly boot or turn off a secure device. It acts as a normal power system. UPSs are commonly used to protect hardware such as computers, data centers, telecommunications equipment, and other electrical equipment. Unexpected power outages can result in injury, death, serious business interruptions, or data loss. The world's largest UPS, 4-megawatt battery electrical storage system (BSES) in Fairbanks, Alaska, empowers the entire city and surrounding rural population in the event of a power outage. The main role of a UPS is to provide short-term power in case of input power failure. However, most UPS units can fix common utility problems like:

- Voltage spike or uninterrupted overvoltage
- Continuous drop in speed or input voltage
- Voltage drop
- Harmonic distortion is defined as a deviation from the expected standard sine wave on the line

Some manufacturers of UPS units categorize their products according to the number of power-related problems they resolved. The three most common categories of modern UPS systems are online, line-interactive, and standby.

Online UPS uses a "dual conversion" method that takes AC input to correct DC via a rechargeable battery (or battery string) and then powers a device at 120 V and returns to 230 VAC. The line UPS puts the inverter on the interactive UPS line and switches the current path of the battery DC to supply when power is lost from normal charging mode. In a standby ("offline") system, the load is driven by the direct input supply, and the backup supply circuit is applied only in case of supply power failure. Most UPS (1 kVA) lines below 1 kW A are of the interactive or standby type and are usually cheaper.

For large power supplies, a dynamic uninterruptible power supply (DUPS) can be used. The synchronous motor/alternator is connected to the mains power supply through a choke. Flywheel stored the energy. In the event of a line failure, the stored current control keeps the load driven until the power of the flywheel is exhausted. The DUPS can be integrated or integrated with the diesel generator. The diesel generator was turned on a little later, making the diesel rotary uninterruptible power supply.

Offline/standby UPS: Offline/standby UPS only provides maximum initial functionality, protection, and battery backup. Welded devices are usually directly connected to incoming commercial power. When the input voltage falls below or above a certain level, the UPS turns on the internal DC-AC inverter circuit powered by the internal battery. The UPS then converts the mechanically connected device to a DC-AC inverter output to switch time up to 25 milliseconds, and it takes a standby UPS to detect lost commercial voltage. UPSs are designed to obtain some special equipment such as personal computers without creating sudden shutdown.

Line-interactive UPS: Line-interactive UPS is similar to the operation of standby UPS, but with the addition of a multi-tap variable-voltage autotransformer. It is a special type of transformer that can add or subtract coil drive. The coils serve to reduce or decrease the magnetic field and output voltage of the transformer. This can also be done with a back booster transformer that is different from an autotransformer. This is because autotransformers can be wired to provide galvanic insulation. This type of UPS can withstand uninterrupted undervoltage drops and overvoltage surges without consuming limited extra battery power.

Online/double-conversion UPS: With an online UPS, the battery is always connected to the inverter, so no power transfer switch is required. When power is lost, the rectifier simply exits the circuit to keep the battery power stable and holding. When the power is restored, the rectifier starts carrying most of the load and starts charging the battery, but the charging current may be limited to protect the high-power rectifier from battery damage. The main advantage of online UPS is that it can provide an "electric firewall" between commercial input power and sensitive electronics. Online UPSs are ideal for devices that require electrical isolation or are highly sensitive to environmental fluctuations.

Multivibrators

A multivibrator is an electronic circuit that oscillates between high and low and produces a continuous output. This circuit operates in a sequential way means in this device, two transistors are connected in such a way so that one controls the state of the other. In other words, multivibrators are amplifiers that provide positive feedback from the output of one amplifier to the input of another. Hence, the ON and OFF states of the whole circuit and the time periods for which the transistors are driven into saturation or cut off are controlled by the conditions of the circuit. Sequential logic circuits that use the clock signal for synchronization are frequency-dependent, so the clock pulse width enables switching operation. Sequential circuits using ascending, falling, or clockwise signals are also seen in previous basic flip-flop circuits using the following list usually related to timing pulses or waveform.

Astable multivibrator

Astable multivibrators are the most commonly used in multivibrator circuits. An astable multivariate is a self-propelled oscillator that does not have a permanent "meta" or "steady" state but constantly shifts its output from continuous low to another state high. This continuous switching operation from High-to-low and low-to-High produces an uninterrupted and stable square wave output that switches between two logic levels, making it ideal for time and clock pulse applications.

Construction of astable multivibrator

Two transistors named Q1 and Q2 are connected in feedback to one another. The collector of transistor Q1 is connected to the base of transistor Q2 through the capacitor C1 and vice versa. The emitters of both the transistors are connected to the ground. The collector load resistors R1 and R4 and the biasing resistors R2 and R3 are of equal values. The capacitors C1 and C2 are of equal values.

Advantages.
- Astable multivibrator does not need any external triggering
- Circuit layout of this multivibrator is very simple and easy
- It is less expensive
- Provide a continuous function

Disadvantages:
- Energy absorption is more within the circuit
- Output signal is of low energy
- Duty cycle less than or equal to 50% cannot be achieved

Monostable multivibrator

The circuit in which a vibrator switches from a stable to an unstable state is called a monostable multivibrator. In this multivibrator, one state is stable, and another is unstable. If we provide an external trigger pulse, the circuit will enter into an unstable state, and when the time is set, the circuit will return to a stable state.

Monostable multivibrators or one-shot pulse generators are usually used to convert short, fast pulses into very wide pulses for time application. When the appropriate external trigger signal or start pulse is applied, the high or low exclusive multivibrator generates a single output pulse.

Construction of monostable multivibrator

Two transistors Q_1 and Q_2, are connected in feedback to one another. The collector of transistor Q_1 is connected to the base of transistor Q_2 through the capacitor C_1. The base Q_1 is connected to the collector of Q_2 through the resistor R_2 and capacitor C. Another DC supply voltage $-V_{BB}$ is provided to the base of transistor Q_1 through the resistor R_3. The trigger pulse is provided to the base of Q_1 through the capacitor C_2 to change its state. R_{L1} and R_{L2} are the load resistors of Q_1 and Q_2.

One of the transistors, when moved into a stable state, an external trigger pulse is provided to change its state. After changing its state, the transistor remains in this quasi-stable state or Metastable state for a specific time period, which is determined by the values of RC time constants and returns to the previous stable state.

Advantages:
- One trigger pulse is sufficient to obtain the output.
- Circuit layout is quite easy and simple.
- This multivibrator is quite a bit cheaper than others.

Disadvantages:
- The time between T (the applications of trigger pulse) must be greater than the RC (time constant of the circuit) is the main drawback of using this.

Bistable multivibrator

The bistable multivibrator has two stable states and maintains a certain output state indefinitely until an external trigger forces the state to change. A bistable multivibrator can move from one stable state to another by applying an external trigger pulse, so two external trigger pulses are required before returning to the original state. Because the bistable multivibrator has two stationary positions, it is commonly known as a Latches.

An isolated bistable multivibrator is a two-state nonregenerative device consisting of two current-connected transistors that act as "ON−OFF" transistor switches. For every two states, one transistor is cut off, and the other is saturated. This means that the bistable circuit can maintain a stable condition. To change the bistable from one state to another, a proper trigger pulse is required for the bistable circuit, and to perform a complete cycle, each of the two trigger pulses requires a stage.

Construction of bistable multivibrator

Two similar transistors Q_1 and Q_2, with load resistors R_{L1} and R_{L2}, are connected in feedback to one another. The base resistors R_3 and R4 are joined to a common source V_{BB}. The feedback resistors R_1 and R_2 are shunted by capacitors C_1 and C_2, known as Commutating Capacitors. The transistor Q_1 is provided a trigger input at the base through the capacitor C_3, and the transistor Q_2 is provided a trigger input at its base through the capacitor C_4. The capacitors C_1 and C_2 are also known as Speed-up Capacitors, as they reduce the transition time, which means the time taken for the transfer of conduction from one transistor to the other.

Advantages:
- Earlier output is stored unless disrupted.
- Design of the circuit is quite easy.

Disadvantages:
- It needs two types of trigger pulses.
- This multivibrator is quite a bit more expensive than others.

Wave-shaping circuits

An electronic circuit that is commonly used to shape a particular wave that changes with time by using transistor, analog, or digital circuits, capacitors, and inductors is called a wave-shaping circuit. The square wave, sine and rectified sine waves, and triangular waves are some of the periodic waveforms. This circuit generates periodic waveforms. The wave-shaping circuit receives waves from AC or DC and emphasizes a certain frequency or a range of frequencies. This circuit can also reduce the effect of a particular frequency or a range of frequencies that is undesired. Wave-shaping circuit is two types, linear and nonlinear.

Linear wave-shaping circuit

In a linear circuit, by changing a certain amount of voltage, the voltage or current in any other part of the circuit can be changed. In this circuit, to shape a signal, some linear elements are used, such as resistors, capacitors, and inductors, etc.

By the filtering process, unwanted signals or a certain portion of a signal are attenuated by an input signal. Filters have certain components that help in this process

1. Capacitors, which allow AC and resist DC
2. Inductors, which allow DC and resist AC

There are mainly four types of filters

- Low-pass filter
- High-pass filter
- Bandpass filter
- Band-stop filter

Filters can be divided into two different types: active filters and passive filters. Active filters have amplifier devices to increase the signal strength, whereas passive filters do not have amplifier devices to increase the signal. The passive filter design has two passive elements, so the amplitude of the output signal is smaller than the corresponding input signal. Therefore, the passive RC filter reduced the signal, and the gain is always less than 1.

Low-pass filter

When a set of frequencies that are less than a specific value are passes through a filter circuit that is called a low-pass filter. This filter allows the passing of the lower frequencies. The capacitor filter or RC filter and the inductor filter or RL filter both act as low-pass filters.

A simple passive RC low-pass filter or LPF can be easily created by connecting a single capacitor and a single capacitor in series. In this type of filter configuration, the input signal Vin is applied simultaneously to a series combination of both a resistor and a capacitor, but the output signal Vout is only received across the capacitor. This type of filter is commonly referred to as a first-order filter or single-pole filter, but the circuit contains only one reactive component, the capacitor.

The response of the capacitor varies with the frequency, but the value of the resistor remains the same as the frequency changes. At lower frequencies, the capacitive reactance Xc of the capacitor is much higher than the resistance value of the resistor. This means that the voltage drop across the capacitor across the potential Vc is much higher than the Vr. The invalid value of the Vc changes as the frequency increases, so when the Vc is smaller, and the Vr is larger as the opposite. The upper circuit is an RC low-pass filter circuit, but it can also be thought of as a frequency-dependent variable voltage divider circuit.

$$Vout = Vin \times \frac{Xc}{\sqrt{R^2 + Xc^2}}$$

where $Xc = \dfrac{1}{2\pi fC}$

f is the AC signal frequency, C is the capacitance and R is the resistance.

By plotting the network output voltage as opposed to the different values of the input and output, you can obtain the frequency response curve or Bode plot function of the low-pass filter circuit.

The Bode plot shows that the frequency response of the filter is almost flat for low frequencies and goes directly through all the outputs of the input signal. The result is about one gain until it reaches the cutoff frequency. This is because the response of the capacitor increases at a lower frequency, blocking the current flowing through the capacitor. After this cutoff frequency, the response of the circuit will be reduced to -20 dB/decade or (-6 dB/octave) with the roll-off operation. Note that the slope angle, this -20 dB/decade roll-off, is always the same for any RC combination. The high-frequency signal connected to the low-pass filter circuit above this cutoff frequency point has been significantly altered. That is, it decreases rapidly. This occurs because the capacitor is so responsive to very high frequencies that it is reacted as a short circuit in the output terminal, and the output is zero. Then, by carefully selecting the right resistor and capacitor combination, we can create an RC circuit that passes through the circuit without affecting the frequency range below the specified value. For this type of low-pass filter circuit, frequencies below cutoff frequency are known to be in the filter passband zone. This passband also represents the bandwidth of the filter. The signal frequencies above this point cutoff point are usually called in the filter stopband zone.

Second-order low-pass filter

So far, we have seen that a simple first-order RC low filter can be made by connecting a single resistor to a single capacitor in the series. This unipolar configuration provides a roll-off slope of -20 dB/decade reduction at frequencies above the -3 dB cutoff point. However, in filter circuits, this -20 dB/decade (-6 dB/octave) slope angle may not be sufficient to remove unwanted signals, which may indicate the use of another low-pass filtering.

This circuit connects or "cascades" two passive first-order low-pass filters to form a second-order filter network. Therefore, we can make that simply by adding another first-order low-pass filter to an additional RC network.

When the n number of this type of RC step is cascaded, the resulting RC filter circuit is called the nth order filter with the nth −20 dB/decade roll-off operation.

For example, the operator of the second filter is −40 dB/deck (−12 dB/octave), the operator of the fourth filter is −80 dB/deck (−24 dB/octave), and many more. This means that as the order of the filter increases, the roll-off operation becomes higher, and the actual blocking band response of the filter comes closer to the characteristics of the ideal blocking band. Secondary filters are important and widely used in filter design. When combined with the first-order filter, it can be designed with a high-order nth order filter. For example, a third-order low-pass filter is made by connecting or cascading first and second-order low-pass filters in series. However, the RC filter stage also has its downside. However, there is no limit to the order in which filters can be created. The gain and accuracy of the final filter decrease as the order increases.

Active low-pass filter

The main drawback of passive filters is that the amplitude of the output signal is less than the amplitude of the input signal. That is, the gain never exceeds 1 unit, and the load resistance affects the properties of the filter. At different stages of using passive filter circuits, this loss of signal amplitude can be fatal. There is a way to recover or control this signal loss by using amplification with an active filter.

As the name implies, active filters have active components such as amps, transistors, and FETs in their circuit design. They draw energy from an external power source and use it to amplify or amplify the output signal. Filter amplification can be used to shape or modify the frequency response of the filter circuit by creating a shorter or wider output bandwidth and more selective output response. Second, the main difference between a "passive filter" and an "active filter" is amplification. Active filters typically use op-amps in their design. Nevertheless, active filters are much easier to design than passive filters, and when used in good circuit design, they provide great performance features with sharp roll-off and low noise with very good accuracy.

This first-order low-pass active filter only provides a low-frequency path for the input of the noninverting op-amp at the passive RC filter level. The amplifier is configured as a voltage follower (buffer) with a gain of 1, as opposed to the previous passive RC filters with a DC gain of less than 1.

The advantage of this configuration is that the high input impedance of the amp prevents additional load on the filter output and the low out frequency point prevents it from being affected by the change in load impedance. This configuration provides great durability for the filter, but its main drawback is that there is no voltage gain greater than 1. However, even if the voltage gain is 1, the power gain is very high because the output impedance is much smaller than the input impedance.

The frequency response of the circuit compliant with the passive RC filter is uniform, except that the output amplitude is increased by the passband gain of the amplifier. For noninverting amplifier circuits, the amount of filter voltage gain is given as the function of the input resistor R1 divided by the response resistor R2.

$$\text{DC gain Af} = \left[1 + \frac{R2}{R1}\right]$$

$$\text{Voltage gain of low pass filter Av} = \frac{Vout}{Vin} = \frac{Af}{\sqrt{1 + \left[\frac{f}{fc}\right]^2}}$$

Af = gain of the filter

where f = Input signal frequency

fc = Cut − off frequency of the filter

$$\text{Magnitude of the voltage gain Av(dB)} = 20 \log_{10}\left(\frac{Vout}{Vin}\right)$$

Second-order active low-pass filter Passive filters allow you to easily convert a primary low-pass active filter into a secondary low-pass filter via an additional RC network of input paths. The frequency response of the second-order low-pass filter is the same as that of the first-order type, with the exception of the blocking band roll-off 40 dB/decade (12 dB/octave), which is twice that of the first-order filter. Therefore, the design steps required for the secondary active low-pass filter are the same.

High-pass filter

If only signals in the form of a low-pass filter can go below the cutoff frequency point, as the name implies, a passive high-pass filter circuit will only pass signals above the cutoff point by removing low-frequency signal waves.

In this circuit configuration, the response of the capacitor is very high at low frequencies, so the capacitor acts like an open circuit, blocking the Vin input signal until it reaches the cutoff frequency point fc. Outside of this cutoff frequency point, the response of the capacitor will be significantly reduced. This is because the capacitor acts like a short circuit, allowing all the outputs of the input signal to pass directly.

The upper Bode plot or frequency response curve for the passive high-pass filter is opposite to a low-pass filter. Here, the signal is damped at low frequencies, and the output is increased to +20 dB/decade (6 dB/octave) until the frequency reaches the cutoff point fc. There is a bend extending from the infinity to the cutoff frequency, and the output voltage amplitude of the input signal value is $1/\sqrt{2} = 70.7\%$ or -3 dB of the input 20 logs (Vout/Vin). We can also see that the phase angle (Φ) of the output signal is led by the input and the frequency is equal to $+ 45$ degrees at fc. The frequency response curve of this filter means that the filter can send all signals to infinity. In reality, the filter response is not infinite, but it is limited by the electrical properties of the components used.

$$\text{Cut-off frequency fc} = \frac{1}{2\pi RC}$$

$$\text{Gain Av} = \frac{\text{Vout}}{\text{Vin}} = \frac{R}{\sqrt{R^2 + Xc^2}}$$

where $Xc = \frac{1}{2\pi fC}$

In case of low frequency $Xc \rightarrow \infty$ and Vout = 0. For high frequency $Xc \rightarrow 0$ and Vout = Vin.

Second-order high-pass filter

Here, two first-order filters are connected or cascaded to form a second-order high-pass network. We can then use the additional RC network to convert the primary filter step to secondary, just as we would for a second-order low-pass filter. The resulting secondary high-pass filter circuit has a slope of 40 dB/decade (12 dB/octave).

For low-pass filters, the cutoff frequency fc is determined by both the resistor and the capacitor.

$$fc = \frac{1}{2\pi\sqrt{R1\ C1\ R2\ C2}}\ \text{Hz}$$

Active high-pass filter

The initial operation of an active high-pass filter (HPF) is similar to that of an equivalent RC passive high-pass filter circuit, but this time the circuit is impeded or included in a design that provides amplification and control.

Technically, there is no such thing as an active HPF. In contrast to passive HPFs, which have an infinite frequency response, the maximum passband frequency response of the active HPF is limited by the open-loop feature or bandwidth that is in use.

The first-order active HPF includes low frequencies and allows high-frequency signals to pass through. It has a passive filter section followed by a noninverting op-amp. The frequency response of the circuit is equal to that of a passive filter, except that the amplitude gain increases the signal amplitude and the value of the passband voltage gain for noninverting amplifiers is (1 + R2/R1).

This first-order HPF is a noninverting amplifier after the passive filter. The frequency response of the circuit is equal to that of the passive filter, except that the amplitude gain increases the amplitude of the signal. For inverting amplifier circuits, the filter voltage gain magnitude is given as the function of the response resistor R2 divided by the input resistor R1.

$$\text{Voltage gain } Av = \frac{Vout}{Vin} = \frac{Af\left(\dfrac{f}{fc}\right)}{\sqrt{1 + \left(\dfrac{f}{fc}\right)^2}}$$

$$Af = \text{the gain of the filter} \left(1 + \frac{R2}{R1}\right)$$

where
$f = $ Input signal frequency

$fc = $ The cut-off frequency.

$$\text{Magnitude of the voltage gain } Av = 20 \log_{10} \frac{Vout}{Vin}$$

Passive filters allow you to easily convert the first-order high-pass active filter into a second-order HPF via an additional RC network in the input path. The frequency response of the second-order HPF is the same as that of the first-order type, with the exception of the blocking band roll-off 40 dB/dc (12 dB/octave), which is twice that of the first-order filter. Thus, the design steps required for the secondary active HPF are the same.

Bandpass filter

Bandpass filters can be used to isolate or filter specific frequencies within a specific band or frequency range. Easy use of these types of passive filters is in audio amplifier applications or circuits, such as speaker crossover filters and preamp tone controls. In some cases, it does not start at 0 Hz (DC) or end at a certain upper-frequency point, but only a narrow or wide fixed frequency within a certain range or frequency band needs to be considered.

By connecting or cascading a single low-pass filter circuit with an HPF circuit, you can create another type of passive RC filter that passes a selected range of bands or frequencies. This new type of passive filter system produces a frequency selection filter commonly known as a bandpass filter.

Unlike low-pass filters, which only pass low-frequency signals, and HPFs that pass high-frequency signals, bandpass filters allow the signal to pass a constant frequency band without distorting the input. This frequency band can be of any width and is commonly known as filter bandwidth. Bandwidth usually consists of two specific cutoff frequency points. Here we can only define the term "bandwidth" BW is known as the difference between low cutoff frequency fc_{Low} and high cutoff frequency fc_{High}. In other words,

$$BW = fc_{High} - fc_{Low}.$$

Obviously, the cutoff frequency of the low-pass filter must be higher than the cutoff frequency of the HPF for the passband filter to work.

An ideal bandpass filter can be used to isolate or filter a specific band frequency, such as the cancellation of noise. Bandpass filters are commonly known as second-order filters because of the two reactive component capacitors in the circuit design.

The upper and lower cutoff frequency points of a bandpass filter are found using the same formula for both low-pass and high-pass filters.

$$\text{Cut-off frequency } fc = \frac{1}{2\pi RC}$$

We can also calculate the Resonance or Center Frequency fr points of the bandpass filter to determine whether the output gain is maximum. This peak value is, as expected, not the mathematical average of the upper and lower limits of the -3 dB cutoff point but a geometric or average value.

$$fr = \sqrt{fc_{Low} \times fc_{High}}$$

Active bandpass filter

An active bandpass filter is a frequency selection filter circuit that is used in electronic systems to separate a series of signals from a signal of a certain frequency or all other frequencies. This band or frequency range is set between two cutoff frequency points, called low frequency and high frequency, to connect the signal outside these two points. The cascade connection of individual low-pass and high-pass passive filters forms a kind of filter circuit with a wide passband. The first stage of the filter is the high-pass phase, which uses capacitors to block the DC bias from the source. This design has the advantage of creating a relatively flat even-odd passband frequency response, representing half the low-pass response and the other half the high-pass response.

Band-stop filter

By mixing a basic RC low-pass filter with an RC high-pass filter, we can create a simple bandpass filter that goes through the frequency range or frequency bands on either side of two cutoff frequency points. However, we can create another RC filter network called band-stop filter (BSF) to combine these low-pass and high-pass filter sections. It is the frequency band between at least these two cutoff frequency points.

The BSF is another type of frequency selection circuit that works in opposite contrast to the bandpass filters we have seen before. BSFs, also known as band-reject filters, pass all frequencies except specific stopband frequencies. If this blocking band is very narrow and greater than a few Hz, a band blocking filter is usually referred to as a notch filter because it exhibits a deep notch (steep side curve) with high-frequency response selection instead of flat broadband. In addition, like bandpass filters, BSFs are secondary (bipole) filters with two cutoff frequencies, usually at 3 dB. Then, the effectiveness of the BSF is that all frequencies pass from zero (DC) to the first low cutoff frequency point band, and all frequencies pass in the second high cutoff frequency.

$$BW = fc_{High} - fc_{Low}.$$

OP-AMP can be used in a BSF design to obtain the voltage gain of the main filter circuit. Two noninverting voltage followers can be easily converted to a basic noninverting amplifier by taking advantage of $Av = 1 + Rf/Rin$ as well as input and response resistors. In addition, if a BSF is required, the cutoff point of -3 dB, like from 1 to 10 kHz, and the stopband gain between them, must be -10 dB. We can easily find low-pass filters and HPF for these requirements and cascades them to create broadband bandpass filter designs.

Nonlinear wave-shaping circuit

A nonlinear wave-shaping circuit transmits through a nonlinear network and alters sinusoidal signals. In this circuit, either the shape of the wave is disabled or the DC level of the wave is altered. Nonlinear elements (like diodes, transistors) in combination with resistors can function as clipper circuits. Nonsinusoidal output waveforms are generated through this circuit from a sinusoidal input; for this procedure, nonlinear elements are used.

For example, a clipper circuit prevents the waveform voltage from crossing the voltage limit, which is already determined, without changing the other parts of the waveform. A clipper circuit is made up of linear and nonlinear circuits, but it does not have any storage component like a capacitor. The main feature of a clipper circuit is that it can remove unwanted noise, and by this circuit, sine waves can also be converted into square waves. In addition, we can obtain a contrast amplitude of the desired wave. The clipper circuit is further divided into two categories.

Positive clipper circuit

When a clipper circuit reduces or disables the positive portion of a wave, the circuit is called a positive clipper circuit. Positive diode clipper circuits can be different types:

- Positive series clipper
- Positive series clipper with positive Vr (reference voltage)
- Positive series clipper with negative Vr
- Positive shunt clipper
- Positive shunt clipper with positive Vr
- Positive shunt clipper with negative Vr

A clipper circuit where a diode is connected to a series of input signals to represent the positive part of the waveform is called a positive series clipper. The following figure shows the schematic of the positive series clipper.

Positive input cycle: When an input voltage is applied, the positive cycle of the input converts to point A of circuit A in agreement with point B. This reverse biases the diode and acts as an open switch. Therefore, no current flows through the load resistor, so the voltage across the load resistor is zero, so $Vo = 0$.

Negative input cycle: Due to the negative cycle of the input, point A in circuit A becomes negative with point B. It causes the diode to move forward and act like a closed switch. Therefore, the voltage across the load resistor is perfectly visible at output Vo and equal to the applied input voltage.

A positive series clipper with positive Vr is a clipper circuit in which a diode is connected to the series with an input signal. A positive reference voltage is biased by Vr, connected to the positive part of the waveform.

During the positive cycle of the input, the diode is inversely biased, and a reference voltage is present at the output. During the negative cycle, the diode becomes more biased and behaves like a closed switch.

A clipper circuit with a diode connected in series with the input signal, a negative reference voltage in Vr, and a positive part of the waveform is called a positive series clipper with a negative Vr.

During the positive cycle of the input, the diode is inversely biased, and a reference voltage is present at the output. Because the reference voltage is negative, the same voltage is displayed with constant amplitude. During the negative cycle, the diode becomes forward biased and behaves like a closed switch. Thus, an input signal output greater than the reference voltage will be displayed.

A clipper circuit where a diode is connected to the input signal by a shunt and represents the positive part of the waveform is called a positive shunt clipper.

Positive input cycle: When an input voltage is applied, a positive cycle of the input is made at point A in circuit A in agreement with the point marriage, making it more biased to the diode so that it behaves like a closed switch. Therefore, no current flows through the load resistor, so the voltage across the load resistor is zero, and Vo is zero.

Negative input cycle: The negative cycle of the input with point B is made at point A of circuit A. It biases the diode in the opposite direction and acts as an open switch. Therefore, the voltage across the load resistor is perfectly visible in the output Vo and equal to the applied input voltage.

A clipper circuit where the diode is separated from the input signal, biased by the positive reference voltage Vr and connected to the positive part of the waveform, is known as the positive shunt clipper positive Vr. During the positive cycle of the input, the diode is more biased, and no reference voltage is displayed at the output. During the negative cycle, the diode is reverse biased and acts as an open switch.

A clipper circuit that separates a diode from an input signal biases it with a negative reference voltage Vr, and a positive shunt clipper with a negative Vr represents the positive part of the waveform. During the positive cycle of the input, the diode is biased forward, and the reference voltage is present at the output. Because the reference voltage is negative, the same voltage is displayed with constant amplitude. During the negative cycle, the diode is reverse biased and acts as an open switch. Thus, an input signal output greater than the reference voltage will be displayed.

Negative clipper circuit

When a clipper circuit reduces or disables the negative portion of a wave, the circuit is called a positive clipper circuit.

Negative diode clipper circuits can be different types:
- Negative series clipper
- Negative series clipper with positive Vr (reference voltage)
- Negative series clipper with negative Vr
- Negative shunt clipper
- Negative shunt clipper with positive Vr
- Negative shunt clipper with negative Vr

A clipper circuit where the negative part of the waveform is connected to the diode series with the input signal to represent it is called a negative series clipper.

Positive input cycle: When an input voltage is applied, a positive cycle of the input is created at point A in circuit A. With the consent of the point marriage, it further biases the diode and acts as a closed switch. Therefore, the input voltage is only displayed to produce the output Vo with the load resistor.

Negative input cycle: The negative cycle of the input with point B is made at point A of circuit A. It biases the diode in the opposite direction and acts as an open switch. Therefore, in the A clipper circuit where the input signal is connected to the diode series, the positive reference voltage is biased into Vr, and the negative part of the waveform is presented, called a negative series clipper with positive Vr.

During the positive cycle of the input, the activity of the diode will start only when the value of the anode voltage exceeds the value of the cathode voltage of the diode.

A clipper circuit with a diode connected in series with the input signal represents a negative reference voltage in Vr, and the negative part of the waveform is called a negative series clipper with negative Vr. During the positive cycle of the input, the diode becomes forward biased, and the input signal appears at the output. During the negative cycle, the diode becomes reverse biased and does not operate. However, the applied negative reference voltage output will be displayed. Thus, the negative cycle of the output waveform is clipped after this reference level.

A clipper circuit where a diode is connected to the input signal by a shunt and represents a negative sequence of waveforms is called a negative shunt clipper.

Positive input cycle: When an input voltage is applied, the positive cycle of the input converts to point A of circuit A in agreement with point B. This biases the diode in the opposite direction and acts as an open switch. Therefore, the voltage across the load resistor is perfectly visible at output Vo and equal to the applied input voltage.

Negative input cycle: The negative cycle of the input is created at point A of circuit A with point B making it forward biased to the diode and acting like a closed switch. Therefore, the voltage of the load resistor becomes zero because no current flows through the load resistor.

A clipper circuit that separates the diode from the input signal is biased with a positive reference voltage Vr and represents the negative part of the waveform called a negative shunt clipper with a positive Vr. During the positive cycle of the input, the diode is reversed biased and acts as an open switch. Thus, the full input voltage, which is higher than the applied reference voltage, is present at the output. The signal stops below the reference voltage level.

During the negative half-cycle, when the diode is biased forward, and the loop is complete, there is no output. A clipper circuit that separates the diode from the input signal biases it with a negative reference voltage Vr and represents the negative part of the waveform is called a negative shunt clipper with a negative Vr.

During the positive cycle of the input, the diode is biased and acts as an open switch. Thus, the full input voltage is displayed at the Vo output. During the negative half-cycle, the diode tends to be forward biased. The negative voltage output to the reference voltage and the rest of the signal is off.

Two-way clipper

It is a positive and negative clipper with a reference voltage Vr.

The input voltage is bidirectional in both the positive and negative parts of the waveform with two reference voltages. For this, two diodes D1 and D2, and two reference voltages VR1 and VR2, are connected to the circuit. This circuit is also known as a combination clipper circuit.

During the positive half of the input signal, diode D1 forms the reference voltage Vr1 and appears at the output. During the negative half of the input signal, the diode D2 forms the reference voltage Vr2 and appears at the output. Thus, both diodes operate alternately and clip the output during both cycles. The output is transmitted through a load resistor.

Signal generator

A signal generator is a device that produces electromagnetic wave signals with a fixed amplitude, frequency, waveshape. Which is commonly used as a stimulus for the electronic measurement purpose of any item; it is also used for musical instruments. Most simple signal generators can be an oscillator with calibrated frequency and amplitude to serve different purposes; there are several types of signal generators.

Function generators

A function generator is a signal generator that produces three or more periodic waves. The function generator can be an analog electronic testing device or software used to produce multiple types of electrical waveforms over a wide range of frequencies. The sine wave, square wave, triangular wave, and sawtooth shapes are some very common waveforms generated by a function generator. General use of function generator is in the development, test, and repair of electronic equipment.

The figure above contains two current sources, an upper current source, and a lower current source. These two current sources are controlled by a frequency–controlled voltage.

Triangular wave

The integrator of the upper block diagram periodically receives an equal amount of AC from the current sources above and below. Therefore, the integrator repeatedly produces two types of output at the same time. The output voltage increases consistently with the time the current is removed from the current source above the integrator. The output voltage of the consistency decreases linearly with the time it takes to bypass the current from the low-current source of the integrator. Thus, the integral in the upper block diagram produces a triangle wave.

As shown in the block diagram above, the output of the integrator, the triangle wave, is applied as the input of the other two blocks by obtaining a square wave and a sine wave, respectively.

Square waves

Triangle waves have the same positive and negative slope alternately and periodically in an equal amount of time. Therefore, the voltage comparator multivibrator in the block diagram produces two types of output at the same amount repeatedly.

During the period of constant (high) voltage at the output of the voltage comparator multivibrator achieves a positive gradient of the triangle wave. Another type of constant (low) voltage occurs at the output of the voltage comparator multivibrator when the multivibrator achieves a negative gradient of the triangle wave.

The voltage comparator multivibrator in the upper block diagram produces a square wave. Voltage Comparator If the amplitude of the square wave generated at the output of the multivibrator is not sufficient, we can use a square wave amplifier to increase it to the required value.

Sine wave

The sinusoidal shaping circuit generates the sinusoidal output from the triangular input waveform. This circuit has a diode resistance network of sine waves. If the sine wave output amplitude at the output of the shaping circuits is insufficient, we can use a sine wave amplifier to increase the required value.

Radio-frequency and microwave signal generators

RF and microwave signal generators can be used in test equipment, receivers, and test systems for various applications such as cellular communication, WiFi, WiMAX, GPS, audio and video transmission, satellite communication, radar, and electronic warfare. RF and microwave signal generators usually have the same features and functionality but depending on the frequency range. RF signal generators typically range from a few kHz to 6 GHz, whereas microwave signal generators cover a very wide frequency range from 1 MHz to at least 20 GHz. Some models can use up to 70 GHz with direct coaxial output and several hundred GHz with external waveguide sources. RF and microwave signal generators can be further classified as analog and vector signal generators.

Analog signal generators: Analog signal generators based on sinusoidal oscillators were common before the introduction of digital electronics and are still in use today. There was a big difference in the purpose and design of RF signal generators and audible frequency signal generators.

The RF signal generator produces an uninterrupted wave radio frequency signal with defined adjustment amplitude and frequency. Many models offer various analog modulations as standard equipment or as an optional feature of the base unit. These may include AM, FM, ΦM (phase modulation), and pulse modulation. A common feature is that the output power of the attenuator signal changes. Output power can range from −135 to +30 dBm, depending on the make and model. A wide range of output power is desirable because different applications require different amounts of signal power. For example, if the signal needs to reach the antenna through a very long wire, a high-power signal may be needed to overcome the losses at the top of the wire and ensure adequate power for the antenna. However, when testing receiver sensitivity, low signal levels are needed to see how the receiver behaves in situations of low signal-to-noise ratios.

The RF signal generator is available in benchtop tools, rack mounting tools, embedded modules, and card-level formats. Mobile, field testing, and onboard applications will benefit from a lightweight battery-powered platform. For automated production testing, web browser access that handles multiple sources and improves output through fast frequency switching speed test time and throughput.

RF signal generator is required for maintenance and installation of radio receivers and is used in professional RF applications. RF signal generators have frequency band, power function (−100 to +25 dBC), single-sideband phase term can be determined by different carrier frequencies, spurs and harmonics, frequency and amplitude switching speed, and modulation function.

Vector signal generator: With the advent of digital communication systems, it is not possible to accurately test these systems with traditional analog signal generators. This led to the development of vector signal generators, also known as digital signal generators. These signal generators can generate digitally modified radio signals that can use many digital modulation formats such as QAM, QPSK, FSK, BPSK, and OFDM. In addition, modern commercial digital communication systems are based on almost all well-defined industry standards, and many vector signal generators can generate signals based on these standards. Examples include GSM, W–CDMA (UMTS), CDMA 2000, LTE, Wi-Fi (IEEE 802.11), WiMAX (IEEE 802.16). In contrast, military communications systems such as the JTRS, which place great emphasis on robustness and data security, typically use very unique methods. To test this type of communication system, users often create their own custom waveforms and download them to a vector signal generator to create the desired test signal.

Arbitrary waveform generators

Arbitrary Waveform Generator (AWG) is an electronic test device used to generate radio waves. These waves can be repetitive or single shots (only once). In that case, we need some kind of trigger source, internal or external. The resulting waves can be injected into the equipment under test and analyzed during the process to ensure the correct operation of the device or to find a fault on the device.

Unlike function generators, AWG can generate an arbitrarily defined waveform as output. Waveforms are usually defined as a series of "way-points" (specific voltage targets that occur at specific times with a wave), and AWG jumps to those layers or multiple layers to differentiate between them. For example, a square wave with a 50% charge cycle easily determines only two points: set the output voltage to 100% at t0 and return the output voltage to 0 at 50%. Set AWG to jump (not interpolated) to the desired square wave between this value and the result. By comparison, you can create a triangular waveform from the same data by setting a line projected between these two points.

Since AWG uses digital signal processing technology to synthesize waves, the maximum frequency is usually limited to a few gigabits or less. The output connector from the device is usually a BNC connector and requires a 50 or 75 Ω termination. AWG, like most signal generators, includes an attenuator, a variety of ways to change the output wavelength, and the ability to automatically sweep the output waveform through a voltage controlled oscillator. This feature makes it very easy to evaluate the frequency response of a particular electronic circuit.

Some AWGs also act as traditional static function generators. These may include standard waveforms such as signs, squares, ramps, triangles, sounds, and pulses. Some units include additional built-in waves such as exponential rise and fall, the function of sinx/x, and ECG signal. Some AWGs allow users to capture waves from multiple digital and mixed-signal oscilloscopes. Some AWG screens may display a waveform graph. Some AWGs have the ability to combine features of both AWG and digital pattern generators to provide a sound pattern output to multiple bit connectors to simulate data transfer.

Some AWG models include different detection schemes to adjust the output wave in real time based on the results of different measurements by triggering signal demodulation, photon counting, or oscilloscope. For example, some implementation of quantum error correction and quantum communication can reduce application response time by integrating signal production and detection.

Digital pattern generators

The digital pattern generator is part of electronic testing equipment or software to generate digital electronic simulation. The excitation of digital electronics consists of electromagnetic waves ("low state" and "high state" "0" and "1") between two conventional voltages combined with two logical states. The main purpose of the digital pattern generator is to stimulate the input of digital electronic devices. As a result, the voltage levels generated by digital pattern generators are often consistent with the I/O standards of digital electronics (TTL, LVTTL, LVCMOS, LVDS, etc.).

Digital pattern generators are sometimes referred to as "pulse generators" or "pulse pattern generators" and can also act as digital pattern generators. Thus, the difference between the two types of devices may not be obvious. Digital pattern generators are the source of synchronous digital excitation. The generated signal is interesting for testing digital electronics at the logic level. This is why it is called a "logical source" Pulse generators are intended for making electrical pulses of various sizes. These are mainly used for testing electrical or analog levels. Another common name for this type of device is "digital logic source" or "logic source."

Digital pattern generators are now available as standalone units, add-on hardware modules on other devices such as logic analyzers and PC-based devices. A standalone unit is a self-contained device that includes everything from the user interface and defines the patterns generated by the electronic interface that produces the output signal. Some test equipment manufacturers are offering pattern generators as an add-on module for logic analysts.

PC-based digital pattern generators are connected to a PC via peripheral ports such as PC, USB, and Ethernet. It uses the PC as a user interface to define and store the digital patterns to be transmitted. Digital pattern generators are characterized by multiple digital channels, maximum rates, and supported voltage values. The number of digital channels determines the maximum width of the generated pattern. This is usually an 8-bit, 16-bit, or 32-bit pattern generator. The 16-bit pattern generator can create any digital sample from 1 to 16 bits. The highest rate defines the minimum time interval between two consecutive patterns. For example, a 50 MHz

(50milion sample per second) digital pattern generator can create new patterns every 20 ns.

Supported voltage values define a set of electronic devices that can ultimately use digital pattern generators. In contrast, the instability and conversion characteristics of the output signal of the digital pattern generator are consistent with these voltage values. Examples of supported voltage values include TTL, LVTTL, LVCMOS, and LVDS.

Most digital pattern generators add features such as repetitive sequences, the ability to generate digital clock signals at specific frequencies, an external clock input, the ability to use trigger options, and pattern generation after receiving an event from the outside.

Signal analyzer

A signal analyzer is a device that measures the magnitude, phase, or any aspect of the input signal at a particular frequency within the intermediate frequency bandwidth. In a modern signal analyzer input signal is converted to an intermediate frequency and then filtered in order to band-limit the signal and prevent aliasing. A signal analyzer can be used as both spectrum analyzers and vector signal analyzers. As a spectrum analyzer, it does spectrum analysis, including phase noise, power, and distortion, and as a vector signal analyzer, it analysis modulation, demodulation quality.

A spectrum analyzer measures the amount of an input signal for frequencies within the entire frequency range of the instrument. Its main use is to measure the spectral power of known and unknown signals. The most common spectrum analyzer measurement of input signals is electrical. However, with the right transducer, we can consider the spectral composition of other signals such as sound pressure waves and light and light waves. Spectrum analyzers also exist for other types of signals, such as optical spectrum analyzers that measure using direct optical techniques as monochromatic.

Spectrum analysis of electrical signals reveals major frequencies, energies, distortions, harmonics, bandwidth, and other spectral components of the signal that cannot be easily detected by waves in the time domain. These parameters are useful for characterizing electronic devices such as wireless transmitters. The performance of the spectrum analyzer shows the frequency of the horizontal axis and the width of the vertical axis. To a casual observer, a spectrum analyzer looks like an oscilloscope. In fact, some experimental instruments can act as oscilloscopes or spectrum analyzers.

The types of spectrum analyzers are distinguished by the method used to obtain the spectrum of the signal. Spectrum analysts based on Swept Tune and Fast Fourier Transform (FFT) are

The Swept Tune analyzer uses a superheterodyne receiver to convert part of the input signal spectrum to the center frequency of the narrow-band passband filter. The instantaneous output power of this filter is recorded as a function of time. It will appear in the form. The frequency (using a voltage-controlled oscillator) at the center of the receiver at multiple frequencies is also a function of the frequency. However, if the sweep focuses on one frequency, it may miss short-term events on other frequencies.

The FFT analyzer calculates the order of time during the period. FFT refers to a special mathematical algorithm used in a process commonly used in conjunction with receivers and analog-to-digital converters. As mentioned above, the receiver input signal decreases the frequency of the part of the spectrum, but this part does not sweep. The purpose of the receiver is to reduce the sample that the analyst must make at a sufficiently low sample rate; the FFT analyzer can process all samples (100% duty cycle), thus avoiding short-term events.

A vector signal analyzer is a device that measures the amplitude and amplitude of the input signal at a single frequency within the device's IF bandwidth. Its main uses are to measure the in-channel of known signals such as error vector magnitude, coded principal intensity, and spectral thickness. Vector signal analyzers are useful for measuring and demodulating digitally modified signals such as W–CDMA, LTE, and WLN. These measurements are used to determine the quality of modulation and can be used for design verification and compliance testing of electronic devices.

CHAPTER 4

Radiological devices

Sudip Paul[1], Angana Saikia[1,2], Vinayak Majhi[1] and Vinay Kumar Pandey[1]

[1]Department of Biomedical Engineering, School of Technology, North-Eastern Hill University, Shillong, Meghalaya; [2]Mody University of Science and Technology, Laxmangarh, Rajasthan, India

Contents

Introduction to Biomedical Instrumentation and Its Applications
ISBN 978-0-12-821674-3
https://doi.org/10.1016/B978-0-12-821674-3.00004-8

Radiography is often used to image bone structures when looking for fractures. Magnetic resonance imaging (MRI) scanners are frequently used to take images of the brain or other internal tissues, especially when high-resolution images are required. Nuclear medicine is used to look within the digestive system or bloodstream for blockage and other medical problems. Ultrasound looks at fetuses in the womb and provides images of the internal organs when high resolution is not required.

Radiography is an image that can visualize the internal appearance of an object with the help of X-rays, gamma rays, or similar ionizing and nonionizing radiation. Radiographic applications include medical ("diagnosis" and "treatment") and industrial radiography. Similar techniques are used for airport security ("body scanners" that typically use backscatter X-rays). In traditional radiography, an X-ray generator produces an X-ray beam to project onto an object and create an image. Depending on the density and structural structure of the object, a certain amount of radiation is absorbed by the object. X-rays that pass through an object contain a detector (photographic film or digital detector) behind the object. Generating flat two-dimensional (2D) images by this technique is called general radiography. In a computed tomography (CT) scan, the X-ray source and related detectors revolve around an object as it passes through a self-generated cone X-ray beam. Any point of content is captured from many directions by different beams at different times. Information about beam fusion is calculated to produce 2D images in three planes (axial, coronal, and sagittal), and further processing can create three-dimensional (3D) images.

The body is a combination of materials at different concentrations, and ionizing and nonionizing radiation reveal this through attenuation. In case of ionizing radiation, the internal body structure to clarify receptor absorption of X-ray photons by high-density substances (such as calcium-rich bones). The study of anatomy using radiographic images is known as radioanatomy. Radiodiagnosis is usually made by a radiologist, who performs the image analysis. Some radiologists also specialize in the interpretation of images. Radiography involves a variety of imaging methods, including various clinical applications.

In *projectional radiography*, X-rays or other high-energy electromagnetic radiation forms are expressed, and latent images capture dormant rays (or "shadows") to create an image called "projection radiography." I would say "shadows" can be converted to light using a fluorescent screen and captured in photographic film, or later "read" by a "laser" with a fluorescent screen. Alternatively, you can activate the matrix directly with a solid-state detector (the charged-coupled device of a digital radiographic camera is quite similar to the larger version). Only bones and some organs (such as the lungs) are suitable for projection radiography. It has a relatively low cost with high diagnostic rates, with cost differences for soft and hard body parts primarily attributable to carbon having far fewer X-ray cross-sectional regions than calcium.

A *CT scan* (formerly known as CAT scan, where "A" stands for "axial") connects ionizing radiation (X-rays) to a computer to capture and generate soft and hard tissue images (hence, "tomography"—"tomo"—means "piece"). CT uses more ionized than clinical X-rays (both use X-ray radiation), but advances in technology have reduced the required CT radiation and scan time. [1] CT scans are usually of less duration, and the media used depend on the tissue to be monitored as long as you hold your breath for a long time. The radiologist can perform these tests with a radiologist (for example, if the radiologist performs a CT-guided biopsy).

Dual-energy X-ray absorptiometry (DEXA), or bone density measurement, is primarily used to test for osteoporosis. This is not standard radiography, because two narrow beams emit X-rays 90 degrees apart that scan the whole patient. Typically, the hip joint (femoral head), lumbar spine (lumbar spine), or calcium are copied to determine the bone density (calcium component) and provide a number (T score). It is not used for bone imaging because the image quality is insufficient to create accurate diagnostic images such as fractures and swelling. It is not uncommon but can also be used to measure total body fat. The amount of radiation emitted by a DEXA scan is much less than that from a standard radiography test.

Fluoroscopy is a term first coined by Thomas Edison during X-ray research. The name comes from the fluorescence of an illuminated plate that had been irradiated with X-rays. This technology provides ongoing, consistent radiographs. Fluoroscopy is primarily performed to visualize movement (tissue or contrast) or to perform a medical intervention in the form of angioplasty, pacemaker insertion, or joint repair/replacement. The latter can be brought into the operating room using a portable fluoroscope called a C-arm. You can rotate around the operating table to create a digital image for the surgeon. Biplanar fluoroscopy works like single-plane fluoroscopy in addition to showing two planes simultaneously. Navigating two planes is important for orthopedics and spinal surgery and can reduce operation time by relocating.

Angiography uses fluoroscopy to visualize the cardiovascular system. An iodine-based contrast agent is observed after injection into the bloodstream. Liquid blood and blood vessels are not very dense, so a higher contrast (such as larger iodine atoms) is used to visualize blood vessels with X-rays. Angiography looks for aneurysms, leaks, thrombosis, and new blood vessel development and determines catheter and stent placement. Balloon angioplasty is often performed by angiography.

Contrast-enhanced radiography uses specific contrast agents to create structures of interest that stand out from the background. Projection therapeutic angiography requires a contrast agent and is called general radiography and CT.

Other medical imaging methods that do not use X-rays are technically not X-ray imaging techniques, but because a hospital's radiology department processes all image forms, diagnostic imaging methods such as positron-emission tomography (PET) and MRI are classified as X-ray imaging. Here, treatment that uses radiation is considered radiation therapy.

Industrial radiography is a nondestructive method for inspecting a variety of manufactured components to verify the internal structure and integrity of the sample. Art radiography can be done using X-rays or gamma rays. Both forms of electromagnetic radiation have different wavelengths for different forms of electromagnetic energy. X-rays and gamma rays have shorter wavelengths, and this property allows them to enter, transfer, and emit various substances such as carbon steel and other metals. Common methods include industrial CT.

The role of the radiographer

The radiologist is responsible for creating many clinical images to diagnose a patient's condition. The patient's body parts must be properly positioned, and the required amount of radioactivity should be applied to create a high-quality clinical image. A radiologist has two important responsibilities. The first involves the proper use of technology, and the second involves taking care of patient needs within the treatment environment. Tests and procedures are performed on patients of all ages, including pediatric and elderly patients. Radiologic technologists work in many areas besides radiology, including surgery, emergency care, cardiac care, intensive care units, and hospital rooms. Radiologists provide specialized talent for diagnosing injuries and illnesses.

Equipment

a. *Sources*: In treatment and dentistry, simple radiography and computer tomography images use X-rays generated from X-ray tubes. Images from a radiograph (X-ray generator/machine) or CT scanner are called, respectively, radiogram/roentgenograms or tomograms.

Many other X-ray photon sources are possible and can be used for industrial radiography and research. These include battens, linear accelerators (Linux), and synchrotrons. Radioactive sources such as ^{192}Ir, ^{60}Co, or ^{137}Cs are used for gamma rays.

b. *Grid*: A Bucky–Potter grid can be placed between the patient and the detector to reduce the number of scattered X-rays, thus improving image contrast resolution but increasing patient exposure to radiation.

c. *Detectors* can be divided into two primary categories, dry image plates and X-ray film (photographic film) image detectors. They are employed to measure dosages or dosage rates—for example, to ensure that radiation protection devices and procedures are in place.

d. *Side markers* of the radioactive structure are added to each image. For example, if the patient's right hand is X-rayed, the radiologist will include a radiopaque "R" marker in the X-ray beam field as an indicator of the hand. The radiologist may add the correct side marker during digital postprocessing if the physical marker has not been included.

e. *Image intensifier and array detector.* As an alternative to X-ray detectors, image synthesizers are analog devices that display readily available X-ray images onto a video screen. The device has a vacuum tube coated with cesium iodide (CSI) on the wide internal input surface. When the X-ray element collides with the phosphor, it emits adjacent photocathode electrons. These electrons focus on using electronic lenses in boosters for output screens coated with phosphorus material. Images can be recorded from the output and displayed through the camera. Digital devices known as array detectors are becoming more common in fluoroscopy. These devices have a distinct pixilated detector, known as a thin film transistor, that indirectly uses a photodetector to detect or capture light emitted from a scintillator material such as CSI and collide it with the detector. Electrons produced by X-ray detectors are not affected by blurring or scattering, either because X-ray photons activate the detector directly or because of phosphorescent scintillators on the film screen.

Table 4.1 briefly describes various imaging techniques:

Table 4.1 Various imaging techniques.

Type	Targeted body region	Advantage	Disadvantage	Use
Radiography	Bony and soft tissue anatomy	Low radiation, inexpensive, readily accessible, quick, no arrangement required	Provides fundamental anatomic data for just a couple of tissue densities	Look at skeleton and chest, investigate intrastomach infection
Fluoroscopy	Anatomic and functional information	Provides images progressively, widely accessible	Radiation portion might be significantly more costly than radiography	Bar vesicoureteric reflux, upper and lower gastrointestinal assessment, therapeutic bowel purges for clogging or intussusception decrease
Sonography	Real-time evaluation of soft tissues	No radiation introduced, painless, portable, widely accessible	Administrator subordinated, pictures obtained are exceptionally reliant on sonographer mastery, more costly than radiography	Assess renal and intracranial pathology, suspected hepatobiliary illness, evaluate stomach or pelvic cycles, hip radiation, formative dysplasia of the hip, scrotal issues, or anomalies of the neck
Computed tomography (CT)	Anatomical data from practically any organ framework in the body	Fantastic portrayal of anatomic detail, very quick test time, with intravenous (IV) contrast can analyze organ upgrade just as veins, helical CT provides multiplanar and three-dimensional (3D) data	Higher radiation portion than radiography, may require oral or IV contrast, relatively costly	Screening assessment of vague indications, evaluate for contamination and stomach and chest injury, diagnose and screen malignancy

Magnetic resonance imaging	Detailed high-contrast resolution images of organs	Considers multiplanar and 3D assessment, does not need routine use of IV contrast material for imaging of midsection and pelvis (in contrast to CT), superior portrayal of delicate tissue and organ contrasts, no radiation introduced, painless	Frequently requires sedation in children younger than 7 years, expensive, scanner is loud, monitoring is restricted, requires patient to be motionless for imaging, picture quality	Assessment of focal sensory system, musculoskeletal assessment, cancer assessment
Nuclear medicine	Shape, structure, and function of organs, soft tissues, and bones	For the most part, conveys lower radiation portion than fluoroscopy or CT, adverse responses uncommon	May take quite some time and require sedation, offers minimal anatomic data contrast and different procedures	Studies identified with provincial perfusion, skeletal pathology, gastroesophageal, or vesicoureteric reflux, gastric purging, tumor observation

Fig. 4.1 shows the generic block diagram of a medical imaging system.

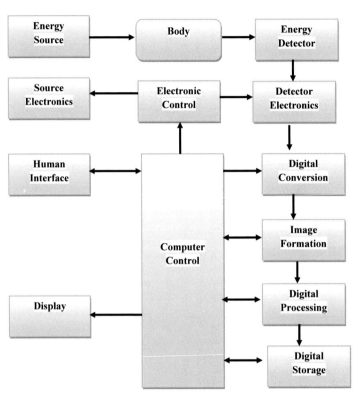

Figure 4.1 Medical imaging system block diagram.

Benefits of medical imaging

X-ray and CT discovery represent great advances in treatment. X-ray imaging is recognized as a valuable medical tool for various tests and procedures:

a. Assists in noninvasive diagnosis and monitoring of treatment
b. Supports treatment plans and surgical treatment plans
c. Guides the healthcare professional when inserting a catheter, a stent, or another device into the body, treating a tumor, or removing a blood clot or other obstruction

Risks of medical imaging

In many medical cases, there is a risk with X-ray images that use ionizing radiation to create body images. Ionizing radiation has enough energy to damage DNA. Risks of exposure to ionizing radiation include

a. X-rays are less likely to cause cancer in later life (general information on cancer detection and treatment for patients and healthcare providers is available from the National Cancer Institute).

b. Effects on tissues such as cataracts, skin redness, and hair loss due to relatively high levels of radiation exposure are rare in many imaging tests. For example, specifically using CT scanners or conventional radiographers should not affect the tissues, but skin administration from long and complicated conventional fluoroscopic procedures may, in some cases, be sufficient.

Another risk of X-ray imaging is a potential reaction associated with intravenous (IV) injection or "dye" that can also be used for vision correction. The risk of developing cancer from radiation exposure in medical imaging is usually very low and depends on

a. Radiation dose: testing patients with higher doses and more X-rays increases the lifetime risk of cancer.

b. Patient age: X-rays are available at a younger age than in patients with a lifetime risk of cancer who receive it at an older age.

c. Patient sex: Women of the same age and exposure have the same risk of radiation-related cancer as men.

d. Body area: Some organs are more radiosensitive than others.

The advantages of medically appropriate radiography generally reduce the risk of exposure to ionizing radiation, but efforts should be made to further reduce this unnecessary risk. To reduce patient risk, tests that use ionizing radiation should only be performed when answering treatment questions, treating an illness, or guiding treatment methods. If treatment with a specific diagnostic imaging procedure is required or other tests using radiation are inadequate, the benefits outweigh the risks, and the doctor will not consider radiation risks or study methods that affect the patient's decision. However, one should always follow the "as low as reasonably achievable" policy when selecting equipment settings to reduce radiation risk to the patient. It is important to consider patient factors in this balance of benefits and risks. For example: Since young patients are sensitive to radiation, special care should be taken to reduce radiation exposure to all types of radiography in pediatric patients. Because of the potential for radiation exposure from developed photons, special care should be taken when imaging pregnant patients. The benefits of potential disease detection should be carefully balanced with the risks of imaging screening studies of healthy, incomplete patients.

Some radiological imaging techniques are described below.

X-ray radiography

X-ray imaging uses an X-ray beam projected toward the body. These rays pass through the frame, and elements of the X-ray beam are absorbed. On the other side of the frame, the X-rays are detected, ensuring an image.

X-rays were discovered in 1895 by Wilhelm Conrad Röntgen, a professor at Wurzburg College in Germany. Agreeing to the Nondestructive Asset Center's "History of Radiography," Röntgen noted that precious stones close to a high-voltage cathode-ray tube showed a fluorescent shine when he protected them with dark paper. A few vital shapes were delivered by the tube entering the paper, causing the precious stones to shine. Röntgen called the obscure vitality "X-radiation." Tests appeared to show that this radiation could enter delicate tissues but not bone and would create shadow images on photographic plates. For this revelation, Röntgen was granted the first Nobel Prize in Physics in 1901.

X-rays, like visible light and radio waves, are a certain type of electromagnetic radiation. X-rays skip through most gadgets; however, variations in density change how they are absorbed. Scientific X-ray machines produce small bursts of radiation that skip through selected body regions, recording images on film, virtual plates, or fluoroscopic screens. Because X-rays are a form of ionizing radiation and can be harmful in excessive doses, exposure ranges are cautiously controlled and monitored. Radiation protection is the first and most important objective for device manufacturers, physicians, and radiologic technologists. The traditional X-ray exam has been used for over a century and still makes up about half of all imaging studies. With its simplicity and low radiation dose, conventional X-ray imaging is frequently the first type of imaging ordered, with applications including chest X-rays, bones, and joint dislocations. DEXA imaging is a specialized exam for assessing bone loss and diagnosing osteoporosis. The block diagram of the X-ray machine is shown in Fig. 4.2.

In forming a radiograph, a patient is positioned so that the portion of the body being imaged is between an X-ray source and an X-ray locator. When the machine is turned on, X-rays travel through the body and are

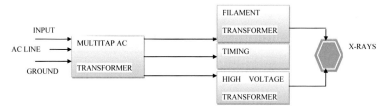

Figure 4.2 X-ray machine block diagram.

retained in distinctly different amounts by diverse tissues depending on the radiological thickness of the tissues they pass through. Radiological thickness is decided by the thickness and the nuclear number (the number of protons in an atom's core) of the materials being imaged. For instance, structures such as bones contain calcium, which incorporates a higher nuclear number than most other tissues. As a result, hard structures show up whiter than other tissues against a radiograph's dark background. Conversely, X-rays travel more effectively through less radiologically thick tissues such as fat and muscle, as well as through air-filled cavities such as the lungs. These structures are shown in shades of gray on a radiograph.

Formation of radiography

X-ray photons have the potential to enter tissue; a portion of them are weakened by the tissue, and a portion pass through the tissue to connect with and expose the radiographic film. Retention of X-rays may be a work of the nuclear number and thickness of the tissues/objects. Tissues/objects with a better nuclear number will retain more radiation than tissues with a lower nuclear number. Thicker tissue/objects will assimilate more X-rays than more slender tissue of comparable composition. The more prominent the sum of tissue retention, the less X-ray photons reach the film, and the whiter the image on the film. The radiograph will show a extend of densities from white, through different shades of dim, to dark. Radiopaque tissues/objects show up whiter, whereas radiolucent tissues/objects show up darker. The resultant design of opacities shapes an image on the radiograph that is recognizable in the frame and can be translated.

Gamma rays

There is no sensation about the definition of the difference between X-rays and gamma rays. One thing they have in common is that they distinguish between two types of radiation based on their source. X-rays are emitted from electrons and gamma rays are emitted from the atomic nucleus. There are some problems with this definition. Other processes can produce higher energy. No method of photon or sometimes generation is known. Isolation of X-rays and gamma rays by a common alternative wavelength (or equivalent frequency or photon energy). Below any wavelength is defined as noncombustion, such as 10^{-11} m (0.1 A), called a gamma ray. This parameter allocates photons in an ambiguous range, but it is possible to know the wavelengths of this (some measurement techniques do not differentiate between wave detection). These two definitions often match. The term is usually used in a specific context for historical instances based on the method of measurement (detection) rather than the wavelength or source, or the purpose of use. Therefore, gamma rays in the range of 6−20 MeV generated for therapeutic and industrial applications such as radiation therapy are sometimes referred to as X-rays in this context.

Properties of X-ray

X-ray photons carry enough energy to ion the atoms and break the molecular bonds. It is a type of ionizing radiation and is therefore harmful to living tissues. Too high a dose of radiation for a very short period of time causes radiation damage, but low doses can increase the risk of radiation-induced cancer. In clinical imaging, the risk of this cancer growth is usually influenced by the convenience of the test. X-ray radiation therapy can kill malignant cells as a cancer treatment method. It is also used for physical features using the X-ray spectrum. The erosion length of X-rays in water shows flattening at higher photon energies due to oxygen absorption edge, light have strength dependence and Compton scattering at 540 LV. The length of the extension is about four times longer for the hard X-ray (right half) than for the soft X-ray (left half). Hard X-rays can be done without absorbing or spreading excessively thick objects. For this reason, X-rays are widely used to image the interior of visible opaque objects. The most common applications are medical radiographs and airport security scanners, but similar technologies are important in the industry (such as industrial radiography and industrial CT scans) and research (such as small animal CT). The depth of penetration varies by different orders of width across the entire X-ray spectrum.

Interaction with matter

X-rays interacts in three main ways: photoabsorption, Compton scratching, and Rayleigh scratching. The energy of these interactions depends on the energy of the X-rays and the initial structure of the material, but the photon energy of X-rays is much higher than the energy of chemical bonding, so it is less dependent on chemistry. Photoelectric absorption is the main communication system in soft X-ray regions and low-rigidity X-ray energy. At high power, Compton dominates the scattering.

Photoelectric absorption: The probability of one photoelectric absorption per unit mass is approximately proportional to Z^3/E^3. Where Z is the atomic number and E is the energy of the photon of the event. This rule is invalid to the underlying shell electron binding energy, however, as it is the general tendency for high-absorption coefficients; thus, low penetration depth is very strong for low photon energy and high atomic number. In the case of soft tissues, photon analysis is at about 26 kV photon energy where Compton is crushed. This limit is higher for substances with higher atomic numbers. The amount of calcium in the bone (Z = 20) is high and dense, so it is clearly visible on medical X-rays. Photons absorbed photons transfer all their energy to interactive electrons, ions the electrons bound in electrons, and in that way create photons that can ion more atoms. These effects can be used for detection by X-ray spectroscopy or further electron spectrum.

Compton scattering

The main contact between X-rays and soft tissues in medical imaging is Compton scratching. Compton scratching is the indescribable scratching of X-ray photons by outer cover electrons. Some photon energy is transferred to the scattered electrons, ions the atoms, and prolongs the wavelength of the X-ray. Scattered photons can go in any direction, but one direction is more likely to be the same as the original, especially for high-energy X-rays. The possibilities of different scattered angles are expressed by the Klein–Nishina formula. Transferred energy is obtained directly from scattered angles by conserving energy and speed.

Production by electrons

X-ray tubes can be produced by X-ray tubes, which are vacuum tubes that use high voltage to accelerate the electrons emitted at high speeds by the cathode. High-speed electrons collide with the anode and the metal target produces X-rays. Treatment X-ray tubes usually target crack-resistant mixtures of tungsten or tungsten (5%) and tungsten (95%), but may be more specific applications with molybdenum, such as soft X-rays. In crystallography, cobalt is most commonly used when copper targets are the most common and can cause problems with regression from the iron content in the sample. The maximum energy event of the X-ray photon produced is limited by the energy of the electron. This is equal to the voltage of an electron-charged tube, so an 80 kV tube cannot produce X-rays above 80 kV. When an electron collides with a target, X-rays are formed by two separate atomic processes:

a. Characteristic X-ray emission (X-ray electrolysis): If the electron has sufficient energy, it can scatter the atomic electron from the inner electron shell of the target atom. Electrons at higher energy levels then emit X-ray photons by filling the voids. This process produces X-ray emission spectra at certain isolated frequencies, sometimes called spectral lines. These are usually changes from the top shell to the K shell (known as the K line), the L shell (L line) and the like. If the conversion is 2p to 1s, it is called $K\beta$, and if it is 3p to 1s, it is called K. The frequency of this line depends on the target material and is therefore called the characteristic line. $K\alpha$ rays are usually stronger than K and are more desirable in isolation tests. Therefore, who. The line is filtered by filter. Filters are usually made of a proton-less metal from anode material (e.g., Ni filter for Cu anode or Nb filter for Mo anode).

b. Bremsstrahlung: Higher Z (number of protons) is the radiation provided by an electron when it is propagated by a strong electric field near the nucleus. These X-rays have an uninterrupted spectrum. The frequency phenomenon of Bremsstrahlung is limited by the energy of the electron.

Therefore, the resulting tube output has a constant Bremsstrahlung spectrum that drops to zero in the line characterized by tube voltage and multiple spikes. The maximum power of X-ray photons is about 20−150 kV because the voltage used in diagnostic X-ray tubes is about 20−150 kV. Both of these X-ray production processes are inefficient and use about 1% of the electrical energy converted to X-rays by the tube, so most electrical energy used by the tube is released as waste heat. When producing an available X-ray stream, the X-ray tube is redundant. Should be made to dissipate heat. One of the specific light sources of X-rays widely used in research is synchrotron radiation produced by particle skin. Its distinctive features are X-ray tube, wide X-ray spectrum, great collimation, and multiple-order X-ray output larger than linear polarized light. The adhesive tape from the backing paper in a medium vacuum without peeling ensures that a small nanosecond burst of pressure is created on the X-ray with an energy peak of 15 kV. This may be the result of a recombination of the charge generated by the triboelectric charging. The intensity of X-ray triboluminescence is sufficient for use as a light source in X-ray imaging.

Projectional radiographs

The practice is to use X-ray radiation to produce 2D images. Bone contains a high concentration of calcium and has a relatively high atomic number, so it absorbs X-rays efficiently. This reduces the number of X-rays that reach the detector behind the bone and makes them more visible on radiographs. Lungs and trapped gases are clearly visible because they absorb less than tissues but the difference in tissue type is confusing. Hypothetical radiographs help to identify pathologies of the skeletal system and some pathological processes in the soft tissues. A few notable examples are quite common: chest X-rays to detect lung diseases such as pneumonia, lung cancer, or pneumonia, and gastric X-rays to detect intestinal diseases. Obstruction, free air can be detected (from intestinal perforation) and free fluid (ascites). X-rays can identify frequent (but not always) treatments (rarely opaque for such rarity) and treatment of kidney stones. Simple X-rays are not very useful for imaging soft tissues such as the brain and muscles. One area where large-scale radiographs are used is to treat orthopedics, such as knee, hip, or shoulder replacements, to determine how they are placed relative to the bones around the body. You can evaluate in 2D from a simple X-ray or in 3D using a technology called "Registration from 2D to 3D." This technique explicitly disables inductive errors related to the placement of implants from plain radiographs. Dental radiography is commonly used to diagnose common oral problems such as tooth decay. In clinical treatment applications, low-energy (soft) X-rays are not necessary because they are completely absorbed by the body and increase radiation levels without contributing to the image. Therefore, a thin sheet of metal, often referred to as aluminum, is called an X-ray filter, usually placed in the window of an X-ray tube that absorbs the low-strength portion of the spectrum. This is called beam stiffening because it shifts the center of the spectrum to higher energy (or harder) X-rays. Preliminary images (angiography) are taken from areas of physiological interest involving arteries and veins to produce images of the cardiovascular system. The second image was taken at the same location after iodized contrast medium was added to the blood vessels in this region. These two images are then digitally subtracted, leaving only the iodized image inside the blood vessel. The radiologist or surgeon then compares the resulting image with a regular physiological image to determine if there is any damage or obstruction.

X-rays have much shorter wavelengths than visible light, so you can examine very small structures compared with using a regular microscope. This property is used for X-ray microscopy and X-ray crystals to obtain high-resolution images for the positioning of crystal molecules.

Computed tomography

CT is a superior imaging technique in which an X-ray source rotates around the patient to supply cross-sectional "slices" that have been reconstructed to show precise images. Regular exams take 10 min to an hour, and many require oral or IV evaluation to increase element. CT provides unique photographs of organs, bones, soft tissue, and blood vessels.

A CT scanner uses a motorized X-ray supply that rotates around the circular beginning of a donut formed structure known as a gantry. At some stage in a CT experiment, the affected person lies on a mattress that slowly moves through the gantry at the same time as the X-ray tube rotates around the patient, generating images using slender of X-rays via the frame. Instead of film, CT scanners use unique advanced x-beam locators, which may be found quickly opposite the x-beam flexibly. Because the X-rays travel from the influenced individual, they are obtained through the finders and sent to a figure. Each time the x-beam flexibly finishes one turn, the CT computer uses state-of-the-art mathematics to construct a 2D photograph slice of the affected person. Intermittently, the filter employs a contrast operator. This contrast operator, sometimes called a dye, makes strides the images by highlighting certain highlights. Our healthcare supplier will either have you drink an extraordinary fluid containing the differentiate specialist or deliver an intravenous (IV) infusion with the differentiate or both depending on the sort of CT filter and the reason for the check. The differentiate specialist is cleared from the body through urine, to begin with quickly at that point more gradually over the following 24 h.

The thickness of the tissue represented in each photo slice can vary relying at the CT device used, but generally tiers from 1 to 10 mm, While a full slice is finished, the image is saved, and the motorized mattress is moved

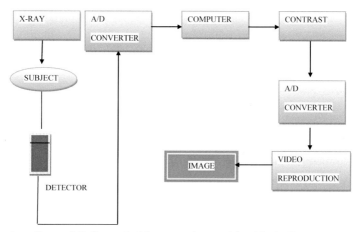

Figure 4.3 Computed tomography machine block diagram.

ahead incrementally into the gantry. The X-ray scanning procedure is then repeated to produce any other photograph slice. The block diagram of a CT machine is shown in Fig. 4.3.

Present day CT machines take nonstop images in a helical (or winding) design instead of taking an arrangement of images of person cuts of the body, as the initial CT machines did. Helical CT (moreover called winding CT) has a few focal points over more seasoned CT procedures: it is quicker, produces way better quality 3D images of ranges interior the body, and may identify little variations from the norm superior.

As an expansion of its use in cancer, CT is broadly used to assist analyze circulatory (blood) framework infections and conditions, such as coronary course infection (atherosclerosis), blood vessel aneurysms, and blood clots; spinal conditions; kidney and bladder stones; abscesses; fiery maladies, such as ulcerative colitis and sinusitis; and wounds to the head, skeletal framework, and internal organs. CT imaging is also used to identify unusual brain functioning or memory in adult patients with cognitive impedance being assessed for Alzheimer's disease and other causes of cognitive decay. Whole-body CT generates images for each zone of the body—from the chin to underneath the hips. This strategy, which is used routinely in patients with existing cancer, can also be used in individuals with no indications of illness. In any case, whole-body CT has not appeared as a compelling screening strategy for sound individuals. Most anomalous discoveries from this strategy do not indicate a genuine well-being issue; the tests required to address an issue can be costly, badly arranged, and awkward, and may uncover unseen additional dangers, such as from an invasive strategy like a biopsy to assess the discoveries. In expansion, whole-body CT can expose individuals to generally enormous amounts of ionizing radiation—about 12 mSv, or four times the normal yearly dosage obtained from characteristic sources of radiation. Most specialists advise against whole-body CT for individuals without any signs or side effects of illness.

CT scans can diagnose life-threatening conditions such as bleeding, blood clots, and cancer. Early detection of these conditions is likely to save lives. However, CT scans use X-rays, and all X-rays produce ionizing radiation. Ionizing radiation can have biological effects on living tissues. It is a risk that increases with the number of risks in a person's life. Moreover, radiation exposure usually carries a risk of cancer.

If there is no area around the abdomen or pelvis of the pregnant woman, there is no risk to the baby from the CT scan of the pregnant woman. Generally, physicians prefer radiation-free tests such as MRI and ultrasound if abdominal and pelvic imaging is required. However, CT may be acceptable as an alternative response option if the first options cannot provide the required response or if an emergency or other time constraint must be addressed. In some patients, contrast media can cause allergic reactions; in rare cases, they may cause temporary renal failure. Ivy contrast media may cause further deterioration in renal function, may sometimes be permanent, and should not be administered to patients with abnormal renal function. Children are more susceptible to ionizing radiation and have longer life spans, so they have a higher risk of cancer than adults. Parents can ask a technician or doctor whether the machine settings have been adjusted for their child.

In contrast to the traditional X-ray, which uses stationary X-ray tubes, CT scanners use an electric X-ray source that revolves around a circular opening called a gantry in a dent-shaped structure. During the CT scan, the patient lies in bed and slowly passes through the gantry. The X-ray tube revolves around the patient and irradiates the body with a thin X-ray beam. Instead of film, CT scanners use a special digital X-ray detector located opposite the X-ray source. When the X-ray leaves the patient, it identifies the detector and sends it to the computer. Each time the X-ray source completes a rotation, the CT computer uses advanced mathematical techniques to create a 2D image of the patient. The thickness of the tissue displayed in each image slice depends on the CT device used but is usually in the range of 1—10 mm. When the whole work is completed, the image is saved, and the electric bed is slowly moved to the gantry. The X-ray scanning process is then repeated to create another image fragment. This process continues until the required process slices are collected. During the scan, the computer has images of two radiologists looking at the CT scan.

Image slices can be viewed or stacked individually on a computer to produce 3D images of patients with skeletal, organ, tissue, and physician identifiable abnormalities. This method has several advantages, such as the ability to rotate a 3D image in space and continuously display fragments to make it easier to find the exact location where problems may occur.

CT scans can identify illnesses and injuries in different areas of the body. For example, CT has become a useful research tool for detecting potential tumors or lesions in the abdomen. If one suspects various heart diseases or abnormalities, a CT scan of the heart can be ordered. Head CT can detect strokes, bleeding, and other injuries resulting from tumors and blood clots. The lungs may reveal tumors, pulmonary embolism (thrombus), excess water, and conditions such as effusions and pneumonia. CT scans are especially effective for imaging complex fractures, severely deformed joints, or bone tumors. This is usually possible with a more detailed traditional X-ray.

Computed tomography contrast agent

As with all X-rays, dense body structures such as bones are easy to image, but soft tissues have different abilities to block X-rays, making them unconscious or hard to see. This has led to the development of intravenous (IV) contrast agents that are highly visible on X-rays or CT scans and can be used safely by patients. Contrast media contain substances that are more readily seen in X-ray images because they are better at X-ray resistance. For example, to test the circulatory system, an iodine-based contrast agent is introduced into the bloodstream to help illuminate the blood vessels. These types of tests look for potential blockages in blood vessels, including the heart. Oral contrast agents, such as barium-based compounds, create images of the digestive system, such as the esophagus, stomach, and gastrointestinal tract.

Magnetic resonance imaging

MRI includes radio waves and magnetic fields to observe the organs and other structures in the body. The process calls for an MRI scanner with a large tube consisting of an enormous circular magnet. This magnet creates a powerful magnetic field that aligns the protons of hydrogen atoms inside the body. The protons are exposed to radio waves that invoke proton rotation. When the radio waves turn off, the protons are released and realign themselves, emitting radio waves in the healing technique that the system can detect to create an image.

At Stony Brook University in 1971, Paul Lauterbur applied 3D pulse field gradients and backprojection techniques to create nuclear magnetic resonance (NMR) images. He first published an image of two water pipes in *Nature* in 1973, then a living animal, a claw, and in 1974, the chest cavity of a rat. Lauterbur called his imaging method zeugmatography, a term that has been replaced by MRI. In the late 1970s, Lauterbur and another physicist, Peter Mansfield, developed MRI-related technologies, including echo-planar imaging technology. Advances in semiconductor technology were essential for developing practical MRIs that required large amounts of computing power, which was made possible by the rapidly increasing number of transistors on a single integrated circuit chip. Mansfield and Lauterbur won the Nobel Prize in 2003.

MRI scanners are particularly well suited to the imaging of nonbony parts or tender tissues. MRI can differentiate between white count and gray matter and provide images to diagnose aneurysms and tumors. Because MRI no longer uses X-rays or other radiation, it is by far the imaging modality of choice when frequent imaging is required for analysis or treatment, especially inside the brain.

A specialized form of MRI, functional magnetic resonance imaging, looks at brain structures and decides which areas are affected. It is mainly used in neurological disorders to detect cognitive impairments.

To acquire an MRI image, an affected person is placed within a large magnet and must stay still for the duration of the imaging process so as not to blur the image. Contrast agents (often containing the element gadolinium) can be administered to a patient intravenously before or during the MRI to increase the speed or proton realignment with the magnetic field. The faster the protons realign, the brighter the photograph will be. The block diagram of an MRI machine is shown in Fig. 4.4.

MRI uses magnets to produce a sturdy magnetic field that forces protons in the body to align with it. When a radiofrequency (RF) current is pulsed through the affected person, the protons are stimulated and spin out of equilibrium, straining toward the pull of the magnetic area. When the RF current is turned off, the MRI sensors can capture the released energy because the protons have realigned with the magnetic field.

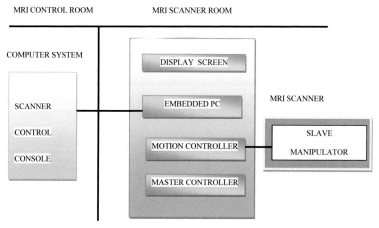

Figure 4.4 Magnetic resonance imaging machine block diagram.

Research opportunities for magnetic resonance imaging hardware

Magnet systems

➤ Development of practical high-temperature superconducting magnets using Nb_3Sn and other high-temperature materials
➤ Development of plans for practical magnets for extraordinary applications including specific anatomic parts—for example, treatment
➤ Development of plans to more prudently use old magnets by retrofitting them into existing demonstration, intervention, and treatment rooms

Pulsed-LED magnetic resonance imaging

➤ Advancement and approval of techniques for a signal readout that limits the display of extra movements and obstructions
➤ Development of means for vitality recuperation during polarization breakdown

Radiofrequency coils

➤ Plan of uniform transmitters and collectors that incorporate dielectric and wavelength effects
➤ Development of techniques for 3D display of RF fields
➤ Design and assessment of high-temperature superconducting loops and related cooled preamplifiers for low-field imaging
➤ Design of signal-to-noise-ratio-efficient rapid image mixing and recreation strategies for multicoil arrays

Gradient systems

➤ Design of extremely short head-slope loops using current return methods at more prominent measurements than those of the essential windings

➤ Design of acoustically quiet inclination coils

➤ Design of head inclination coils with great subject access for visual and auditory assignment introduction

➤ Optimization of head immobilization gadgets viable with head gradient coils

➤ Development of strategies for electrical decoupling of RF loops and angle curls specific to strategies for electrical decoupling of RF, particular loops, and angle curls

Construction of magnetic resonance imaging

In most therapeutic applications, a proton-containing hydrogen nucleus generated in the tissue forms a signal that is processed to form an image of the body about the density of the nucleus in a particular region. Considering that protons are affected by other atoms in the field where they are bound, it is possible to separate the reaction from the hydrogen of a particular compound. In a study, individuals are placed inside an MRI scanner that creates a strong magnetic field around the magnetic field being copied. First, energy from the vibrating magnetic field is temporarily applied to the patient at the appropriate resonant frequency. Scans with X and Y gradient coils create the exact magnetic field that the patient's selected magnetic field had required to absorb energy. The excited atomic output emits an RF signal measured by the coil. Gradient coils process RF signals and reduce position information by observing RF levels and phase changes due to local magnetic field fluctuations. These coils rapidly change the excitation and response time for performing flow line scans, causing the windings to flow slightly because of the magnetic material and resulting in the repeated sound characteristics of MRI scans. An external contrast agent may be incorporated to sharpen individual images. An MRI scanner's main components are a primary magnet that polarizes the sample, a bean coil that corrects the symmetry of the original magnetic field, a gradient system to scan the magnetic field, and an excitation RF system. The system, controlled by one or more computers, takes a sample and identifies the resultant NMR.

A strong and uniform magnetic field of several ppm is required during the MRI scan. The strength of a magnetic field is measured with a telescope. Most systems operate at 1.5 T, but commercial systems are available at 0.2—7 T. Most diagnostic magnets contain superconducting magnets and require liquid helium to keep them very cool. Lower magnetic field strength can be achieved with permanent magnets often found in "open" MRI scanners for patients with claustrophobia. A low-field-power, FDA-approved portable MRI scanner was also used in 2020. MRI is usually performed within ultralow magnetic field strengths ranging from microtesla to millitesla where appropriate signal quality is possible. The Larmor precession field measured approximately 100 μT using prior permission (on the order of 10—100 mT) and a sensitive superconducting quantum interference device.

T1 and T2

Each tissue returns to equilibrium after stimulation by an independent T1 (spin-lattice) relaxation process. This is magnetization in the same direction as the static magnetic field and T2 (spin—spin; direction across the static magnetic field). Changing the repeat time to create a T1-weighted image allows it to recover before measuring the magnetic resonance (MR) signal. The weight of this figure is effective for cerebral cortex diagnosis, adipose tissue detection, localized liver lesion detection, general morphology information, and subsequent contrast imaging. A T2-weighted image can be created by changing the echo time to reduce the magnetic field before measuring the MR signal. This figure is useful for identifying edema and inflammation from weightlifting, detecting white matter lesions, and determining the zonal anatomical structure of the prostate and uterus.

Diagnostic

a. Use by organ or system: MRI has a wide range of uses for diagnostic treatment, and it is estimated that more than 25,000 scanners are used worldwide. While MRI affects diagnosis and treatment in many disciplines, its health impact can be challenged.

MRI is the decision-making process for identifying, staging, and following up on tumors, the prestages of rectal and prostate cancers, and determining tissue samples through biobanking.

b. Neuroimaging: MRI is the best diagnostic tool for neurological cancer in CT because it better visualizes progressive fossa, including the brain and cerebellum. Because of differences between gray and white matter, MRI is ideal for many central nervous system conditions, including demyelinating diseases, dementia, neurological disorders, infections, Alzheimer's disease, and epilepsy. The ability to study functional and structural brain abnormalities using MRI in guided stereotactic surgery and radiosurgery, aneurysm dysfunction, and other surgical conditions is known as N-localizer treatment of intracranial tumors.

c. Cardiovascular: Cardiac MRI complements other imaging techniques such as echocardiography, cardiac CT, and nuclear medicine and can evaluate heart structure and function. Its applications include assessing myocardial ischemia and survival, cardiomyopathy, myocarditis, iron overload, vascular disease, and congenital heart disease.

d. Musculoskeletal: Applications of the musculoskeletal system include spinal imaging, joint disease, and soft tissue tumor assessment. In addition, MRI technology can be used for clinical imaging of systemic muscle diseases.

e. Liver and gastrointestinal: Hepatobiliary MRI identifies lesions of the liver, pancreas, and bile ducts. Local or diffuse liver diseases can be assessed using expansion-weighting, antiphase imaging, and dynamic contrast-weighting sequences. Extracellular contrast media are widely used in liver MRI, and new hepatobiliary contrast media offer the opportunity for effective bile imaging. Anatomical imaging of the bile duct is achieved using magnetic resonance cholangiopancreatography using the T2-enhanced sequence. Effective imaging of the pancreas occurs after secretin administration. MR enterography provides an invasive assessment of inflammatory bowel disease and tumors of the small intestine. MR colonography may play a role in identifying large polyps in patients at risk of colorectal cancer.

f. Angiography: Magnetic resonance angiography (MRA) takes images of the arteries to assess them for stenosis (abnormal stenosis) or aneurysm (vessel wall dilation, risk of rupture). MRA is often used to assess cervical and cerebral arteries, chest and abdominal arteries, renal arteries, and legs (known as "runoffs"). Images can be created using a variety of techniques, including the administration of paramagnetic contrast media (gadolinium) and "flow-related magnification" (such as the sequence of 2D and 3D time-of-flight). The signal in the figure was due to the blood that had recently entered the aircraft. A technique of phase accumulation (known as phase-difference angiography) can easily and accurately map the flow velocity. Magnetic resonance venography is a similar procedure used in vein images. This method causes tissue depression, but signals are collected on the surface just above the excitation surface. Therefore, venous blood transferred from the recently stimulated surface has been photographed.

Magnetic resonance spectroscopy

The MR spectrum measures different metabolic levels in body tissues. This can be achieved with a variety of single-tone or imaging-based techniques. The MR signal, including the isotopes "excited," produces a resonance spectrum corresponding to different molecular sequences. This signature diagnoses specific metabolic disorders, particularly those affecting the brain, to provide information on tumor metabolism.

The MR spectrum combines the spectrum and image (MRSI) to create a patient's spatial or spatially local spectrum. Although the spatial resolution is very low (limited by the available signal-to-sound ratio), each vowel spectrum contains some metabolic information. MRSI must achieve higher SNR at higher field strengths (3 and above) because the available signals encode spatial and spectral information. High field strength, combined with high collection and maintenance costs of MRI, hampers the popularity of MRI. Recently, however, compressed sensing-based software algorithms have been proposed to achieve superresolution without the need for such high field power.

Real-time magnetic resonance imaging

Real-time MRI is a series of images of moving objects (such as the heart) in real time. One of the various techniques developed in the early 21st century is based on radial flash MRI and repetitive reconstruction. It provides an image with an in-plane resolution of 1.5−2.0 mm with a resolution of 20−30 ms. Balanced and stable free precision imaging has better image contrast between blood pool and myocardium than flash MRI, but an architect will occur if B0 differentiation is stronger. Real-time MRI can add important information about heart disease and joint disease, often making MRI tests easier and more comfortable for patients, especially those who cannot hold their breath or have arrhythmias.

Interventional magnetic resonance imaging

MRI is ideal for interventional radiology, where images created by an MRI scanner guide perform minimally invasive surgery with no adverse effects on patients and operators. No such procedures use ferromagnets. A growing subset of interventional MRIs is intraoperative MRI, which uses MRI in surgery. Imaging can be done simultaneously with the surgical procedure with the help of some specialized MRI systems. Surgery is usually temporarily interrupted so that MRI can evaluate the success of the surgery or guide for later surgery.

Magnetic resonance-guided focused ultrasound

In guided treatment, high-intensity ultrasound rays are focused on the tissue and controlled using MR thermal imaging. Because of the high energy in focus, the tissue is completely destroyed above 65°C (150°F). This technique can properly lift the injured tissues. MRI provides a 3D view of the target tissue, allowing the accurate focus of ultrasonic energy. MRI provides quantitative, real-time thermal imaging of the treatment area. This ensures that the temperature generated during each cycle of ultrasonic energy is sufficient to create pyrolysis in the tissues of interest, otherwise ensuring effective treatment. We can customize the parameters to create it.

Multinuclear imaging

Hydrogen is the most frequently depicted nucleus in MRI because it is abundant in living tissue and its high gyromagnetic ratio provides a strong signal. However, the nucleus with pure nuclear spin may be able to combine with MRI. These nuclei include helium-3, lithium-7, carbon-13, fluorine-19, oxygen-17, sodium-23, phosphorus-31, and xenon-129. 23Na and 31P are naturally abundant in the body and can be directly depicted. Gaseous isotopes such as 3He and 129X must be hyperpolarized, in which case their atomic concentrations are too low to receive useful signals under normal conditions. Adequate 17O and 19F can be administered in liquid form (e.g., 17O-water) where no additional growth is required. Using helium or xenon has the advantage of reducing background noise and improving image contrast. This is because these elements are not usually present in living tissues. In addition, the nucleus of an atom with pure atomic spin and bound with a hydrogen atom can be visualized instead of a low gyro magnetic ratio nucleus through heteronuclear magnetization transfer MRI. The hydrogen nucleus is involved with a hydrogen atom, and theoretically, heteronuclear magnetic transfer MRI can detect the presence or absence of certain chemical bonds. Multinational imaging is currently essentially a research method. However, possible applications include functional imaging and imaging of organs seen on 1H MRI (such as lungs and bones) or the inclusion of images as alternative contrast agents. Respiratory hyperpolarization can be used to illustrate the distribution of space within the lungs. Injections containing static bubbles of 13C or hyperpolarized 129X have been studied as agents in contrast to angiography and perfusion imaging. Information on bone density and structure and functional images of the brain can be provided by 31P. Multinational imaging can graph the human brain's distribution of lithium, an ingredient used as an important drug in situations such as bipolar disorder.

Molecular imaging by magnetic resonance imaging

MRI has the advantage of much higher spatial resolution and is suitable for metaphorical and functional imagery but also has some drawbacks. First, the sensitivity of MRI ranges from about 10^{-3} mol/L to 10^{-5} mol/L, which can be very limited compared with other types of imaging. This problem occurs because the population difference between the spin states of atoms at room temperature is very small. For example, in 1.5 T, the intensity of a specific field of clinical MRI, the difference between high- and low-energy states is about 2 million in nine molecules. Improvements from increasing MR sensitivity include increased magnetic field strength and hyperpolarization due to optical pumping or dynamic nuclear polarization. There are also various signal amplification projects based on chemical exchanges to increase sensitivity.

A molecular image of a disease biomarker using MRI can be achieved with a target MRI contrast agent with high specificity and high rest (sensitivity). To date, several studies have devoted themselves to the development of MRI contrast agents aimed at achieving molecular imaging with MRI. Small protein domains such as peptides, antibodies, or small ligands and HER-2 affibodies are usually applied to achieve the target. These targets are usually combined with a high-piled MRI contrast agent or a high-reflective MRI contrast agent to increase contrast agent sensitivity. A new class of MR contrast agents has been introduced to show the unique mRNA of the gene and the activity of the gene transcription factor protein.

These new contrast agents can detect unique mRNA, microRNA, and viral cells. Tissue response to inflammation of the living brain. MRIs report changes in gene expression positively related to Taqman analysis, optical, and electron microscopy.

Safety

MRI is usually a safe technique, but it can lead to injuries due to failed safety procedures or human error. Improvements in MRI include most cochlear implants and cardiac pacemakers, debris, and metal foreign bodies. MRI during pregnancy seems to be protected if there is a slight contrast between the second and third trimesters. MRI does not use ionizing radiation, but it is used more than CT if the moderator can receive the same information. Some patients experience claustrophobia and may require sedation. Because MRI uses strong magnets, it can transfer magnetic materials at high speeds with the risk of projectile loss. Millions of MRIs are performed worldwide each year, but deaths are extremely rare.

Overuse

If a physician recommends using MRI on a patient, the treatment committee will issue guidelines. Although MRI can diagnose health problems and confirm a diagnosis, clinicians suggest that MRI is not the first step in developing a plan to identify or manage patient complaints. For example, although it is common to use MRI to determine the cause of back pain, the American College of Physicians indicates that the procedure is less likely to provide good patient results than some alternate approaches.

Artifacts

MRI artifacts are visual abnormalities in visual architecture and visual representation. MRI can create various artifacts, some of them affected by clinical quality and others confused with pathology. Arts can be classified as patient-related, signal processing-dependent, and hardware-related (machine).

Nonmedical use

MRI is primarily used industrially for routine analysis of chemicals. Atomic MR technology is also used, for example, to measure the ratio of fatty foods in food, to monitor the flow of corrosive fluids in pipes, and to study molecular structures such as catalysts.

MRI is noninvasive and can be used to study plant anatomy, water transport processes, and water balance, and this also applies to clinical veterinary radiology. MRI's high cost limits its use in zoology. However, it can be used in many species. In paleontology, MRI is used to investigate fossil compositions. The forensic imaging manual provides graphic documentation of the autopsy investigation rather than the autopsy. CT scans provide rapid systematic imaging of skeletal and paleontology changes, and MRI imaging provides a better presentation of soft tissue deformities. However, MRI is more expensive and time-consuming 10°C.

Ultrasound

Ultrasound uses high-frequency sound waves to create real-time fixed and transferring photos of internal frame structures. The examination is usually painless and takes 30—60 min. A sonographer or radiologist moves a special device on the skin over the region of interest, generating real-time video images. Commonly it includes imaging pregnant women and reviews of organs, blood vessels, and approaches together with needle biopsies and aspirations.

Ultrasound images are constructed from the electricity of contemplated acoustic waves, wherein the reflecting source's range (or intensity) is determined clearly by timing. Because of the incredibly small sound space in tissue (1.5 mm/μs or $1.5*10^6$ mm/s), various precise measurements may be received from easy timing measurements. Hence, ultrasound snapshots fundamentally show local, microscopic mechanical residences of tissue. Lateral resolution in ultrasound snapshots is generally decided through diffraction. Maximum imaging structures work near the diffraction limits. Consequently, the acoustic lens layout is critically vital for ultrasound imagers. Acoustic lenses can be synthesized electronically using phased-array transducers (i.e., sampled apertures). Directed ultrasound beams can be generated at digital costs using phased arrays. Thus, ultrasound imagers are inherently real-time.

Ultrasound makes use of high-frequency sound waves bounced off tissue to create images of organs, muscular tissues, joints, and various gentle tissues. It is like a light shining within the inner part of the frame, gently traveling through layers of skin that might only be considered with digital sensors.

An ultrasound professional called a sonographer applies an extraordinary lubricating jelly to the skin during the ultrasound process to reduce friction so the ultrasound transducer can be rubbed on the skin. The transducer's appearance is comparable to that of a mouthpiece. The jelly also affects sound wave transmission. The transducer sends high-frequency sound waves through the body. The waves resound as they hit a thick area, such as an organ or bone, at which time the echoes are reflected to a computer. The pitch of the sound waves is too high for the human ear to hear. The constructed image can be deciphered by a specialist. Depending on the region being inspected, the patient may need to change position to provide the specialist with better access. After the process is complete, the gel is cleaned from the skin. The full process is usually completed in less than 30 min, depending on the region being inspected. One may be free to go almost the typical exercises after the method has wrapped up.

An essential ultrasound machine has the following accompanying parts:

➤ transducer probe—sends and transforms sound waves
➤ central processing unit—performs the calculations and contains the electrical force supplies
➤ transducer pulse controls—change the length of the beats transmitted through the transducer probe
➤ display—detects the ultrasound images
➤ keyboard/cursor—inputs the information
➤ disk storage device (hard, floppy, CD)—stores image results
➤ printer—prints results from the displayed information

Ultrasound has been used in various clinical settings, including obstetrics and gynecology, cardiology, and malignant growth discovery. The fundamental bit of leeway of ultrasound is that exact structures can be seen without using radiation. Ultrasound can likewise be performed much more quickly than X-ray beams or other radiographic procedures. The following are certain ultrasound uses:

➤ obstetrics and gynecology
➤ estimating the size of the baby
➤ monitoring the baby to check whether it is in the ordinary head-down position or breech
➤ checking the position of the placenta to determine whether it is inappropriately placed over the opening to the uterus (cervix)
➤ monitoring the number of embryos in the uterus
➤ determining the sex of the infant
➤ monitoring the baby's development rate

➤ distinguishing ectopic pregnancy
➤ deciding whether the level of amniotic liquid padding the infant is appropriate
➤ observing the child during specific methodology
➤ seeing tumors of the ovary and bosom

Cardiology

➤ determining unusual structures or capacities
➤ estimating blood flow levels

Urology

➤ estimating blood flow through the kidney
➤ detecting kidney stones
➤ recognizing malignant prostate growth early

Fig. 4.5 shows the ultrasound machine block diagram.

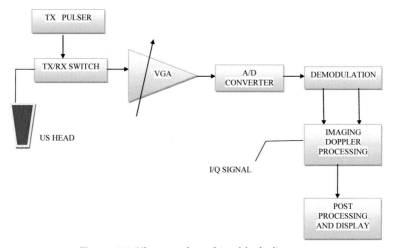

Figure 4.5 Ultrasound machine block diagram.

External ultrasound

The sonographer applies a lubricating gel to the patient's skin and places a transducer over the gel-covered skin. The transducer is moved over the portion of the body to be inspected. Cases incorporate ultrasound examinations of a patient's heart or a baby within the uterus. The patient ought not to feel inconvenience or pain. They will feel the transducer over the skin. During pregnancy, slight inconveniences may occur in the case of a full bladder.

Internal ultrasound

If internal regenerative organs or the urinary framework must be assessed, the transducer may be placed within the intimate organs. To assess portions of the stomach-related framework—for illustration, the esophagus, the chest lymph nodes, or the stomach—an endoscope may be used. A light and an ultrasound gadget are joined at the end of the endoscope, which as a rule is embedded into the patient's body through the mouth. Before the procedure, patients are administered medicines for pain. Internal ultrasound filters are less comfortable than external ones, and there is a slight chance of internal bleeding.

Ultrasound in anesthesiology

Anesthetists frequently use ultrasound to direct a needle with an anesthetic close to nerves. An ultrasound can be done at a doctor's office, at an outpatient clinic, or within the hospital. Most scans take between 20 and 60 min. It is typically not difficult, and there is no noise. No unusual arrangements are necessary in most cases, but patients may wish to wear loose-fitting and comfortable clothing. If the liver or gallbladder will be influenced, the patient should fast for a few hours before the procedure. For a scan during pregnancy, particularly early pregnancy, the patient should drink plenty of water and remain in a strategic location that allows for urination a few times before the test. When the bladder is full, the check produces an improved image of the uterus. A specialist or specially trained sonographer will carry out the test.

Positron-emission tomography

A PET experiment is an imaging technique to check for illnesses in the body. The test uses a unique dye containing radioactive tracers. Those tracers are swallowed, inhaled, or injected into a vein in the arm, depending on what part of the frame is tested. Certain organs and tissues soak up the tracer. When detected via a PET scanner, tracers help a doctor see how well the organs and tissues are working. The tracer gathers in areas of higher chemical activity, which are beneficial because positive tissues of the body, and certain diseases, have a better stage of chemical interest. Regions of disorder will display as bright spots on the PET test.

A PET scan works by using a scanning tool (a system with a large hole at its center) to hit upon photons (subatomic particles) emitted by using a radionuclide within the organ or tissue being examined. The radionuclides used in PET scans are made by attaching a radioactive atom to chemical substances used by a particular organ or tissue at some stage in its metabolic manner. For example, in PET scans of the brain, a radioactive atom is implemented to glucose (blood sugar) to create a radionuclide called fluorodeoxyglucose (FDG) because the brain uses glucose for its metabolism. FDG is widely used in PET scanning.

Different substances can be used for scanning, depending on the reason for the test. If blood flow and perfusion of an organ or tissue are of interest, the radionuclide may be a form of radioactive oxygen, carbon, nitrogen, or gallium. The radionuclide is run right into a vein via an intravenous (IV) line. Next, the PET scanner slowly actions over the part of the frame being examined. Positrons are emitted through the breakdown of the radionuclide. Gamma rays are created during the emission of positrons, and the scanner then detects the gamma rays. A computer analyzes the gamma rays and uses the information to create an image map of the organ or tissue being studied. The amount of radionuclide collected within the tissue influences how brightly the tissue appears on the image and shows the organ or tissue feature level.

The figure below shows the block diagram of a PET detection mode. Acquisition by 2D versus 3D continues to assume a function in image recreation, with 3D increasingly gaining in routine use for all PET reproductions. In prior periods of PET use, 2D imaging was the most attractive of the possible imaging methods because too many events would be detected within the PET crystal array with excessive dead time and image degradation in adjacent PET detector rings. This was overcome by putting septa containing tungsten or lead in the middle of the finder rings. Alongside these septa, the scanner gadgets were arranged to only distinguish incident occasions from inside a restricted plane to reject noncollinear occasions. This additionally diminished the sensitivity of occurrence location and the compared image goal.

With enhancements of innovation and indicator hardware, it became conceivable to eliminate the septa that isolated PET rings and identify collinear occasions in the nearby PET rings. This could happen without corresponding dead time influences and considered an almost fourfold increment in sensitivity. The figure below shows the PET modes.

A PET check is a compelling way to look at the chemical movement in parts of our body. It may offer assistance distinguish an assortment of conditions, counting numerous cancers, heart infections, and brain clutters. The images from a PET scan provide data distinct from those revealed by other sorts of scans, such as CT or attractive reverberation imaging (MRI). A PET check or a combined CT-PET scan empowers the specialist to thoroughly analyze sickness and evaluate conditions.

Images from a PET scan show shining spots where the radioactive tracer was collected. These spots reveal higher levels of chemical action and points of interest about how our tissues and organs are working. A specialist with specific training to translate scan images (radiologist) will report the discoveries to the doctor. The radiologist may also compare PET images with images from other tests experienced of late, such as CT or attractive reverberation imaging (MRI). Or the images may be combined to supply more detail almost our condition. PET scans show metabolic changes happening at the cellular level in an organ or tissue. Usually critical because illness frequently starts at the cellular level. CT scans and MRIs cannot determine issues at the cellular level. PET scans can distinguish exceptionally early changes in cells. CT scans and MRIs can only distinguish changes afterward, as illnesses modify the structure of organs or tissues.

When either of these scans is performed in conjunction with a PET scan, they result in image fusion. A computer combines the images from the two scans to form a 3D image that provides more data and permits a more exact diagnosis. Gallium filters are comparable to PET scans in that they include the infusion of gallium citrate, a radioactive tracer. Gallium scans are performed 1 to 3 days after the tracer is employed, so it is a multiday process. These scans are not performed as commonly for the location of cancer, even though a few shapes of the gallium scan are combined with more up-to-date tests such as the PET check.

The concept of excretory and infectious tomography was introduced by David E. Kuhl, Luke Chapman, and Roy Edwards. Kuhl subsequently designed and manufactured several tomography instruments at the University of Pennsylvania. Tomography imaging technology was developed by Phelps, Edward J. Hoffman, and others. Gordon Brownell, Charles

Burnham, and Massachusetts General Hospital made significant contributions to the development of PET technology in the early 1950s and included the first exposure to apocalyptic radiation for medical imaging. His innovations include Using light pipes and volumetric measurements. Analysis was important in the case of PET imaging deployments. In 1961, James Robertson and his colleagues at the Brookhaven National Laboratory created the first single-aircraft PET scan called the "Head Compressor." One of the most responsible factors for adopting positron imaging is the development of radiopharmaceuticals. In particular, the development of 2-fluorodeoxy-D-glucose (2FDG), labeled by the Brookhaven group under the direction of Alfred Wolf and Joanna Fowler, was a major factor in expanding the field of PET imaging. In August 1976, Abas Alvi donated the campus to the first two general volunteers at the University of Pennsylvania. Images of the brain obtained with a simple (non-PET) nuclear scanner show the concentration of FDG in that organ. This component was then used in modern process-specific PET scanners. The logical extension of the positron measure was a design using a 2D array. The PC-1 was the first device to use this concept, designed in 1968, and reported in 1972. In 1970, the first application of PC-1 was described as separate from mathematical tomography mode. Instantly it became clear. Many people involved in the development of PET believe that a circular or cylindrical array of detectors is the logical next step in PET devices. Many researchers have adopted this method, but James Robertson and Zhang-Hi Cho first proposed a ring system prototype of the current size of PET. The PET-CT scanner, which originated from David Townsend and Ronald Newts, was nominated by TIM in 2000 as a medical invention.

Initially, PET methods were performed in committed PET centers because the hardware to create the radiopharmaceuticals, counting a cyclotron and a radiochemistry lab, had to be accessible in expanding to the PET scanner. Presently, the radiopharmaceuticals are delivered in numerous regions and are sent to PET centers, so the scanner is required to perform a PET check. Further expanding the accessibility of PET imaging may be an innovation called gamma camera frameworks (gadgets used to scan patients who have been infused with small numbers of radionuclides and right now are used with other atomic medication methods). These frameworks have been adjusted for use in PET filter methods. The gamma camera framework can total a filter more rapidly and at less effective, than a conventional PET scan (Fig. 4.6).

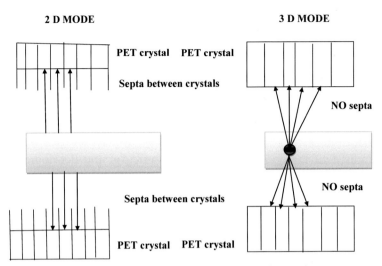

Figure 4.6 Positron-emission tomography mode.

Oncology

PET scans using Tracer 18F-FDG are widely used in clinical oncology. FDG is a glucose analog absorbed by glucose-using cells and phosphorylated by hexokinase, which is significantly enhanced in mitochondrial morphology in the case of rapidly growing defects. Metabolic traps of radioactive glucose molecules allow the use of PET scans. The concentration of the imaged FDG tracer depends on local glucose and, therefore, reflects the tissue's metabolism. 18F-FDG is used to detect the possibility of cancer spreading to other parts of the body (cancer metastasis). These 18F-FDG-PET scans are the most common in traditional medicine (equivalent to 90% of current scans) for detecting cancer metastases. You can use the same tracer to determine the type of dementia. Rarely, other radiotracers, usually not labeled with fluorine 1, characterize the tissue density of different types of molecules in the body.

The effective radiation dose of the normal dose of FDG used in oncological scans is 7.6 mSv. All cells require the hydroxy group replaced by fluorine-1 to produce FDG for the next stage of glucose metabolism, so no further response is observed in FDG. In addition, most tissues (except the liver and kidneys) cannot remove the phosphate added by hexokinase. This means that phosphorylated sugars cannot leave the cells due to ionic charges and are trapped in the cells that carry them until their death. Intense radio-balancing of tissues with high glucose uptake contributes to most cancers, including common cancers of the brain, liver, and kidneys. The Warburg effect causes these tissues to have higher glucose levels than most normal tissues, resulting in FDG-PET can be used for cancer diagnosis, staging, and monitoring, especially for Hodgkin's lymphoma, non-Hodgkin's lymphoma, and lung cancer. A 2020 review of research on PET for Hodgkin's lymphoma found that negative results of intermediate PET scans were associated with higher overall and progression-free survival. However, the reliability of the available evidence was very low for medium-to-moderate and progress-free survival. A few other isotopes and radiotracers have been gradually introduced into oncology for specific purposes. For example, 11C-labeled metomidate (11C-metomidate) has been used to detect adrenocortical tumors. In addition, FDOPA PET/CT (or F-18-dopa PET/CT) has been proven to be a more sensitive alternative to pheochromocytoma detection and detection than MIBG scans.

Operation

The radionuclides used in PET scans are usually small particles like Carbon-11 ($\sim$20 min), Nitrogen-13 (10 min), Oxygen-15 (min 2 min), Fluorine-18 ($\sim$110 min) I Period Gallium-68 ($\sim$67 min), zirconium-89 ($\sim$78.41 h), or rubidium-22 ($\sim$1.27 min). These radionuclides are commonly found in compounds used in the body, such as glucose (or glucose analogs), water, ammonia, or molecules that bind to receptors or other drug activity. Such labeled compounds are known as radiotracers. PET technology can identify the biological pathways of living humans (and many more species) if they can be radioactively stable in PET isotopes. Thus, the precise processes that can be investigated with PET are virtually unlimited, and radiotracers are synthesized for new target molecules and processes. As of this writing, dozens of treatments have been used and applied in hundreds of studies. By 2020, the most widely used radiotracer in clinical PET scans was 18F-FDG, an analog of fluorine-18-labeled FDG glucose. Because this radiotracer is used for all oncology scans and most neuroscience scans, it makes up the majority (>95%) of the radiotracers used in PET and PET-CT scans. Most positron-emitting radioisotopes have short half-lives, and thus, radiotracers have traditionally been created using cyclotrons in the vicinity of PET imaging facilities. Fluorine-11 has a long half-life, so fluorine-1-labeled radiotracers can be commercially prepared off-site and sent to imaging centers. Recently, rubidium-22 generators have come on the market. They contain strontium-22, which has been determined to produce positron-emitting rubidium-22 by electron capture.

Limitations

The radiation dose is an interesting feature of using radionuclide, at least for this subject. In addition to its established role as a diagnostic strategy, PET has a broad role in evaluating treatment responses, especially in cancer treatment, where a lack of knowledge about disease progression far outweighs the risk of a radiological examination. Because PET is radioactive, the elderly and those who are pregnant cannot use it because of radiation risk.

The high cost of cyclotrons to produce short-term radionuclides for PET scans will limit the wide use of PET stem, requiring a particularly optimal site chemical synthesis process to prepare and produce radiopharmaceuticals. The molecules of organic radioactive tracers with positron-emitting radioisotopes cannot be synthesized at first, and the biological carrier is destroyed when a cyclotron is struck to prepare a radioisotope, so radioisotopes are included. Cannot be prepared with. Instead, the chemical reaction before the isotope is made and then the organic radiotracer (e.g., FDG) should be done very quickly and very quickly before the isotope decays. Some hospitals and universities may maintain such a system, and most clinical PET radiotracers are supported by a third-party supplier that can deliver multiple sites simultaneously. This restriction essentially prohibits clinical PET for fluorine-18 labeled trailers. It has a half-life of 110 min and can be kept at a reasonable distance before use or transported as Rubidium-22 (used as Rubidium-Chloride 62). A 1.27 min half-life is included in a portable generator to study myocardial perfusion. Nevertheless, in recent years, some site-cyclotrons equipped with integrated shields and "hot laboratories" (automated chemical laboratories capable of processing radioisotopes) have begun to travel to remote hospitals equipped with PET units. In response to the high cost of isotope transport, remote PET machines are expected to increase the presence of small site-cyclotrons in the future by compressing cyclotrons. The shortage of PET scans in the United States has been overcome in recent years as the deployment of radiopharmaceuticals for the supply of radioactive material has dropped to 30% annually. Since the half-life of fluorine-13 is about 2 h, the final dose of a radiopharmaceutical that affects this radionuclide falls after several half-lives in a working day. This requires frequent repetition of the remaining dose (determination per unit volume) and careful planning of the patient's schedule.

Lasers

Lasers have been used in medication for medical procedures for quite some time, with applications running from the searing of veins to penetrating gaps through the heart. However, laser-based indicative gadgets are multiplying in some areas—for example, biomedical imaging and fundamental organic exploration. Ultrafast lasers are credited with making a large number of these applications conceivable.

The benefit of knowledge of the mechanisms governing the operation of laser requires the submission of aspects of physics that are the substratum for the study of the laser—matter interaction. A consequential parameter for the study of the effects of macroscopic and microscopic interaction with the matter and that characterizing the laser light is represented by the intensity, which is defined as the ratio of the emitted beam power and the unit of the irradiated area. It is possible to distinguish the laser intensity as high ($>10^{16}$ W/cm^2), medium ($\sim 10^{10}$ W/cm^2), and low ($>10^6$ W/cm^2).

The last of these is particularly used in medicine for diagnostic use, intermediaries for surgical use, and those of high power mainly for research purposes. The capacity to be concentrated in a small solid angle of an enormous power lends the laser to many applications. The peculiarities of laser (high directionality and high spatial and temporal precision, an excellent hemostatic effect, the reduction in pain, and postoperative complications) make them a valuable and indispensable tool for the care and the intervention on certain pathologies otherwise not curable.

The wavelength plays a fundamental role in the laser—matter interaction, as it is inherent in the value of the absorption coefficient and its reverse, or rather in the depths of the absorption of the laser light. Likewise, the length of the laser beat related to high or low energies guarantees a restricted bar statement with less or higher vitality.

Benefits of medical imaging

➤ Analysis of the ailment, and the seriousness or generous nature of that procedure, are made rapidly and precisely.

➤ Obtrusive symptomatic techniques—for example, exploratory medical procedure or angiography or cardiovascular catheterization—may not be fundamental.

➤ At the point when a child has a constant infection or a type of malignancy, clinical imaging is basic as a beginning.

Surgical removal of tissue with a laser is a physical cycle like modern laser boring. Carbon-dioxide lasers working at 10.6 µm can consume tissue with smoldering heat as the infrared pillars are emphatically consumed by the water that makes up the heft of living cells. A laser bar sears the cuts, halting seeping in blood-rich tissues—for example, gums. Essentially, laser frequencies close to 1 micrometer (Neodymium-YAG Laser) can enter the eye, welding a withdrawn retina back into the spot, or cutting inward films that regularly develop darkly after waterfall medical procedures. Less-serious laser heartbeats can wreck unusual veins that spread over the retina in patients experiencing diabetes, deferring the visual impairment frequently connected with the infection. Ophthalmologists precisely right visual imperfections by eliminating tissue from the cornea, reshaping the straightforward external layer of the eye with exceptional bright heartbeats from Excimer Lasers.

Laser light can be conveyed to places inside the body that the beams could not reach through optical filaments like the little strands of glass that convey data in phone frameworks. One significant model includes stringing fiber through the urethra and into the kidney with the goal that the finish of the fiber can convey serious laser heartbeats to kidney stones. The laser vitality parts the stones into sections sufficiently little to go through the urethra without requiring careful entry points. Filaments likewise can be embedded through little cuts to convey laser vitality to exact spots in the knee joint during an arthroscopic medical procedure. Another clinical application for lasers is in the treatment of skin conditions. Beat lasers can blanch some tattoos as well as dim red pigmentations called port-wine stains. Corrective laser treatments incorporate eliminating undesirable body hair and wrinkles.

Some laser components are as follows:

➤ Gaining medium capable of sustaining stimulated emission

➤ Enough energy components to increase the gain medium

➤ A reflector to reflect energy

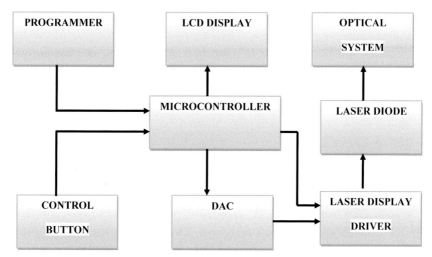

Figure 4.7 Block diagram of a laser biostimulator.

➤ Partial reflector
➤ Laser beam output

The gain medium and resonator decide the wavelength of the laser pillar and control of the laser.

Fig. 4.7 shows the block diagram of a laser biostimulator.

Indeed, even today, most clinical lasers sold are applied in ophthalmology. Dentistry was the second clinical order to which lasers were presented. In any case, albeit significant examinations have been conducted, the outcomes have not been exactly as promised much of the time, and the conversation on the helpfulness of dental lasers continues despite everything. Today, the significant focus of clinical laser research is zeroing in on different tumor medicines, for example, photodynamic treatment and laser-prompted interstitial thermotherapy. These assume extraordinary function in numerous other clinical controls such as gynecology, urology, and neurosurgery. Because of later advances in instrumentation for a negligibly intrusive medical procedure—for example, the advancement of smaller than normal catheters and endoscopes—novel methods are under present examination in angioplasty and cardiology. Exceptionally fascinating laser applications were found in dermatology and orthopedics. In addition, fruitful laser medicines have been accounted for lately in gastroenterology, otorhinolaryngology, and pulmonology, as discussed toward the end of this section. In this manner, it tends to be inferred that—right now—laser medication is a quickly developing field of both exploration and application. This is not surprising by any stretch of the imagination, since neither the improvement of novel laser frameworks nor the plan of suitable application units has yet come to stagnation. Additionally, laser medication is not limited to one order or a couple of orders. Rather, it has been acquainted with practically every one of them in the interim, and, typically, additional clinical applications will be created soon.

Some types of laser surgery are

➤ refractive eye surgery
➤ tooth whitening
➤ cosmetic scar, tattoo, or wrinkle removal
➤ cataract or tumor removal

Lasers are more exact than customary technical instruments, and cuts can be made more limited and shallower. This results in less harmed tissue. Laser tasks are typically more limited than customary medical procedures. On the off chance that overall sedation is required, it is typically used for a more limited time. Individuals likewise will generally mend faster with laser tasks. One may have less agony, expanding, and scarring than with customary medical procedures.

Some applications of lasers are

➤ finding laser range
➤ processing information (DVDs)
➤ reading bar codes
➤ laser surgery
➤ holographic imaging
➤ laser spectroscopy
➤ laser material processing (cutting, engraving, drilling, marking, surface modification)

Laser design

Some lasers are a means of gain, a method to activate it, and optical feedback. Gain media content must be amplified by the emission of excited light. Light of a certain wavelength that goes through the gain is amplified (power is increased). The reaction initially allows the excited emission to increase the optical frequency at the top of the gain frequency curve. As the excited radiation increases, a frequency ultimately dominates everything. This means an integrated beam is formed. The stimulus emission process, for example, relates to an audio pendulum with positive feedback that can occur if the loudspeaker system is located near the microphone—scratch audio vibration at the top of the gain frequency curve of the hearing amplifier. To increase light by gain, energy must be supplied through a process called pumping. Energy is usually supplied as an electric current or light of a different wavelength. Pump lights can be supplied by flash lamps or other lasers. The most common lasers use optical resonator feedback (a pair of mirrors on either side of the gain). The light bounces back and forth in the center of the mirror, goes through the gain, and widens each time. Typically, the output coupler, which is one of the two mirrors, is partially transparent. Some light escapes through this mirror. Depending on the design of the cavity (whether the mirror is flat or curved), the light emitted by the laser can scatter or form a thin beam. Like electronic pendulums, this device is sometimes referred to as a laser pendulum. Most practical lasers have additional components that affect the characteristics of the emitted light, such as polarization, wavelength, and beam size.

CHAPTER 5

Analytical instruments

Sudip Paul[1], Angana Saikia[1,2], Vinayak Majhi[1] and Vinay Kumar Pandey[1]

[1]Department of Biomedical Engineering, School of Technology, North-Eastern Hill University, Shillong, Meghalaya, India; [2]Mody University of Science and Technology, Laxmangarh, Rajasthan, India

Contents

Introduction to Biomedical Instrumentation and Its Applications
ISBN 978-0-12-821674-3
https://doi.org/10.1016/B978-0-12-821674-3.00009-7

Introduction

It may seem like an odd statement to say that some healthcare providers are unfamiliar with the workings of a pathology laboratory. But in real life, the delivery of equipment to the laboratory is initiated by a complex series of events starting with demonstration and resulting in a pathologic diagnosis/ interpretation. This chapter provides information about pathological analytical instruments. Analytical instruments are an integral subset of medical instruments. Vast categories of medical instruments include those used in audiology, ophthalmology, dentistry, prosthetics, and surgery.

Analytical instruments are used in research and diagnostic applications in chemical, medical, life sciences, and other fields. Traditionally, such instruments have typically been found only in medical laboratory environments. But as medical diagnostics has grown and continues to move from laboratories to points of care, the demand for analytical instruments increases daily, as they are integral to hospital settings.

The market size for analytical instruments continues to increase with the healthcare growth trend, increasing the scope for system designers, lab technicians, and skilled healthcare professionals. Patient care delivery systems include many different personnel, professionals, and specialties in the technology era. Caregivers should have a thorough understanding and good practical knowledge, including the role of diagnostic evaluation and other necessities, in pursuing professional endeavors. Medical professionals use laboratory facilities, and diagnostic tests are tools in themselves. In correlation with a pertinent medical history and physical inspection/examination, these tests can confirm a diagnosis and provide useful information about the disease and a patient's health status, and later, responses to medication. In addition, laboratory diagnoses are essential for epidemiological surveillance and research findings. An entire laboratory network can provide better service to society if it is used effectively to contribute to health care and disease prevention. To obtain the desired result, every health professional must

❖ know and understand the roles and responsibilities of their job and laboratory and their contributions in service to other human beings
❖ appreciate the need to involve all members in healthcare service
❖ follow professional ethics and codes of conduct
❖ experience job satisfaction and have professional loyalty.

Medical laboratory science is a complex field embracing many different disciplines: microbiology, clinical chemistry, hematology, urinalysis, serology, immunology, histopathology, immune–hematology, molecular biology, and others. Medical laboratory technology is a basic course for medical students that equips them with the essential knowledge, techniques, and skills of the medical laboratory, such as the

✔ importance of laboratory services
✔ role of the medical laboratory technologist
✔ use of laboratory wares, instruments, and sterilization techniques
✔ prevention and control of laboratory accidents
✔ process for instituting a quality control system.

Moreover, laboratory science is an extremely important course for students, as it paves the way to easily understanding and absorbing various professional courses: hematology, bacteriology, urinalysis, parasitology, and others. Hence, more emphasis should be given to teaching and training to produce skilled, competent, task-oriented, and efficient medical laboratory technologists. It is difficult to say exactly when and where medical laboratory science began. However, early historical references tell us that bodily fluids were examined around the era of Hippocrates. One of the most important events contributing to the development of this profession was the discovery of the microscope by German scientist Antonie van Leeu-wenhoek. In the early days, no one talked about the medical laboratory science field without discussing the medical specialty of pathology. Earlier medical laboratory practitioners were physicians or pathologists, but laboratory professionals achieved new heights with a separate discipline, medical laboratory technology, and educational requirements and standards. In India, the English were the first to establish health laboratories with 19th century standards. During the colonial period, the ruling country established medical infrastructure for the public within their colonies. Immediately after independence, these countries started developing medical infrastructure with the help of the global community, which assumed health laboratory activity worldwide. After some time, World Health Organization (WHO) took greater responsibility for developing basic medical infrastructure in the least developed and less wealthy economic countries. WHO developed a team to review healthcare conditions in various countries. This organization also supported the establishment of facilities to produce vaccines as well as other diagnostic activities. Meanwhile, the laboratory technician training program was financially supported for skill development.

A laboratory is equipped with various instruments, equipment, and chemicals (reagents) for performing experimental work, research activities, and investigative procedures. Specifically, the medical laboratory is equipped with various biomedical instruments, equipment, materials, and reagents (chemicals) for performing laboratory investigative activities using biological samples/specimens (whole blood, serum, plasma, urine, stool, etc.). WHO classifies medical laboratories according to four levels based on biosafety.

Basic laboratory (level 1)

Basic laboratory level 1 is the simplest and is adequate for work with organisms that pose a low risk to individual laboratory personnel and community members. Such organisms are categorized under Risk Group 1 by WHO. These organisms are unlikely to cause human diseases—for example, food spoilage bacteria, common molds, and yeasts.

Basic laboratory (level 2)

Basic laboratory level 2 is suitable for work with organisms predisposing moderate risk to laboratory workers and limited risk to community members. Such organisms are categorized under Risk Group 2 by WHO. They can cause serious human diseases but not serious hazards, because effective preventive measures and treatments are available. Examples of these organisms are staphylococci, streptococci, enterobacteria except *Salmonella* Typhi, and others. Such laboratories should be clean, provide enough space, have adequate sanitary facilities, and be equipped with an autoclave.

Containment laboratory (level 3)

A containment (level 3) laboratory is more advanced and used when working with infectious organisms presenting a high risk to laboratory personnel but a lower risk to the community. Such organisms are categorized under Risk Group 3 by WHO. Examples are tubercle bacilli, *Salmonella* Typhi, HIV, *Yersinia,* and others. The principle is to remove particularly hazardous organisms and activities from the basic laboratory. They are easily transmitted through air, by ingesting contaminated food or water, and parenterally. Such a laboratory should comprise separate rooms with controlled access by authorized staff. It should also be fitted with a microbial safety cabinet.

The maximum containment laboratory is intended for work with viruses that predispose a high risk for both laboratory personnel and the community. Such organisms are categorized under Risk Group 4 by WHO. Examples are smallpox, Ebola, Lassa fever, COVID-19, and others. Most of these organisms cause serious diseases and are readily transmitted from one person to another. These laboratories usually comprise separate buildings with strictly controlled access.

Laboratory organization

A laboratory is an orderly, structured system in which things are kept together in a procedural order that supports undertakings involving cooperation and teamwork. The emphasis is on arrangements that provide a good working environment to accomplish common objectives that are efficient, planned, and economical. The organization of a single medical laboratory has two interlocking components: laboratory heads and other staff with duties and responsibilities for better service to society. Size, power consumption, cost, and ease of use are critical considerations for penetrating nontraditional markets. Instruments must be small enough to fit on benchtops. If portable, they must be optimized for battery operation.

Additionally, they should be inexpensive enough for purchase by point-of-care providers, with enough versatility for use in a range of tests to improve cost efficiency. Lastly, because operators may no longer be trained technicians, tools should be easy to use and intuitive. The medical instrument company's advanced process technologies can bring new levels of precision, integration, and power savings to medical instrument designs.

Flow injection analysis

Flow injection analysis (FIA), a type of continuous flow analysis (CFA), was unveiled for the first time by Ruzicka and Hansen in 1974 in articles explaining the novel approach and analytical changes. The idea of performing chemical tests was published in early 1975. Previous CFA methods had been based on a complete mix of samples and reagents (physical identity) to develop chemical reactions to achieve chemical homogenization. Demonstrating stable conditions under the test, FIA exploded with this way of thinking. However, FIA measures transient signals or inputs by injecting them. This system results in axial and radial distribution processes formed for an infinite concentration gradient. A regular liquid portion shows all concentrations; each section of the time delay can be used for analysis. If the network itself responds, there will be a distribution process with a side window (if added). The chemical reaction is then allowed to occur at the example reagent interface and continue to the scattered front and tail section sample areas. Offspring product specifications may also be considered with the right detector. The system moving from the injection site to the reception site is precisely controlled. In addition, the concentration gradient is controlled by the blood vessels. The pattern I will provide has existed for a long time but can be recycled to combine features such as a note detector that continuously absorbs the energy of electrodes and other objects. Other parameters are clear when modified cell flow patterns indicate that achieving chemical equilibrium (stable state) is unnecessary. All patterns continue to run as long as the signal is repeated until the condition is stabilized. In addition, it allows quick chemical production diagnosis and thus facilitates sampling prices; it also makes it possible to use more powerful methods than the general method. Many people say this is difficult or impossible because of the differences discussed in the sections that follow. This system has been used since the inception of scientific and analytical chemistry and has been the subject of more than 13,500 scientific articles and 20 available monographs worldwide through 2003.

The chapter applies to three generations of FIA characteristics, using typical specific FIA through preselected samples rather than F-specific CFA. The following sections highlight the unique possibilities that will be clarified in next-generation FIA with various benefits (and limitations compared with FIA). Most importantly, in the present context, FIA must demonstrate that chemical testing can be performed automatically at high sampling rates with sampling and reuse within 1 min. It must provide solutions, innovation, and opportunities for unique implementation.

Inductively coupled plasma optical emission spectrum spectrometry

Many sources provide specific and detailed information about inductively coupled plasma optical emission spectroscopy (ICP-OES) strategies. This section discusses the role of ICP-OES technology with some references. Necessary and basic information about ICP-OES is being written, and the language is easy for those who know it. Knowledge of other spectral chemical technologies—for example, pressure and spark emission spectrometry and the atomic absorption spectrum (AAS)—can be used in the ICP-OES strategic area. It is written for beginning spectral chemical analysis and analytical cases in general chemistry.

Introduction to inductively coupled plasma optical emission spectroscopy

When confronted by the question, "Is the material present and at what level?" an analyst will resort to the most likely strategy based on the atomic spectrum. As the name suggests, electromagnetic radiation (such as light) from detection, measurement, and analysis is also absorbed in the atomic spectrum technique. It is emitted by the atom or ion of the element or elements being analyzed in the test. Quantitative data (key level) comes from the amount of electromagnetic radiation absorbed or released by atoms or ions.

It is interesting to learn what qualitative information (which material) from the wavelength electrons is absorbed or emitted by the magnetic radiation atoms or ions of interest. It is necessary to mention a third technique, atomic mass spectrometry—in most cases, mass spectrometry of inductively coupled plasma (ICP-MS)—where quantitative data refers to the number of ions observed with a mass spectrometer, where qualitative information is the correlated mass/ratio (m/z), or the discovered charge of the ions.

Particle concentrations of different elements can be determined using a spectrometer. Even with limited linear dynamic range (LDR), this analysis is still used in some foundations in the United States when samples are not taken from the conductive electrodes and a standard content library. Arc and spark irradiation methods are used for metal detection, and fire radiation spectroscopy alkali is widely used for naming and other elements that can be easily moved. The atomic spectrum separated by fire is much simpler. But more successful commercial nuclear spectroscopy devices have emerged in many markets. Manufacturers are still balancing these devices. Because these tools are inexpensive today, the lab is relatively easy to maintain. Where alkaline analysis is required, a fire analysis is also required.

This technique was still very popular in the 1960s and 1970s—optical emission spectrometry (OES) had both fire and arc in Walsh's first dissertation after publishing on AAS in 1955. OES assets will soon be replaced by AAS. Although about half the elements can be explained by periodic OES, this technology can no longer compete with AAS. Because it absorbs light, the atom is at prime time, and high temperature is required. Historically, flame atomic (nuclear) absorption spectroscopy (FAAS) has been used primarily to analyze metal alloys. This method requires the dilution of solid samples.

Introduction to flame atomic absorption spectroscopy

In addition to the linear calibration range, there is FAAS. Because of graphite FAAS (GFAAS) procedures, each item must be a separate hollow lamp with various terms and conditions—these methods are never compatible. They can easily correct several factors with an immediate analyte. Atmospheric pressure was used for basic inductively coupled plasma (ICP) analysis performed using OES.

Testing and treatment

Check the following results to create a curve; knowing the visibility (value solution) provides the possibility to represent everything. With this power, you can then set the location to the normal package and select the measurement curve for each item. Explain the difference by trying an unknown model and comparing the strengths. The basis of the measurement scale is determined by the object the viewing power currently equals. This can be done with the front computer and is not required for the program. This is the key to building a curved arm to evaluate item sizes within the model.

Analytical and general indicator features are one of the main advantages of ICP-OES's technology-wide breadth. All ingredients are complete, and this method can describe current programming. Most elements can be identified, but it is often impossible to identify some low-grade ICP-OES components at the fingerprinting level. It belongs to the first group samples were taken from sources other than them. An obvious example is the inability to identify argon—argon will have similar restrictions on ICP if C and H are water or organic mills serve as lubricants. Air current plasma discharge determines H, N, O, and C.

When possible, halogen composition is very difficult and usually cannot be determined at the follow-up level, even with great motivation. There are no rest steps or people who can never be identified, with artificial constructions of extremely radioactive or very short-lived substances.

Gamma ray spectrometry is the best method. The second advantage of ICP-OES technology is this ideal width of the linear fit, which is usually four in nine sequences.

The diagnostic limit of the substance can be 104—109 times special emission lines. For the concentration in ICP-OES from the display area to the upper limit of the object, the download line is called LDR.

Unlike AAS, such as arcade-lightning technology, there are usually only one or two LDR sizes. ICP spectroscopy can technically use two solutions to organize space. The high-level calibration curve when the radius is its linear range is longer, of course. Extended LDR is also common. Normal model analysis requires no dilution. Methods for working with narrow LDRs are often more demanding to dilute the sample to maintain the analyte standard linear range concentration. As indicated in the extensive article coverage of large-scale LDRs, the third advantage of ICP-OES is that it covers many aspects. It can be easily identified in a single analysis cycle (e.g., at the same time). It is characteristic that hot plasma emissions occur due to all the quantitative quality codes required for production to distribute information from plasma. This variety is colorful, while the nature of the ICP can be strengthened. When ICP-OES was first introduced as a technology in modern electronics and computer capabilities, some fundamentalists viewed it as fundamental analysis.

Subject matter experts perform this technique without interference. In fact, in ICP-OES, the tool is not blocked. If it were, it would be the subject of the most serious intervention into one of the most commonly used analytical atoms spectrometric methods. It is more or less plasma.

It overcomes chemical barriers. Spectrometry through interference is the most common cause of ICP-OES wrongdoing. The ability to choose from various emissions complex lines of spectral interference with a calibration program allows one to measure without any interruption in many ways. Matrix barriers, instead of weakening the sample, are the stability of the substrate use of internal standards and methodology standard supplement.

Nebulizer

In ICP-OES, all types are usually converted; the liquid is formed first and then pumped directly through the pump. The liquid in the instrument becomes a device called aerosol or fog. Nebulization is the process of an important stage of ICP-OES. A perfect nebulizer should be able to convert all fluids so the aerosol sample will emit plasma reproduction, evaporation, atomization, ionization, and excitement. Two antifog pills have been used successfully in ICP-OES. Most companies use an ultrasound nebulizer (USN) for lung-type pneumonia. It uses high-speed gas flow to build an air pressure. Another type of USN uses one piezoelectric converter for aerosol manufacturing. A common pulmonary tuberculosis USN will be briefly discussed here. The most common pulmonary tuberculosis is addressed with a concentrated nebulizer, usually glass or quartz.

This type of vacuum cleaner is soaked in a sample solution. The capillaries produce rapid pressure through low-voltage circuit argon flows at the end of the capillary. Very fast argon gas is combined with a low-pressure range reproduction of air-fluid samples. In concentrated pneumonia, fog is known for its sensitivity and stability. However, the small hair follicles are weak obstacles, usually with small solutions such as 0.1% total dissolved solids (TDS) and some new shapes. In this regard, the central nebula has improved; thus, some model solutions can be eliminated if 20% of the height of soluble hardware is unbroken. In the second type of pneumonia, vertical argon gas flow is vertical to samples of blood vessels (compared) with high-speed gas central fog parallel to the blood vessels. Again, the solution model or low-pressure hair includes blood vessels in an area created or revitalized by a fast-moving fluid vein through the bladder. It produces liquid samples with the high-pressure argon gas aerosol required by ICP-OES. Generally, crossflow fog is not as effective as density foggers to make fine drops or airdrops. However, this type of fog resistance to obstruction is because a large blood vessel tube can be used to make materials other than glass or quartz:

They are relatively strong and resistant to corrosion with central fog. The first two lungs are not as popular as tuberculosis; the third type is the Babington Nebulizer.

The sample fluid flows smoothly through a little hole, from which gas flows faster. The hole is cut through the sample fluid at the air-passing system. Such fog is not very sensitive to obstruction and uses a larger sample line or hole in a high-viscosity TDS model. Notice the difference between the Babington Nebulizer and the V-Groove Nebulizer.

Instead of flowing on a flat surface, a ditch with a small hole in the middle crosses it with high-speed arginine gas. This fog is also used for fogging by the sample solution, with high salt levels, such as seawater and urine samples. The last type of nebulizer is the USN, in which liquid samples are sprayed with piezoelectric shock. The swivel vibration is interrupted with a very fine liquid sample at the airport. USN efficiency is usually 10%−20%, and a 1%−2% thick foam is obtained with a fog remover or solution. Because most aerosol models achieve ICP release, an ICP-OES is usually equipped with USP ICP-OES-centered fog.

Sprayer

Plasma conversion is 10 mm or less in the middle of the atomizer spray, and only 1%−5% of specimens may reach plasma with others. Samples are sent to the river as waste. In the spray room, they are made of quartz and other things, and samples are made of corrosion-resistant materials. High-density acids, especially hydrofluoric acid, are affected by plasmapheresis and other presentation systems. The ICP version has amazing features; these are special sources of desertification.

Atomization, ionization, and excitation restrictions and their analytical complex applications are perfect. They must be in liquid form for analysis functions. This model comes naturally, and with much effort, spec-trophotologists require that it be confused with the liquid form so these samples can be used as nebulizers. Plasma spray for ICP-OES analysis was used in several display systems made to replace fog pesticides and ICP-OES spray chambers. The most common option is hydrate production. This technology has not yet changed the design in liquid form; after dilution, acid samples are usually mixed with a reducer, a sodium borhohydride solution in sodium hydroxide. Sodium borohydride reacts with acids of volatile hydride with a limited number of elements.

These compounds as gases and are then separated from the rest with the reaction compound and flow into the purple plasma gas. The dramatic development of sensitivity and cognition to 1000 elements obtained the limiting factor from hydride formation techniques more than reducing different compounds. The matrix components of the patterns are not represented from sample analysis against the sample entry system with sprays and spray rooms. Second, the electric heating source (ETV), a graphite furnace, is a single flat GFAAS with some modifications. ATV is commonly used at today's utility research facilities with liquid or solid samples, and the resulting steam of transferred plasma discharge with argon flow with twice-higher sensitivity is possible with other decisions. This device is a discount case condition (because there is often more than one factor) and a more limited number of matrix effects and requests for this technique. In addition, it offers devices of persistent roles with the ICP-OES speed meter for processing transient signals. The bound ATV was not very popular for ICP-OES commercial success, whereas ICP-MS is a companion. In addition, it runs many programs on the ETV system application model.

The third is a laser imaging system, a new and versatile technology that is a solid example for ICP-OES. There is considerable energy in that technology from the brightness of the focused laser shape highlights that are consumed. A paralyzed sample is then taken as plasma discharge by Hodgkin. This technology takes different sample materials: nonconductive, organic, solids, and powders. In addition to mass analysis, the concentrated laser beam is as small as today's area to conduct microanalysis and solve the gap. The main disadvantage of this technology is that it is perfect. A quality that happens with a matrix is often impossible to go on, and it is a completely quantitative definition. This is quite difficult in many instances

There were many other alternatives to enrollment. This can be checked using the ICP-OES system over the past 40 years. Here, a few more rigid observational models are also used. The use of plasma arc and spark sources is outdated. It was used in the early days of development. Many researchers have formed or dried liquid plasma discharge center samples using a special computer-controlled mechanical device. This technique is called a good record of samples. A carbon electrode is used in this technique. The solid sample is filled with the correct sample blank or dried liquid salts, which are caught in the open plasma and dispersed. Ionization and aroma technology have never worked commercially as an example of a presentation method. Practical research related to ICP will continue in the near future.

Compound microscope working principle

The standard compound microscope is considered the common microscope for general purposes.

It has two lens system arrangements: (1) an objective lens to observe the object and (2) ocular lens/eyepiece through which the eye sees the image. Light from a mirror or electric lamp passes through a transparent object. A magnified image of the real object is formed by the objective lens. The image is then further magnified with the ocular lens to achieve a magnified "virtual image" that is visible through the eyepiece.

Compound microscope parts

Mechanical parts are the first category of parts in a microscope. They provide physical support to the optical lenses for smooth adjustment and focusing on the object. Following are the mechanical components of a compound microscope:

1. Base: The metallic base that bears the whole microscope. A mirror is inserted into it.
2. Pillars: The microscope's body is fixed to the base through a pair of elevated pillars.
3. Inclination joint: A movable joint of the microscope's body to the base of the pillars. The body makes an arc/angle with this joint to form the inclined position required by the technician for easy observation. Permanently fixed base bodies are available in the market with an inclined position; thus, they do not need a pillar or joint.
4. Curved arm: The microscope pillars hold the body, stage, tube, adjustment arrangement, curved structure, and coarse adjustment.
5. Body tube: Vertical tube that holds the eyepiece at the top and movable nosepiece with objectives at the bottom.
6. Draw tube: The upper part of the body tube, slightly narrow, into which the eyepiece slides during observation.
7. Coarse adjustment: A knob with a rack-and-pinion technique to move the body tube up and down to focus the object on the visible field. Rotation of the knob via a small angle moves the body tube a long distance compared with the object to perform the coarse adjustment.
8. Fine adjustment: A small knob rotates through a large angle to move the body tube just a small vertical distance. It is used for fine adjustment to obtain a clear image.

9. Stage: A horizontal platform projecting from the curved arm. It has a hole at the center, which is where the object on the slide is viewed. Light comes from a light source passed through the object into the objective.

10. Slide mover: It consists of two knobs in a rack-and-pinion manner. The slide contains the object, which is moved in two dimensions with rotation of the knobs to focus the required portion.

11. Revolving nosepiece: A rotatable disk at the bottom of the body tube. Objectives have various magnifying powers. The required magnification can be achieved as the nosepiece is rotated.

Optical parts are the second category of parts in a microscope. The optical components of the microscope include the following:

1. Light source: A built-in electric light source in the base is an integral part of the microscope. The light source is supplied through a regulator controlling the brightness of the field. Earlier, a mirror was used as a light source. The plane mirror is on one side, and the concave mirror is on the other end. It is used in the following manner: (1) Condenser present: Plane mirror is used, as it converges the light rays. (2) Condenser absent: (a) Daylight: Plane or concave.

2. Diaphragm: Light coming through the light source is passed to the object through the condenser, the object is illuminated, and an infusion of bright light makes the unclear object visible. To reduce the amount of light entering the condenser, the iris diaphragm is fixed below the condenser.

3. Condenser: The condenser is located between the light source and the stage with a series of lenses for converging direct light rays from the source onto the object. The light rays travel through the object and finally enter into the objective. The numerical aperture of the condenser, which indicates the capacity of light condensing, light converging, or light gathering in the same manner as the light capture capacity of an objective, is called the "numerical aperture of the objective." The numerical aperture becomes higher or lower in accord with wider convergence of light in the condenser. A condenser with such a numerical aperture sends light through the object with an angle sufficiently large to fill the aperture back lens of the objective

with the highest numerical aperture. The smaller the numerical aperture compared with that of the objective, the lower the illumination of the peripheral portion of the back lens will be, and the image will have poor visibility. A greater numerical aperture of the condenser relative to the objective may result in decreased contrast. There are three types of condensers as follows:

(a) Abbe condenser (numerical aperture = 1.25). It is used extensively.

(b) Variable focus condenser (numerical aperture = 1.25)

(c) Achromatic condenser (numerical aperture = 1.40). It has been corrected for both spherical and chromatic aberration and is used in research microscopes and photomicrographs.

4. Objective: It is the most important lens in a microscope. Usually, three objectives with different magnifying powers are screwed to the revolving nosepiece. The objectives are

(a) Low-power objective (×10): It produces 10 times magnification of the object.

(b) High dry objective (×40): It provides a magnification of 40 times.

(c) Oil-immersion objective (×100): It provides a magnification of 100 times, wherein immersion oil fills the space between the object, and the objective scanning objective (×4) is optional. The primary magnification (×4, ×10, ×40, or ×100) provided by each objective is engraved on its barrel. The oil-immersion objective has a ring engraved on it toward the tip of the barrel.

The resolving power of the objective is the ability of the objective to resolve each point on a minute object into widely spaced points so that the points in the image can be seen as distinct and separate from one another, thus resulting in a clear, unblurred image. It may appear that very high magnification can be obtained by using more high-power lenses. Though possible, the highly magnified image obtained this way is a blurred one. That means each point in the object cannot be found as a widely spaced distinct and separate point on the image. A mere increase in size (greater magnification) without the ability to distinguish structural details (greater resolution) is of little value. Therefore, the basic limitation in light microscopes is not of magnification but of resolving power, the ability to distinguish two adjacent points as distinct and separate, i.e., to resolve small components in the object into finer details on the image.

5.   Eyepiece: The eyepiece is a drum that fits loosely in the draw tube that magnifies the real image formed by the objective. Usually, a microscope has two eyepieces with different magnifying powers of $\times 10$ and $\times 25$. Depending on the magnification requirement, one of the two eyepieces is inserted into the draw tube before viewing.

6.   Spectrophotometer: The spectrophotometer is used for measuring the intensity of electromagnetic radiation at various wavelengths. Spectral bandwidth and the range of absorption are important parameters of spectrophotometers.

The measured diffusivity operational light ranges in EM spectrum normally cover around 200 nm 2500 nm with different controls and calibrations. Reflection or transmission properties of a material measured through a spectrophotometer are quantitative and a function of wavelength. The term electromagnetic spectroscopy is very specific and deals with visible light within ultraviolet and near-infrared ranges. Time-resolved spectroscopic techniques do not cover anything that allows the measurement of temporal dynamics and kinetics processes.

Its measurement principle is simple and easy to comprehend A spectrophotometer has two major classifications, single beam and double beam. A double-beam spectrophotometer compares the light intensity between two light paths, one route containing the reference sample and the other the test sample. A single-beam spectrophotometer measures the relative light intensity of the beam before and after the test sample is introduced.

Colorimeter

A device used for measuring colors is known as a colorimeter, and the process is known as colorimetry. It works on the principle of absorbance of different light wavelengths by a solution. A known solute concentration is measured by this technique. A colorimeter compares the amount of light passing through a solution with the amount that can pass through a pure solvent sample.

Digital colorimeter.

A colorimeter contains a photocell to detect the amount of light that passes through the solution under investigation. It is a light-sensitive instrument that measures the absorbing capacity of an object or substance. Color determination is based on the red, blue, and green part of light absorbed by the sample, much as the human eye does. When light passes through a medium, a fraction of the light is absorbed, and as a result, there is a decrease in how much of the light is reflected by the medium. The colorimeter measures that change so users can analyze the concentration of a particular substance in that medium. The working principle of this device is based on Beer–Lambert's law, which states that the absorption of light transmitted through a medium is directly proportional to the concentration of the medium.

X-ray

X-ray analytical methods are very common and the most powerful techniques for internal body organ analysis. This section reviews the X-ray techniques used for determining and probing cell, molecular, and supramolecular structures. Additionally, it includes instrumental aspects from a characterization point of view. When a beam of high-energy electrons, protons, deuterons, a-particles, or heavier ions are bombarded on a matter, X-rays are emitted. These rays lie within the wavelength range of 0.05 to 100 A°. X-rays propagate from light velocity and in straight lines without being deflected by electric or magnetic fields. When X-rays encounter matter, several processes take place, and every process may be used in the research and study of particular properties of materials. This includes reflection, refraction, diffraction, scatter, absorption, fluorescence, and polarization.

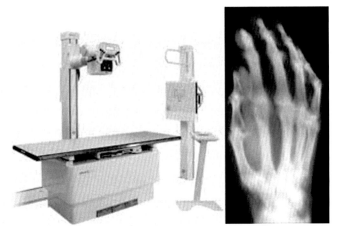

X-ray machine and generated X-ray image.

The overall choice of analysis and appropriate analytical method in materials characterization depends on the physicochemical properties of the material under investigation and the length scale to be probed. In general, the length scale to be probed in a scattering experiment is according to the wavelength of the scattered radiation. Further, the penetration depth of a particular type of radiation in a sample depends on the energy of the radiation. X-rays are not suitable for studying light elements and are not found in applications to distinguish isotopes of the same element.

Electromagnetic flowmeter

According to Faraday's law of electromagnetic induction, a conductor moving in a magnetic field generates an induced current. According to Fleming's right-hand rule, which defines the direction the current flows, the right thumb, index finger, and middle finger give the direction of the conductor motion, magnetic field, and induced current.

Electromagnetic flowmeter.

Fleming's right-hand rule applies to several types of electrical equipment in which an electric generator produces a current. An electric motor needs an external current supply, in which case, Fleming's left-hand rule also applies. An electromagnetic flowmeter uses a probe that cuffs around a blood vessel. The probe contains a magnetic coil that produces a magnetic field B. Two electrodes are arranged in such a manner that both are in the direction perpendicular to the direction of magnetic field "B" to obtain the induced potential e, according to blood movement through the blood vessel at a flow velocity v. $e = B L v$, where L is the distance between the two electrodes. The volume flow rate is the integration of the velocities over the cross section of the lumen of the blood vessel, which can be approximated by the average flow velocity times the cross-sectional area. Because the cross-sectional area inside the flow probe does not change, the instantaneous volume flow rate can be measured. However, a calibration process is required that involves measuring the blood volume through the flow probe at a constant flow rate during a known period.

Cardiac output monitoring

Cardiac output monitoring is essential for a patient in critical condition who needs continuous monitoring to ensure tissue oxygenation. This is generally accomplished with a pulmonary artery catheter (PAC). However, some researchers and scientists have questioned using PAC and believe it to be unnecessary and potentially harmful. There are two types of cardiac output monitoring systems, invasive and noninvasive, with decreased availability of invasive cardiac output-measuring devices reducing the widespread use of PAC. Today, many devices are available to measure cardiac output with various methods. Some are invasive devices to continuously track stroke volume and provide dynamic indices of fluid responsiveness, whereas others allow assessment of volumetric preload variables, and still others provide continuous measurement of central venous saturation via catheters attached to the same monitor. All the variables together with cardiac output provide improved results in the hemodynamic assessment of critically ill patients.

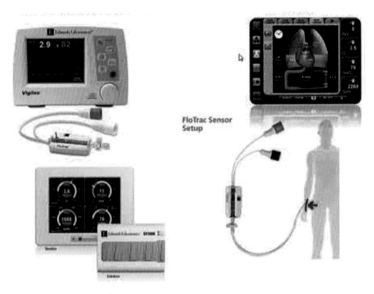

Cardiac output monitoring system.

It is important to know that each device has inherent limitations. Cardiac output monitoring devices can change patient outcomes unless their use is coupled with an intervention that by itself has been associated with improved patient outcomes. Therefore, the concept of hemodynamic optimization is increasingly recognized as a cornerstone in the management of critically ill patients and is associated with improved outcomes in preoperative and intensive care unit settings.

Pulmonary function analyzer

Today's pulmonary function analyzer (PFA) is a collection of several connected, discrete respiratory system diagnostic instruments to provide objective data about the ventilation, diffusion, and distribution of gases in the patient's lungs, as well as valuable data on oxygen exchange and carbon dioxide levels in the lungs.

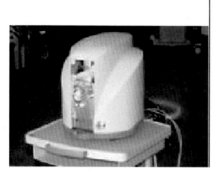

Various pulmonary function analyzers.

Data such as lung volume, flow rate measurements, flow volume testing, muscle strength assessment, and diffusing capacity measurement provide the pulmonologist with a clinical picture to assess the patient's respiratory system. Besides aiding in diagnosing respiratory dysfunction, such as asthma, bronchitis, or emphysema, a PFA can provide baseline data for preventive medicine/industrial hygiene purposes and assess patient treatment progress. Current PFA technology has parametric components such as forced vital capacity, peak flow, forced expiratory volume in 1 s, and airway resistance (Raw). An exhaled gas analyzer provides additional information such as functional residual lung capacity and lung diffusing capacity for various gases such as carbon monoxide. These otherwise standalone instruments are connected to a personal computer (with the usual peripheral devices) that provides output to a printer. Of course, proprietary software is involved, but placing all these devices together on a single cart creates a basic pulmonary function analyzer system.

Pulse oximetry system

Nowadays, various personal health devices are becoming popular as people become more conscious about their health. The necessity for healthcare technologies within home premises is a demand in today's world. Healthcare devices measure data on user health—for example, a pulse oximeter, activity monitor, or medication reminder. Among healthcare devices, the pulse oximeter to measure pulse oximetry data (SpO_2 and pulse rate) has become important for the early diagnosis of heart disease. Existing pulse oximeters are equipped with communication capabilities.

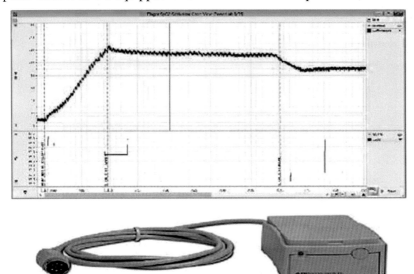

Pulse oximetry system.

Modern oximetry systems come with network capabilities for transmission and reception of data, focusing on exchanging pulse oximetry data with limited device management functions. Network-capable healthcare devices are being applied to various e-health services, and the standardization and calibration of such devices are important to guarantee interoperability.

Blood cell counters

Automatic blood counters are fancy at present. Three basic principles operate in detecting blood counts: image analysis, the Coulter principle, and flow cytometry. The image analysis method counts blood cells by first taking optical images of cells and then extracting cellular feature information (cell size, nucleus shape, etc.) from those images. Blood samples are normally treated with staining reagents beforehand to enhance the discrimination of different cell types. The image analysis method is no longer popular in modern automated blood counters due to limited counting throughput and accuracy. For example, the ambiguity of classifying white blood cells (WBCs) by image information alone could introduce substantial inaccuracy to the WBC differential count result.

Blood cell counter machine.

The Coulter principle detects the electrical impedance of blood cells by flowing the sample through a narrow aperture to ensure the measurement of individual cells. This method can evaluate cell properties such as cell size (direct current impedance) and six-cell electrical opacity (alternating current impedance). It can deliver a three-part WBC differential (lymphocyte, monocyte, and granulocyte) and is still widely used in automatic blood counters because of its low reagent costs and label-free detection. However, the Coulter principle alone is not sufficient to distinguish all five major types of WBCs (lymphocyte, monocyte, neutrophil, eosinophil, and basophil). Therefore, it is often used in parallel with other detection methods, such as light-scattering detection, to extend the capability of distinguishing WBC types. The flow cytometry principle is a general platform technology where single cells could be accurately quantified in suspension with high throughput, and the quantitative methods mainly involve optical detections such as light scattering, light absorption, and fluorescence detection. It is widely appreciated as a powerful technique with a significant impact in various fields of biology and medicine.

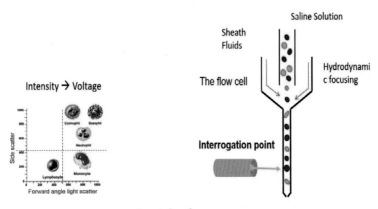

Principle of cytometry.

However, most flow cytometers are quite complex to operate and need careful calibration. Some relatively simplified versions have been developed specifically for blood analysis. For measurement by the flow cytometer, blood samples must usually be treated with labeling reagents (fluorescent dyes, fluorescent conjugated antibody assay, etc.) beforehand to introduce the cell labels for optical detection. Above all, the state-of-the-art instruments for automatic blood cell counting are mostly based on the Coulter principle, the flow cytometry principle, or a combination of both. During the last 2 decades, the automated blood cell counter has undergone a formidable technological evolution owing to the introduction of new physical principles for cellular analysis and progressive software evolution.

Computer-aided biopsy

Imaging-guided biopsy is a primary method to obtain tissue specimens. Various imaging techniques, including computed tomography (CT) fluoroscopy and ultrasound, can be used to guide biopsies, but CT is most frequently employed because of its high spatial and contrast resolution as well as its 3D imaging ability; in many cases, CT is preferred based on the localization of the nodule or other patient-related factors.

High-performance liquid chromatography techniques

High-performance liquid chromatography (HPLC) is an analytical method used to separate organic compound mixtures based on their volatile thermal energy and high molecular weight.

A liquid support called a mobile pad sends the injected sample to the column and the probe for separation. In the separation column, the individual components are separated based on a physicochemical interaction, and the upper layer array is based on such an interaction. Subdivided elements are subject to changes in light absorption or reflection. Discovery through exploration based on electrochemical/electrochemical transformation or distribution of dissociated molecules.

The detector responds to peaks with a constant current proportional to the volume. The output is monitored and verified by an operating program that provides not only the necessary calculations but also the infusion volume, infusion sequence, detection wavelength, and output cycles, including the necessary calculations.

HPLC is a common and flexible technique that provides cost-effective solutions for sorting, identifying, and quantifying components of complex organic samples. In the first place, you may be afraid to mix compounds and tubing. However, once you understand the role of each component and start solving the puzzle, the HPLC system is an easy-to-use device that ensures the highest accuracy and reliability when properly maintained results are analyzed after analysis.

This PowerPoint presentation, "Introduction to High-Performance Liquid Chromatography," is a brief and concise introduction to the components of HPLC. It is important to understand how each component contributes to the overall reliability of the analysis. For convenience, HPLC is an easy-to-use, unlimited system.

On the last slide of this HPLC presentation, you can find a link to the HPLC extended certificate. We invite you to take part in the program, participate in periodic discussion forums and seminars, and interact with experts. See the HPLC PowerPoint slide presentation below for more detail.

The advanced software features available today make it user-friendly. Most times, the user goes to the mobile updates for intermediate and regulatory recording and drafting of documents.

These analytical science techniques support the development and understanding of analytical chemistry in food and drink. The polymer allowed better quality control of consumer products such as environmental monitoring.

Chromatography is a commonly used analytical method. Chromatograms distribute complete feature penalties but with the advancement of technology. Scientists have increased the amount of waste. It was found to be believable that separate sections are considered internally. The current mixture may be a tortoise species but cannot be identified using 1D chromatography.

Storage time is not significant and must consider the complex patterns to be measured. This requires individual complementary and complex ingredients as we attempt to learn more about sample size. Analytical methods are gaining popularity, and improved analytical efficiency can be analyzed in two ways. Multidimensional approaches have been further developed, particularly in analyzing complex chromatographic systems. Specifically, gas chromatography (GC) is multidimensional, but in addition, other analytical methods have been considered recently, and there has been growing interest in regional development separation for liquid chromatography. The second method is evolving toward additional analytical methods; will this chromatography succeed in the workplace? Other analytical and species identification methods have been developed, providing a means to ask questions about analysis with liquid chromatographic ablation (LC reduction).

GC is also associated with many neurological disorders; alternatives analytical tools may include Fourier transformation into liquid chromatography, magnetic resonance imaging, and ICPs. In addition, different types of chromatography can be combined, such as LC–GC, and a question arises here about multidimensional separation or printing technology.

Different combinations of the abovementioned parts are as follows:

pump

detector

injectors offering unlimited application-based customization

To obtain the most reliable data, you should be familiar with the HPLC system and its principles. Understanding how each component works will increase your comfort with the HPLC system, be highly dependent on release, and ensure long-term use.

Mobile phase

The liquid phase is pumped to the stationary phase column at a constant speed. Before entering the column, the sample is introduced into the carrier flow.

When they reach the column, the sample components are received and stored based on the physicochemical interactions between the analyte molecules and the stationary phase.

The moving phase sends the sample to the system. An important criterion of the mobile phase is the inertia of the selected components. A combination of pure solvent or buffer is generally used. The moving phase must be free of solids and removed before use.

Mobile phase pools

Mobile phase pools are inert containers for mobile storage and transport. Clear glass containers are often used to facilitate visual inspection of the moving object within the container. A stainless-steel particulate filter can be placed to remove particles from the moving phase.

Pump

Fluctuations in mobile phase flow affect sample component flexibility and cause errors. The pump has constant phase pressure flow to the column.

Injector

A fixed volume is injected into the mobile phase flow. Injection inefficiency and repetition must be accounted for to maintain high precision.

Column

The column is a stationary phase stainless steel tube. This is important and should be properly maintained according to the supplier's instructions to increase insulation repeatability and efficiency.

Column oven

If the temperature changes during the test, the shelf life of the isolated components may change. The column oven maintains the temperature of the column with the help of circulating air. This ensures constant flow through the mobile phase column.

Detector

The detector responds to the isolated components of the column and provides the required sensitivity. It must be independent of any change in the composition of the displacement phase. Many applications require UV detection, but detectors based on other detection methods are also widely used.

Data collection and control

Modern HPLC systems comprise computers and software that control operating parameters such as roll phase composition, temperature, flow rate, volume, and the sequence of injection and output.

New period analytical method

Analytical chemistry is a popular career choice for chemistry students. An obvious reason for this is that chemical knowledge is used to understand the chemical properties of substances. This is a valuable contribution to determining the quality of natural resources and ready-made industrial products intended for consumption. The quality and consistency of these sources depend on the precise definitions used by analytical chemists in their laboratories.

Chromatography and HPLC have made significant contributions, especially in product research and development. It should be noted that HPLC has found many applications in pharmacology, food, biochemical research, synthetic polymers, biomolecule research, and environmental monitoring. The number of HPLC applications is growing fast, and many new programs appear in popular magazines nearly every day.

The following sections describe recent developments that have led to widespread recognition of HPLC as a common technology in modern laboratories.

Operating area

HPLC allows sample analysis over a wide range of molecular strengths and weights. Sample sizes are usually distinguished from diagrams in grams by different column sizes and methods. The evolution of synthetic polymers and large biomolecules has been due to the ability of HPLC to process samples from low to several million molecular weights.

Technological advances

Silicon columns are used to form a stationary base. In addition, columns with a homogeneous phase, basic zirconium phase, and shell-based core contribute to the analysis rate by increasing the working temperature and pressure range to study the expanded pH range.

Sensor selection

HPLC offers a variety of identification options based on the properties of eluted compounds, including the following sensor types: UV, IR, light code array, fluorescence, conductivity, light distribution, and selective mass. LC-MS-MS is a highly sensitive technology that offers analytical solutions in biological research and pesticide residue analysis in agricultural products to analyze complex impurities and set lower limits.

Choosing mobile phone operating steps

The most common method of operation is reverse phase chromatography, where poly solvents are used as the mobile phase. However, using a mixture of solvents, buffers, and isocratic/gradient stages with different polarities can be useful for dissolving complex mixtures of molecules with spikes. The diagnostic capabilities of HPLC have increased its popularity. In modern laboratories, several HPLC systems may operate continuously.

Gas chromatography

In the early 1920s, Mikhail Semyonovich Tsvet developed GC as a compound separation technology. In organic chemistry, liquid and solid column chromatography are often used to separate organic compounds in solution. Among the various types of GC, gas-liquid chromatography (GLC) is the most common method for separating organic compounds. The combination of chromatography and MS is a valuable means of identifying molecules. Typical GC consists of an injection port, a column, a gas—air circulator, a furnace, and a heater to maintain the gas temperature integrated connector and injection column, recorder, and sensor.

Gases and liquids within compounds are separated by injecting a sample solution containing the organic compounds of interest into the sample, which in turn is evaporated. The inert gas then drives injected vapor patterns, often using helium or nitrogen. This inert gas flows through a glass column filled with liquid containing silica. Less soluble substances and liquids increase the yield faster than more soluble substances. This unit aims to provide a better understanding of the measures and technology used in applying GC.

In GLC, the hydrostatic phase is adsorbed on an inert package that is fixed to the capillary walls or inactive. A column is considered full when the glass or metal pipes of the column are lined with a small spherical inertial support. The liquid phase on the surface of these grains moves in a thin layer. In the capillary column, the tubular walls are coated with a stationary or adsorbent phase that can carry a liquid phase. However, the GLC method can be used only to a limited extent in the laboratory, and then only rarely because of the wide finish and semipermanent storage of the polishing compound in the column. Therefore, the discussion here on the gas—liquid method in GC is brief. The purpose of this unit is to provide a better understanding of the measures and technology used in its application.

Sample injection

A sample angle connection is required to place the sample in front of the column. In modern injection methods, heated sample wells are often used, whereby the sample can be injected and evaporated almost simultaneously. A calibrated microsyringe is used to transport a sample of a few microns through the rubber barrier into the steam chamber. Most parts require just a small fraction of the initial sample volume, and part of the sample is used to remove too many samples. Commercial chromatography often allows separate injections without inoculation when packaged columns and capillary columns are packed in series. The evaporation chamber is usually heated to 50°C above the lowest boiling point in the sample and mixed with the carrier gas to transfer the sample to the column.

Carrier gas

Carrier gas plays an important role and is characterized by the use of GC. The carrier gas should be a dry kinetic phase, free of oxygen, chemically inert, and used in chromatography. Helium is used primarily because it is safer but more efficient, is suitable for hydrogen, has a wider flow range, and is compatible with many sensors. Depending on the desired sensor efficiency, nitrogen, argon, or hydrogen may be used. Hydrogen and helium are commonly used in most conventional sensors, such as isolation (flame ionization, or FID), thermal conductivity (TCD), and electron sampling (electronic capture detector, or ECD), and have shorter detection times and lower sampling temperatures due to higher current velocity and lower weight molecules. For example, hydrogen or helium as a carrier gas is more sensitive to TCD because the difference in thermal conductivity between organic vapor and hydrogen/helium is greater than that of other carrier gases. Other sensors, such as spectroscopes, use nitrogen or argon, which have higher molecular weights than hydrogen or helium and thus have a tremendous advantage in terms of increased vacuum pump efficiency.

All gases are available in pressure vessels; controllers, manometers, and flow meters are used to precisely control gas flow. Most gas reserves employed should be 99.995%—99.9995% pure and contain small amounts (<0.5 ppm) of oxygen and total carbohydrates in the tank. The carrier gas system contains a molecular sieve to remove water and other toxins. Another option is to use traps to keep the system as clean and sensitive as possible while removing traces of water and other contaminants. Two-stage pressure balancing is required to reduce water hammer and control gas flow. A gas flow controller is also required on the side and inlet of the chromatography gas to control the gas flow. This applies to the different types of gas used by various regulators. The carrier gas is heated and filtered through a molecular sieve to remove toxins and water before entering the steam chamber. In a GC system, a gas dryer is usually required to guide the injection and apply the gas sample portion to the GC column associated with the sensor.

Column oven

The temperature of the column can be adjusted in tenths using a thermostatic oven for precise operation. The oven can be operated by either isothermal or temperature programming. During isothermal programming, the temperature of the column is kept constant during separation. The optimal column temperature for isothermal operation is in the middle of the boiling point of the sample. However, uniform temperature programming works best when the range of sampling sources is narrow. When a low boiling temperature is used with a high-temperature isothermal column, the low boiling fractions dissolve well, but the high boiling fractions are slowly washed with a wide bandwidth. When the temperature approaches higher boiling points, the elements with higher boiling points are eluted as sharp spikes, but the elements with lower boiling points are washed away so quickly that no separation occurs.

In temperature programming mode, the column temperature rises continuously or gradually as separation continues. This method is well suited for isolating a mixture with a high boiling point. The analysis begins by dissolving the low-boiling components and continues with subtraction to dissolve the less volatile portion of the sample. A typical temperature programming speed is 5°C—7°C/min.

Detection system

The sensor at the end of the column makes it possible to combine the components of the bulk washing mixture with the properties of the carrier gas. The object has different real expectations from assets other than the gas mixture in practice. These funds are divided into two categories, special and property. It is of good size, so it is a special or high-quality analytical petrol line even in such situations. Sensors can use phosphorus levels such as nitrogen to compensate for increased sensitivity.

The two main parts of each sensor, if they must fill their cabinets to detect fraud, are signaling electrical changes in the sensors. However, you do not remember it in a color sense; the first part should be the sound, if it was supposed to start from the optimization column came close to the sensor according to the electronic equipment used to scan the analog computer signal for color analysis. This is larger than the return of a faster analog-digital signal because the analog is so simple that this component becomes a sound signal for various forms of interference.

Bulk analyzers

Bulk analyzer sensors are the most powerful of all GC sensors. In the GC/MS system, the substance is constantly searched for in the mass analysis during isolation. As the sample exits the chromatography column, it passes through a transmission line to enter the MS. The sample is usually ionized and decomposed by ions that decompose the electron. During this process, the bomb explodes, ionizing high-energy electrons in the molecule and causing electron loss due to the electron beam. Additional bombardment will result in ion decay. The ions are then transferred to a mass analyzer to be sorted by m/z or mass and charge. Most ions have just one charge.

Chromatography shows the retention time, and peaks are used in the mass analysis to determine the types of molecules in the mixture. The figure below shows a typical spectrum of water masses with absorption peaks at a suitable m/z ratio.

Gas chromatography/mass spectrometry unit

One of the most common GC/MS analysis units is the four-pole ion analyzer, which allows long-term trapping of gas anions or cations by electric and magnetic fields. One square ion trap consists of a ground electrode shield with two grounded solenoid valves. The ions are introduced into the cavity at the upper end of the network. Corresponding ions with m/z values pass through the radio electrode at different radio frequencies within the cavity. As the radio frequency increases linearly, ions with a constant m/z value are released in order of mass. Excess or light ions are unstable and have a neutral charge when they collide with the electrode wall. The emitted ions then enter an electronic multiplier that converts the detected ions into electrical signals. The computer then receives these electronic signals through various programs. The result is chromatography showing the m/z ratio and richness of the sample.

GC/MS devices are useful because they allow the mass of smart material to be determined immediately and can be used to identify incomplete components. They are reliable, are easy to use, and can detect the sample almost immediately after washing. The disadvantages of mass analyzers are the tendency of the samples to dissipate heat for detection and the result of removing the decomposition of the entire sample.

Ion flame sensor

FID sensors are the most commonly used sensors. In FID, the sample is sent to the air-hydrogen flame after leaving the column. The air temperature flame takes a sample or decomposes chemically when heated at high flame temperatures. Disinfected carbohydrates release ions and electrons that carry current. A high impedance manometer measures this current to monitor sample removal.

FID should be used sparingly, as the sensor does not affect nonflammable gas and water quantities. These features allow for high FID sensitivity and low noise. The device is both reliable and relatively easy to use. However, this technique requires a flammable gas and destroys the sample.

Thermal conductivity sensor

The first sensors developed for chromatography were TCD sensors. TCD sensors work by measuring the change in the thermal conductivity of the carrier due to the presence of a sample with a thermal conductivity other than the carrier gas. Their structure is relatively simple and consists of an electrically heated source maintained at constant power. The source temperature depends on the thermal conductivity of the surrounding gas. The source is usually platinum and gold. The resistance of the conductors depends on the temperature, which depends on the thermal conductivity of the gas.

The TCD typically uses two sensors, one used as a carrier gas reference and the other to monitor the thermal conductivity of the carrier gas and sample mixture. Carrier gases such as helium and hydrogen have a very high thermal conductivity, so impurities can be easily detected in a small number of samples.

Electronic capture detector

ECDs can detect selected versions of Clostridia and common compositions like organic halogen, such as peroxides, quinone with niter, and groups, and they react to all the others with little or no compounding. Hence, it cannot be detected by suitable use of pesticides that are narrow because of color of the whole economy.

The gaseous form is cast into the burning proportions of the ECD of electron radiators in the electric field. The GC analysis moves like a train and withdraws in order, which is usually the more severe nickel 633 or tritium. The ionizer emitter of nitrogen carrier gas causes the electron to explode, with a constant current between two electrodes in the absence of organic compounds. The addition of organic compound functional groups significantly reduces the electron current to trap electron functional groups.

Selection ECDs have higher sensitivity and specific organic electronic resource service groups. But this sensitivity is countered by limited detector range, possibly because of radiation risk. The reason for attempting to sign from the right of those who had the oversight of shear repeatedly and destruction is cast into the detector.

Nuclear emission sensor

An important device, the gas chromatograph outfitted in the emission detector feed, is selective in the first part of the ionizer gas to the detector molecule in a sample of all the elements like their nuclear program, the write emission. Micromechanical sensor is applied in very efficient and widely used options to detect nuclear items. There are three ways to use search-produced plasma, microwave-induced plasma (MIP), ICP, or direct current plasma. MIP is the most widely used form and monitors the use and emissions of nuclear components with a diode array.

Chemiluminescence detector

Chemiluminescence spectroscopy (CS) is a procedure by which properties can be determined by focusing on qualitative and quantitative optical species radiation. It is not like AES; it uses the light emitted by atoms and molecules to become excited. Because the time for impeding gas or solution in luminescence can make use of the conditions, the light source of the gecko's reaction from the chemical industry produces rays of light in work. This light source is used in place of the strip, as it is a separated light beam.

There are other ways to work with the CS's main limitations and photoconductor tube (PMT) limitations, and the public has become aware of CS's limitations. This is a PMT issue; it must detect the darkness to light emitted from the analyzer.

Photoionization detector

Using GC, the photoionization detector (PID) takes advantage of CS. A PID is a portable gas detector of steam aromatic hydrocarbons, simple organic heteroatoms, mixed species, and other organic compounds. PIDs detect ultraviolet lights of the photon emission composite ionization that leave the GC column and are absorbed into the bedroom. In fact, a small, ionized molecule as part of the analyzer is more pernicious and would not permit otherwise determined analytical results to be confirmed by the detector. In addition, the PID is available as a portable PDA with few lights. The result is almost instantaneous. PID is placed in the soil to detect volatile organic compounds; silt, air, and water are always used to detect contaminants in the soil and the air around it. It does not find fault with heavy downward quality, and the PID is peculiar to hydrocarbons of molecular compositions that may be a metaphor for the ethane.

DNA analysis techniques

In the 1980s, people increasingly became involved in searching for the human genotype. Laboratory tests for gene products—protein or enzyme—and AB and O Rhesus blood samples are classic examples of collecting genotypes from the reaction of genetic products with certain chemicals. Genetic engineering was a major success in the mid-1980s. The genotype leaps forward with the possibility of DNA. Now genetics can be studied without going through the tedious process of analytical development; particular differences in proteins and enzymes can be detected directly. Direct DNA analysis provides the advantage of detecting alleles in unencrypted DNA fragments and polypeptide chains. As a result of these new developments, the number of discovered genetic sites has increased rapidly, and this has quickly led to the identification of gene anomalies that had been mysteries for much of the 20th century.

This section describes the main methods of molecules. That is the point that provides a quick and clear guide for a sociologist. The content of this section does not require an understanding of genetics, what genes are, and how they are related to human behavior. In fact, this part can be studied without too much loss. Thus, only the necessary elements are provided, and the reader's interest is in the laboratory. These methods apply to modern textbooks of scientific molecular genetics.

Electrophoresis

Electrophoresis is an electronically short process that separates charged particles from liquids using an electrically charged field. In biology, it is mainly used to separate protein molecules or DNA and can be obtained by various processes depending on the type and size of the molecules. The methods vary, but all require an electric charge source, support medium, and buffer. It is used in electrophoresis laboratories to separate molecules by size, density, and purity.

Working principle

The molecules are subjected to an electric field. The force of the electric field is as great as the charge of an atom, so the front support in the center of the atom moves according to its mass.

Some examples include applications of DNA and RNA analysis and electrophoresis, as well as protein electrophoresis, a treatment used to analyze and isolate molecules found in fluid samples (mainly blood and urine samples).

Types of electric fields

Various gels are often used as a substrate for electrophoresis and are more effective depending on whether they are in the form of plates or tubes. A gel plate can contain several samples simultaneously, so it is often used in the laboratory. However, tubular gels provide better resolution results, so they are often chosen for protein electrophoresis.

Argos gel is widely used for DNA electrophoresis. It has a large porous structure that allows large molecules to move easily but is not suitable for indexing small molecules.

Polyacrylamide gel electrophoresis has a clearer resolution than early gels, making it more suitable for quantitative analysis. This allows one to determine how proteins bind to DNA. Plasmid analysis provides insight into how bacteria become resistant to antibiotics. Two-dimensional electrophoresis separates molecules along the x-axis and y-axis, one by charge and the other by size.

Cloning

In popular fantasy, "cloning" is associated with scientific literature. "Cloning" means to copy any part of the DNA. Historically, cloning began with the isolation of a small fragment (i.e., several thousand pairs) of human DNA that, after detailed steps, is added to the DNA of another body. The vector is needed to create multiple copies of a human DNA segment so that new genetic material can reproduce quickly. Therefore, they are the most common carriers with the smallest organisms and high reproductive potential—plasmids, viruses, bacteria, and yeasts. There are several motivations for human DNA cloning. Multiple human DNA sequences could be used in other forms for research methods in molecular genetics. Another reason is to create arguments from DNA libraries. Therefore, cloning is also used in therapeutic genetics technology. The hope is that a vector here has a cloned copy of the "good" gene. The loss of a protein can be placed in the patient's cells, which can lead to the loss of protein or enzyme.

The test is a series of DNA or RNA segments with a specific nucleotide order. The test is synthesized or cloned in the laboratory in large quantities that are physically exposed to human DNA under treatment. Due to the additional base pair, the study is linked to a DNA segment containing a specific nucleotide sequence that completes the study.

Purified DNA and RNA are comparable to viscous water. When many single-stranded pieces of DNA undergo electrophoresis, the joints move to the other side of the gel depending on the size but are invisible. If you add research, they will link to other single-stranded DNA fragments suitable for a double strand. However, nucleotide samples are invisible and designed for analysis after DNA binding and therefore carry the biological equivalent of a lightbulb. There are two main types of lightbulbs used in molecular genetics. The first type comes from the investigation mark of radioactive isotopes, whereas other types of research use special fluorescent dyes and have a photo taken under special lighting conditions. However, radioactive use of fluorescence technology is rapidly replacing it.

CHAPTER 6

Cardiac pacemakers and defibrillators

Sudip Paul[1], Angana Saikia[1,2], Vinayak Majhi[1] and Vinay Kumar Pandey[1]

[1]Department of Biomedical Engineering, School of Technology, North-Eastern Hill University, Shillong, Meghalaya, India; [2]Mody University of Science and Technology, Laxmangarh, Rajasthan, India

Contents

Introduction to Biomedical Instrumentation and Its Applications
ISBN 978-0-12-821674-3
https://doi.org/10.1016/B978-0-12-821674-3.00007-3

The heart is a muscular organ that provides continuous blood circulation throughout the human body. In normal situations, heartbeats are regular, ranging from 60 to 100 beats per minute. The heart rate can become much faster for physiological reasons—physical activity, pain, stress, etc. The heart comprises two upper chambers, the left and right atria—and two lower chambers, the left and right ventricles. Circulation of the blood for one heart cycle is performed in two loops.

Pulmonary loop

a. Oxygen-poor blood enters the right atrium via superior and inferior vena cava.
b. The right atrium pumps the blood into the right ventricle through a tricuspid valve.
c. The right ventricle contracts, pumping blood through the right and left pulmonary arteries to the lungs for oxygenation.

Systemic loop

a. The oxygen-rich blood from the lungs passes through the pulmonary veins and enters the left atrium.
b. The left atrium pumps the blood into the left ventricle through a mitral valve.
c. The left ventricle contracts, pumping oxygen-rich blood through the aorta to the rest of the body.
d. Regular contractions of the heart are managed by the heart's electrical system, which has three important components:
   • Sinoatrial (SA) node—known as the heart's natural pacemaker, leading the rhythm in a healthy heart
   • Atrioventricular (AV) node—the bridge between the atria and ventricles, passing electrical signals from the atria to the ventricles
   • His—Purkinje system—carries electrical signals throughout the ventricles and includes the His bundle, right bundle branch, left bundle branch, and Purkinje fibers

Problems with the heart's electrical system can disrupt its normal rhythm. A pacemaker is a small battery-powered medical device that distributes electrical impulses to ensure regular heart contractions and normal blood circulation to the heart muscles. Therefore, cardiac arrest and death can be prevented. Modern pacemakers excite the heart muscle and store various data from internal cardiac activity, sensation signals, and pacing systems. These include a pulse generator and one or more pacing leads that transmit electrical impulses from the pulse generator to the heart before returning to the pulse generator. Depending on the underlying heart disease, a pacemaker can be programmed to move the upper chamber, the lower chamber, or both. The American College of Cardiology and the American Heart Association have published clinical guidelines for permanent pacemaker implants, selecting the type of pacemaker according to the patient's relevant characteristics.

Pacemakers are embedded to control the pulse. They can be embedded incidentally to treat a slow heartbeat after a coronary failure, medical procedure. Or on the other hand, they can be embedded for all time to address a slow or irregular heartbeat or, in certain individuals, to help treat cardiovascular breakdown. Surgery is required to embed a pacemaker in one's chest. There are two sections: a generator and wires (leads). The generator is a little battery-fueled unit. It creates the electrical motivations that invigorate the heart to pulsate. The generator might be embedded under the skin through a little cut and is associated with the heart through tiny, simultaneously embedded wires. The driving forces move through these prompts to coordinate the stream at regular intervals as a normal pacemaker would do. A few pacemakers are external, brief, and not precisely embedded.

A pacemaker replaces the heart's imperfect natural pacemaking capacities. The SA hub, also called the sinus hub, is the heart's natural pacemaker. It is a small mass of cells in the upper chamber of the heart. It delivers the electrical driving forces that cause the heart to thump. A chamber of the heart contracts when an electrical motivation or sign moves across it. For the heart to thump appropriately, the sign must go down a particular way to arrive at the ventricles (the heart's lower chambers). When the heart's regular pacemaker is inadequate, the heartbeat might be excessively quick, excessively slow, or irregular. Rhythm issues can additionally happen because of a blockage of the heart's electrical pathways. The pacemaker's heartbeat generator sends electrical driving forces to the heart to assist it with siphoning appropriately. Most pacemakers have a detecting mode that prevents them from sending driving forces when the heartbeat is higher than a specific level. It permits the pacemaker to fire when the heartbeat is excessively slow. These are called demand pacemakers. Pacemakers work just when required. On the off chance that one's pulse is excessively slow (bradycardia), the pacemaker imparts electrical signs to the heart to address the beat. Likewise, more up-to-date pacemakers have sensors that distinguish body movement or breathing rate and signal the pacemaker to increase or vary the pulse during exercise. Two small, leadless pacemakers, which can be embedded straightaway into the heart, have been endorsed for use in the United States. Since no lead is required, this gadget can limit certain dangers and speed recovery. Although this sort of pacemaker seems to function admirably and securely, longer-term study is required.

The development of the cutting-edge pacemaker is a great story in the progression of innovation and our comprehension of how the heartbeat functions. The first pacemaker was embedded in an individual in 1958. It did not have a long operating life, as the patient lived to age 88 and had 26 pacemakers over the course of his life.

Pacemakers reached the cutting edge in 1969 with the primary lithium battery. A man arrived on the moon the same year that pacemakers discovered batteries.

The first pacemakers paced just the ventricle, or base chamber of the heart. "Physiologic" was first used in pacing to portray a pacemaker that could detect the patient's hidden beat and not rival it. It was known as a "demand" pacemaker. It would possibly produce a heartbeat if the pacemaker detected that the patient's heart had no beat of its own.

However, that was not adequate. It did not represent the chamber or upper chamber of the heart in patients with a hidden mode in the chamber. The subsequent stage in "physiologic" pacing was a double-chamber pacemaker that could pace the atria, or upper chambers, and the ventricles, or lower chambers, in succession like an ordinary heart. Double-chamber pacing was then called "physiologic."

The next stage of development was the "rate-responsive pacemaker." This pacemaker used a piezoelectric crystal (a precious stone similar to that found in a phonograph stylus). The crystal would twist or disfigure with body action. A pacemaker could use this electrical sign to build the pulse action as an ordinary heart would. The crystal was supplanted by "accelerometer" a long time ago to more precisely reflect movement. There are several available sensors for doing this.

A standard pacemaker has a lead that paces the right ventricle. A few patients have diminished heart work, and a pacemaker that paces the two ventricles simultaneously has been demonstrated to improve heart work. This pacing, designated "biventricular pacing" can pace the chamber, the right ventricle, and the left ventricle. The leadless pacemaker is nevertheless another advance in pacemaker development. It does not have a double chamber yet can build a pulse with action. It might be the correct decision for patients with an irregular atrial mode (for example, atrial fibrillation).

Pacemaker pulses

Pacemakers typically supply current between two electrodes; one has an electronic lead, and the other does not. When current flows between the two lead electrodes of the pacemaker, it is called bipolar pacing. When the current is flowing, the space between the tips of the pacemaker generator is called unipolar pacing. Bipolar and unipolar pacing differ in their surface ECG presence. The current flows over a large body area between the pulse and the vibration generator in unipolar pacing. Therefore, unipolar pacing produces large stimulus artifacts (hundreds of millivolts). In bipolar pacing, the distance of the electronic card to the surface is very small (about 1 cm) between the two poles carrying current and excitation. The ECG is very small on the surface and nearly invisible. In pulse pacing with average patterns, the skin surface may be smaller than the width and as small as 25 μs (e.g., height 1 mV, width 500 μs). Most pacing pulses have edge spans with growth that is very fast but changes less than 10 μs?

False-pace pulse detection

Motion artifacts, muscle noise, microventilation pulses, telemetry signals, and noise from the patient's other treatment devices may cause false-movement pulse detection. Speed is a well-known source of artistic and EMG sound errors and ECG analysis. Microventilation pulses (used to control the pacing rate of rate-response pacemakers) are always less than 100 wide and vary by about 15 degrees. This is 100 μs. Hence, they can lead to false pulse speed detection. Another major source of noise is the H-field telemetry scheme used with most superposed cardiac devices. Avoiding such errors requires a reusable form and an appropriate time interval between pulses. Multiple ECGs are at the top, and sound and pulse rate can prevent false motion while identifying the more complicated true motion pulses.

Types of pacemakers

Pacemakers vary according to the number of leads they have.

*   *Single-chamber pacemaker*

    This pacemaker uses one of the electrodes in the right atrium or the right ventricle of the heart. When the sinus node generates irregular pulses, the electrode is placed in the right atrium. However, to use this stimulation method, the rest of the heart's conduction system must function normally. The single electrode is often placed in the right ventricle to help correct a slower or irregular heartbeat. It is most often the case when the electrical current is slowed or blocked in the region of the AV node and the normal impulses from the atria cannot reach the ventricle, thus producing too much slow-beating heart.

*   *Dual-chamber pacemaker*

    This pacemaker uses one terminal in the heart's upper chamber and one anode in the lower chamber. This pacing most intently mirrors the heart's ordinary conduction design by pacing consecutively from the atria to a ventricle, thereby amplifying the heart's pumping capacity. By having terminals in both the atria and ventricle, the beat generator is ready to constantly control the heart's electrical activity in the two chambers. These are the most regularly used pacemakers these days.

*   *Biventricular pacemaker*

    This pacemaker uses three leads—first into the right atrium, second into the right ventricle, and third into the left ventricle. It is applied when the two ventricles do not contract at the same time and are not synchronized with the atria. In cardiovascular breakdown, the right and left ventricles do not pump together. When the heart's compressions become unsynchronized, the left ventricle cannot pump enough blood to the body. In such cases, biventricular pacing (aka CRT) is applied.

    Pacemakers vary according to their programming.

*   *Fixed-rate pacemaker*

    This pacemaker produces a consistent pace. A fixed-rate pacemaker cannot distinguish characteristic pulses and simultaneously puts out electrical driving forces while the heart's own pacemaker fires, causing competing beats.

- **"*On-demand*" *pacemaker***

  This pacemaker monitors heart rhythm and generates electrical pulses only if the heart beats too slowly or misses a beat. The benefit of an on-demand pacemaker over a fixed-rate pacemaker is that it generates pulses only for serious events has greater battery life.

- **Rate-responsive pacemaker**

  This pacemaker accelerates or hinders the pulse contingent on the dynamics of the individual. The ideal pulse is controlled by an extra sensor for an individual's movement or a breathing sensor that identifies the breathing rate.

Pacemaker operating modes

The working method of the pacemaker is coded with four letters. The first letter shows the chamber being paced—A for atrium, V for ventricle, D (double) for both. The second letter represents the chamber detected—it can be one of the letters in the first position (i.e., A, V, or D), or an additional selection, 0 for none. The third letter identifies the reaction to a detected electrical sign and shows whether a pace is restrained or activated. Hindrance of pace is done when an inherent beat is detected in the chamber, and another planning cycle begins. In activated mode, an upgrade is transmitted when an event in one chamber is detected in the other. The predetermined codes are I (inhibited) for restraint of pace in the detected chambers, T (triggered) for activating in the detected chambers, D for double (inhibition + trigger), and 0 for none. The fourth letter identifies the rate balance, with R (responsive) for balanced rate and 0 for no rate regulation.

Fig. 6.1 shows the block diagram of a pacemaker.

Table 6.1 summarizes the advantages and disadvantages of a cardiac pacemaker.

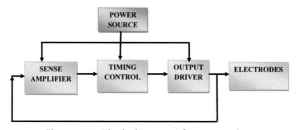

Figure 6.1 Block diagram of a pacemaker.

Table 6.1 Advantages and disadvantages of a cardiac pacemaker.

Advantages	Disadvantages
Mimics natural action: The pacemaker performs electrical actions to work at a faster rate when the normal functioning of the heart fails to manage heartbeat pacing.	Procedure risks: Pacemaker implantation may expose patients to site diseases. In addition, they can become oversensitive to sedation and become susceptible to swelling and bruising.
Heart efficiency: A biventricular pacemaker works to guarantee the ventricle and atria cooperate and improve the heart's pumping effectiveness.	Lifestyle adjustment: • Pacemaker users must be cautious around electrical apparatuses—for example, mobile phones, microwaves, and high-tension
Atrial fibrillation: Used to correct a heart ventricle that vibrates rather than beats in a normal mode.	wires. Completely changing one's lifestyle may protect one's life and well-being.
Heart failure: Cardiovascular breakdown can be forestalled using a pacemaker to facilitate electrical signs between the ventricles of the heart.	• Pacemaker users should also be careful about clinical hardware, for example, MRI with exceptionally high magnetic forces and shockwave lithotripsy to eliminate kidney stones. • All these factors can disturb a pacemaker's functioning, preventing it from working and possibly causing death.

Types of pacing

a. Percussion pacing, also known as transthoracic mechanical pacing, usually uses a closed fist on the lower-left end of the sternum above the right ventricle to induce ventricular vibration. (The *British Journal of Anaesthesia* suggests that ventricular pressure should be increased by 10−15 mmHg to induce electrical activity.) This is an old method used only as a life-saving measure until an electric pacemaker is procured.

b. Transcutaneous pacing (TCP), also known as percutaneous or external pacing, is recommended to stabilize all types of hemodynamically important bradycardia at an early stage. The procedure is performed in the anterior/posterior position by placing two pacing pads on the patient's chest. The rescuer chooses a pacing rate and gradually expands into a wide QRS complex (characterized by a wide QRS complex) with a long, wide T wave pacing current (measured by MAA) in the ECG. Vibration. Putting the artwork on the ECT and causing severe muscle tremors can make this decision difficult. External pacing should not be a long-term solution; it is an emergency procedure that acts as a bridge until the procedure has passed or other treatments are provided.

c. Temporary epicardial pacing is used during open-heart surgery. The electrodes are kept in contact with the outer wall of the ventricle to maintain satisfactory cardiac output until a temporary transvenous electrode enters.

d. Transvenous pacing is an alternative to TCP. The pacemaker wires are cleanly placed in the vein and pass through the right atrium or right ventricle. Then the pacing wire is connected to a pacemaker outside the body. Pacing is often used as a bridge for the permanent placement of pacemakers. It can be kept until the permanent pacemaker is embedded or removed after the pacemaker is needed.

e. Permanent transvenous pacing involves inserting one or more pacing electrodes into a permanent pacing chamber or chamber with an implantable pacemaker and implanting the pacemaker into the skin under the wrist. The procedure is performed by an appropriate venous cutdown in which an electronic lead is inserted and placed in the chamber through the heart valves along the vein. This procedure is facilitated by fluoroscopy that allows the physician to observe the passage of electronic leads. After verifying that the electrons are good for the electronics, the opposite end is taken to the pacemaker generator.

f. Leadless pacing uses leadless pacemakers small enough to hold the generator to the heart, thus eliminating the need for pad reeds. Pacemaker leads can fail over time, so avoiding these components gives a pacing system a theoretical advantage. Leadless pacemakers make an incision in the groin and can be implanted in the heart using a maneuverable catheter inserted into the femoral vein.

Biventricular pacing

Cardiac resynchronization therapy (CRT) is used in patients with failure who do not simultaneously contract the left and right ventricles (asynchronous ventricles), which occurs in about 25%–50% of patients with heart disease. CRT uses a biventricular pacemaker (BVP) that can trigger the middle and side walls of the left ventricle. By pacing on both sides of the left ventricle, the pacemaker can resynthesize ventricular contractions. The

CRT device has at least two leads, one through the vena cava and the right through the atrium and left ventricle. Rhythmic patients with normal sinuses often have a lead in the right atrium to facilitate synchronization with atrial contractions. Therefore, optimal cardiac function can be achieved by adjusting the time between atrial and ventricular contractions and the middle septum and sidewall of the left ventricle. CRT devices have been shown to reduce mortality and improve the quality of life for patients with symptoms of heart failure. The LV ejection fraction is less than 35%, and the QRS duration at EKG is more than 120 mm. Biventricular pacing alone is called CRT-P (for pacing). CRT may be combined with an implantable cardioverter-defibrillator (ICD) for certain patients at risk of arrhythmias in a device known as a CRT-D (for defibrillation) for life-saving arrhythmias. It also provides effective protection.

His bundle pacing

RV apical pacing around the upper right ventricle or in traditional ventricular placement can adversely affect cardiac function. It is associated with an increased risk of atrial fibrillation, heart failure, myocardial infarction, and reduced life expectancy. Bundle pacing (HBP) leads to more natural or completely natural ventricular activation, which has generated strong research and clinical interest. HBP uses specialized reed and placement techniques to directly stimulate the Harken fiber network, resulting in synchronized and consequently more efficient ventricular activation and avoiding long-term myocardial disease. HBP may also change the bundle branch block pattern in some cases.

Functional advancements

The main step in pacemaker function is to create a rate-response pacemaker with parameters using different inputs such as the QT interval, pO_2 to pCO_2 (dissolved oxygen or carbon dioxide level). In imitating nature, create a stable heart rate or simultaneous control over the accelerometer, body temperature, ATP levels, possible physical activity such as adrenaline, and expected breathing potential. The first dynamic pacemaker was invented in 1972 by Anthony Rickards of the National Heart Hospital in London, England. Dynamic pacemaking technology could be applied to the artificial heart of the future. Advances in transition sublayers support these and other synthetic/joint/tissue exchange efforts. Stem cells may be interested in the transitional epithelium. When replaced, some progress was made to improve pacemaker control. Many of these have been replaced by pacemakers made with microprocessor-controlled pacemakers that control not only ventricles but also small cells. A pacemaker that controls both the atrium and the ventricle is called a dual-chamber pacemaker. Dual-chamber models are usually expensive, but some ventricular extruders can improve the heart's ventricular pumping efficiency and aid in heart failure response. The rate-response pacing allows the device to understand the patient's physical activity and increase or decrease the basic transit rate through appropriate response algorithms. David's experiments suggest that unnecessary passage of the right ventricle may accelerate heart failure and increase the incidence of vertigo. The new dual-chamber device reduces the amount of proper ventricular passage and prevents the severity of heart disease.

Pacemaker insertion

Pacemakers can be implanted locally with anesthetics to cripple the skin or sleep with or without general anesthetics. Antibiotics are usually given to reduce the risk of infection. Pacemakers are usually implanted in the left or right shoulder area in front of the chest. The skin is prepared by shaving or shaving at the implant site before cleaning the skin with skin disinfectant. A hole is made under the collarbone to create space or pockets under the skin to hold the pacemaker generator. This pocket is usually made just above the pectoralis major, but in some cases, the device can be placed under the muscle (below the muscle). Lead or lead is delivered to the heart through large veins operated by X-ray images (fluoroscopy). The lead tip can be placed in the right ventricle, right atrium, or coronary sinus, depending on the type of pacemaker needed. Surgery is usually completed within 30—90 min. After transplantation, the healed wounds should be kept clean and dry until they heal. To reduce the risk of a pacemaker lead disconnection, one should be careful not to move the shoulder excessively for the first few weeks. Pacemaker generator batteries usually last 5—10 years. As the battery nears the end of its life, the generator is usually replaced in a simpler process than for a new implant. Replacement involves removing the existing device, disconnecting the lead from the old device, reconnecting the lead to a new generator, installing a new device, and skinning.

Periodic pacemaker checkup

After installing the pacemaker, one should regularly confirm that the device is turned on and working properly. The device can be tested as often as necessary depending on the number of times prescribed by the following physicians: Regular pacemaker tests are usually performed in the office every 6 months depending on the condition of the patient/equipment and the availability of remote monitoring. The new pacemaker model can also be searched remotely, allowing patients to transmit pacemaker data using their home transmitter connected to a geographic cellular network. Technicians can access these data through the device manufacturer's web portal. During internal follow-up, tools are specified for performing diagnostic tests. These tests include the following:

a. Sensing is the ability to "see" the internal cardiac activity of the device (displacement of the atrium and ventricle).

b. An obstacle is a test to measure reading integrity. A large or sudden increase in obstruction may indicate a reed break, and a large or sudden decrease in obstruction may indicate a reed insulation violation.

c. Marginal width is the minimum amount of energy (usually 1/100 of a volt) required to guide the atrium or ventricle.

d. Marginal period is the time required for the device to be given a width to increase the speed of the atrium or ventricle attached to the lead.

e. Percentage of pacing determines how much the patient depends on the device. The pacemaker actively paces after searching the previous device.

f. Estimated battery life at current rates depends on device life and use, as modern pacemakers are "on demand" and only accelerate when needed. Other factors that affect the life of the device include programmed outputs and algorithms (features) that receive elevated levels of current from the battery.

All events, especially arrhythmias such as atrial fibrillation, have been archived since the last follow-up. These are usually set by a physician and stored based on specific patient-specific criteria. As soon as the event starts, some devices have availability to display potential intracardiac maps of the event. This is especially effective for determining the cause or source of an event and making necessary programming changes. An embedded electronic pacemaker mimics the behavior of a natural electrical system. The pacemaker consists of two parts.

Pulse generator: This small metal container has a battery and an electrical circuit that controls the speed of electrical pulses transmitted to the heart.

Lead (electronic): One to three flexible insulating wires are placed in one or more heart chambers to supply electrical pulses to control the heart rate.

Pacemakers only work when needed. If the heartbeat is too slow (bradycardia), the pacemaker sends an electrical signal to the heart to correct the heartbeat. Also, the new pacemaker sensors detect body speed and breathing rate, indicating that a pacemaker can increase one's heart rate as needed during exercise. Two small lead-free pacemakers that can be implanted directly into the heart are approved in the United States. Since no lead is required, this device can reduce some risk and speed recovery.

Complications of pacemaker use

The pacemaker may be defective by
a. oversensing events
b. reducing incidence
c. failing at speed
d. not being captured
e. pacing at unusual speeds

Tachycardia is a particularly common complication. Speed-modulated pacemakers can increase excitation as a response to voltage during magnetic field-induced vibrations, muscle activity, or magnetic resonance imaging. In pacemaker-mediated tachycardia, a functional dual-chamber pacemaker sensitizes the ventricular premature or ash nodes (transmitted through the atrioventricular node or retrograde axillary pathway) and the ventricle of a strong repetitive cycle. Additional complications associated with extracurricular devices include cross talk suppression by sensing atrial pacing impulses through dual-chamber pacemaker ventricular channels, ventricular pacing, and pacemaker syndrome induced by ventricular pacing AV, which may cause fluctuations, hallucinations (e.g., the presence of light), neck disorders (for example, neck beats), or respiratory symptoms (e.g., dyspnea). Pacemaker syndrome is controlled by atrial-synchronous atrial pacing (AAI), single-lead atrial sensing ventricular pacing (VDD), or dual-chamber pacing (DDD). Environmental interference comes from electromagnetic sources such as surgical electrodes and MRI, but MRI can be safe if it is not inside the pacemaker generator and lead magnet. Mobile phones and electronic security devices are potential sources of interference. One's phone should be kept away from the device but only when it is used for conversation. Passing a metal detector does not cause a pacemaker to malfunction unless the patient is bending over.

Procedures before pacemaker insertion

a. Electrocardiogram (ECG pacemaker): In this optional test, a sensor pad attached to a wire called an electrode is placed in the chest and can be placed on a limb to measure the heart's electrical impulses.

b. Holter monitoring is the portable version of an ECG. It is especially effective in diagnosing rhythmic diseases that occur at unexpected times. When one wears the monitor, it records information about the heart's electrical activity and normal activity for a day or two.

c. Echocardiogram. Harmful sound waves are used in this noninvasive test to allow the doctor to see the heart's movements. A small device called a transducer is placed on the chest. The sound waves (echoes) collected from the heart are transmitted to the machine, which uses the sound wave pattern to create images of the beating heart on the monitor.

d. Stress test. Some heart problems occur during exercise. The stride test takes an electrocardiogram before and after walking on a treadmill or riding a practice bike. In some cases, echocardiography or atomic imaging is applied.

Precautions for cardiac pacemaker use

a. Cell phones. It is safe to talk on one's cell phone, but it should be kept at least 6 inches (15 cm) from the pacemaker. Do not leave the smartphone in a shirt pocket. When talking on the phone, hold the pacemaker close to the ear where it is embedded.

b. Security systems. Passing through an airport metal detector will not interfere with the pacemaker, but the metal in it may sound an alarm. However, the metal does not lean near or tilt the detection system. To avoid potential problems, one should carry an ID card that indicates the use of a pacemaker.

c. Medical equipment. Make sure all doctors and dentists know when a person has a pacemaker. Some treatment modalities, such as magnetic resonance imaging, CT scans, cancer radiotherapy, electrification to control bleeding during surgery, and shock wave lithotripsy to destroy large kidney stones and gallstones, are likely to interfere with the pacemaker.

d. Power-generating equipment. Stand at least 2 ft (61 cm) from the welder, high-voltage transformer, or motor-generator system. If one wants to avoid such a device, the doctor should be consulted to arrange a test to see whether the device affects the pacemaker.

Pacemaker indications

a. sinus node dysfunction
b. acquired atrioventricular (AV) block
c. chronic bifascicular block
d. after the acute phase of myocardial infarction
e. neurocardiogenic syncope and hypersensitive carotid sinus syndrome
f. postcardiac transplantation
g. hypertrophic cardiomyopathy (HCM)
h. pacing to prevent tachycardia

Pacemakers can be implanted using an endocardial or epicardial approach.

The most common method is the endocardial (transvenous) approach. Local anesthetics (analgesics) are given to numb the area. A slit is made in the chest where the reed and pacemaker will be inserted. The lead is inserted into a vein through a hole, and the heart is imaged with fluoroscopy. The tip of the reed is attached to the heart muscle, and the other end of the tube (attached to the pulse generator) is placed in a pocket made under the skin at the top of the chest.

Epicardial procedures are less common in adults but more common in children. General anesthesia is used for this surgical procedure. The surgeon attaches the tip of the skull to the heart muscle and places it in a pocket made under the abdominal skin at the other end of the rim (which connects to the pulse generator). Although recovery from the epicardial approach is longer than recovery from the transvascular approach, minimally invasive strategies have made hospital status and recovery time more active.

After replacing the lead, it is checked that the straw is working properly and can increase one's heart rate. The name of this reading function test is "Pacing". A small amount of energy is distributed to the bone muscles through lead. This energy makes the heart contract. After testing the lead, the doctor attaches it to the rear machinery and determines the pacemaker's speed and settings. After the pacemaker embedding procedure, the physician uses an external device (programmer) to program the final pacemaker settings.

Pacemaker troubleshooting

a. Battery: the battery may be defective, life of the battery indicates that the battery has no work.
b. Diagnosis of lead: pacemaker is programmable for increased speed and leads in bipolar mode ring or unipolar mode (reed tip of pulse generator). Bipolar sensing is preferred because of its small sensitivity, with an "antenna" that reduces the processing of extracardiac signals. There is an advantage to reducing bipolar pacing stimulation of pectoral muscles; however, in bipolar marginal lift or sensory impairment (e.g., ring electrode due to conductor loss), unipolar pacing or sensing can be seen. The programmable sensitivity is the minimum signal and can be identified as a cardiac event. So a setting of 0.5 mV is twice as sensitive as 1.0 mV. In general, a two-sensing protection margin is desirable. If the potential atrial map is 1.0 mV, set it to 0.5 mV. Sensitivity is adjusted to improve oversensing and undersensing. Some devices can have their sensitivity automatically adjusted based on measurement amplitude awareness-raising events. The two most common device-related arrhythmias are tachycardia through pacemaker and high-rate movement with high and low ventricular rates.

Pacemaker operating modes

a. VOO mode: pace ventricle, sense nothing, no response. Ventricular asynchronous pacing is the simplest mode without sensing or mode opinions. Ventricular pacing occurs at programmed low-rate intervals (LRIs) regardless of internal heart activity.
b. AOO mode: pace atrium, sense nothing, no response. This mode is similar to VOO, with the only difference being that instead of the ventricle, the atrium is paced after every LRI.
c. DOO mode: pace atrium + ventricle, sense nothing, no response. The DOO timing cycle consists of defined AV and ventricular-atrial (VA) intervals. The VA interval is a function of the AV and ventricular—ventricular (VV) intervals. The intervals do not vary, because no activity is sensed.

d. VVI mode: pace ventricle, sense ventricle, inhibit pacing. In this mode, the pacemaker paces the ventricle and senses the electrical activity in the ventricle. Initially, LRI is set for a programmed interval that starts with a sensed or paced ventricular event. During this LRI, when the pacemaker senses an intrinsic ventricular beat, it inhibits the pacing pulse, and LRI is reset. In the opposite case, if the ventricular beat is not sensed until the end of LRI, the pacemaker paces the ventricle, and again LRI is reset. VVI pacemakers are refractory after a paced or sensed ventricular event—a period known as the ventricular refractory period (VRP). Any ventricular event occurring within the VRP is not sensed and does not reset the LRI. VVI is the most commonly used pacing mode. It offers protection from lethal bradycardias but does not restore or maintain atrioventricular synchrony and does not provide rate responsiveness.

e. AAI mode: pace atrium, sense atrium, inhibit pacing. AAI pacing incorporates the same timing cycles as VVI with the obvious difference that a sensed atrial event inhibits pacing and sensing from atrium and pacemaker output. In this mode, LRI starts from an atrial event. An atrial paced or sensed event initiates an atrial refractory period (ARP) during which the pacemaker senses nothing.

f. DDD mode: pace atrium + ventricle, sense atrium + ventricle, inhibit + trigger pacing. In this mode, the pacemaker paces both the atrium and ventricle and senses the electrical activity in the atrium and ventricle. Some intrinsic activity in the atrium inhibits the pace but triggers a pace in the ventricle after the AV interval. If there is an intrinsic activity in the ventricle, the pace in the ventricle is also inhibited.

g. DVI mode: pace atrium + ventricle, sense ventricle, inhibit pacing in atrioventricular sequential (called sequential demand) mode, both the right atrium and the right ventricle are paced, but only the ventricle is sensed. The DVI timing cycle consists of defined AV and VV intervals. The VA interval is a function of the AV and VV (LRI) intervals. Both chambers are paced at the same rate, separated by a fixed AV sequential interval. The ventricles are paced if the defined AV interval expires and there is no sensed intrinsic ventricular beat. If there is some intrinsic ventricular beat within the AV interval, the pace of the ventricle is inhibited, but the VA interval is not reset. With no sensing in atria, even with some intrinsic atrial beat in the VA, the pace is not inhibited, and the pacemaker paces atria when the VA interval expires.

h. DDI mode: pace atrium + ventricle, sense atrium + ventricle, inhibit pacing. In this mode, sensed atrial activity inhibits the atrial pacing but does not trigger a ventricular pace pulse—i.e., the atrial sensed event does not produce a physiological AV delay, and if the intrinsic ventricular beat is not sensed, the ventricles are paced when the VV interval has elapsed.

i. VDD mode: pace ventricle, sense atrium + ventricle, inhibit + trigger pacing. In this mode, the pacemaker paces only the ventricle, senses both atrium and ventricle, and responds by inhibiting the ventricular pace when intrinsic ventricular activity is sensed and triggering a ventricular pace in response to an intrinsic P wave.

j. VAT mode: pace ventricle, sense atrium, trigger pace in ventricle. In this mode, the ventricle is paced after every LRI even if there is a ventricular beat in the LRI (no ventricular sensing, no inhibition). The atrium is sensed, and if intrinsic atrial activity is found, ventricular pace after AV interval is triggered in the ventricle, and a new LRI is started. VAT mode can be used in patients with complete heart block with normal sinus and atrial electrical function, but it should be avoided when an abnormality of sinus or atrial function is present.

k. VVT mode: pace ventricle, sense ventricle, trigger pace in ventricle. In this mode, the ventricle is paced and sensed. If there is any intrinsic activity sensed in the ventricle, ventricular stimuli are triggered. When no spontaneous electric activity is sensed in the ventricle outside the VRP, the ventricle is paced at the LRI cycle. In VVT, the pacemaker delivers a pulse when no ventricular event has been sensed within a preset LRI and generates pace pulses triggered by any spontaneous ventricular event within the preset LRI. This mode was invented to overcome difficulties that the earlier VVI pacemaker encountered when exposed to magnetic interference and falsely inhibited the delivery of pacing pulses. VVT is rarely used—only when a patient is routinely exposed to electromagnetic interference. Its general drawback is that it may cause tachyarrhythmias.

l. AAT mode: pace atrium, sense atrium, trigger pace in atrium. In this mode, the atrium is paced and sensed. If there is any intrinsic activity sensed in the atrium, atrium stimuli are triggered. When no spontaneous electric activity is sensed in the atrium outside the ARP, the atrium is paced at the LRI cycle. In theory, this mode is used when the AAI mode cannot be used because of symptomatic skeletal muscle sensing.

Defibrillation

Defibrillation is a system used to treat hazardous conditions that influence the cadence of the heart, for example, cardiovascular arrhythmia, ventricular fibrillation, and pulseless ventricular tachycardia. The methodology includes conveying an electric shock to the heart that causes depolarization of the heart muscles and restores ordinary conduction of the heart's electrical drive. The machine used to convey this remedial shock to the heart is known as a defibrillator. The various kinds of defibrillators used include external defibrillators, transvenous defibrillators, and embedded defibrillators. Defibrillation was first introduced by Prévost and Batelli, two physiologists from the University of Geneva, Switzerland, in 1899. In animal studies, they saw that little electric shock conveyed to the heart could trigger ventricular fibrillation, while the conveyance of enormous electrical charges could switch the fibrillation.

In 1947, the method was used without precedent for a human patient. Claude Beck, a professor of medical procedure at Case Western Reserve University, rewarded a student experiencing surgery for a chest deformity and determined how to reestablish an ordinary sinus mode in the child's heart. The early types of defibrillators conveyed a charge of somewhere in the range of 300 and 1000 V to the heart using "paddle" type terminals. In any case, the units had significant disadvantages, such as the requirement for open-heart medical procedure, the transformers were enormous and hard to move, and postmortem assessment indicated the method was harming the heart muscles. Moreover, the method was regularly ineffective in really switching ventricular fibrillation. During the 1950s, an elective technique for conveying an electric shock to the heart was spearheaded by V. Eskin and colleague A. Klimov from the USSR. Instead of the paddle electrodes used in open-heart medical procedures, the closed chest gadget could apply a charge of more than 1000 V through nodes applied outside the chest cage. In 1959, Bernard Lown and architect Baruch Berkovitz built up a method of conveying the charge using protection from making a less solid sinusoidal wave that would last 5 ms using paddle anodes. The specialists also settled the ideal planning regarding when stuns ought to be conveyed, which empowered use of the strategy in different instances of arrhythmias, for example, atrial fibrillation, atrial shudder, and one type of tachycardia. This strategy was named the Lown—Berkovitz waveform, and it turned into the standard defibrillation treatment to be used into the late 1980s. From there on, the biphasic shortened waveform (BTE) was embraced as a similarly powerful waveform that necessary less charge to accomplish defibrillation. The unit was likewise lighter to move. The BTE waveform related to programmed transthoracic impedance estimation frames the premise of the cutting-edge defibrillator. The present compact defibrillators were presented in the mid-1960s by Prof. Frank Pantridge in Belfast. Today, these instruments are part of the basic gear found in an emergency vehicle. A further advancement was the ICD, which was created at Sinai Hospital in Baltimore by a group of individuals that included Stephen Heilman, Alois Langer, Jack Lattuca, Morton Mower, Michel Mirowski, and Mir Imran at Sinai Hospital in Baltimore. The gadget was made by Intec Systems of Pittsburgh.

Table 6.2 summarizes some defibrillation milestones.

Defibrillation is often an important step in cardiopulmonary resuscitation (CPR). CPR is an algorithm-based intervention aimed at restoring cardiac and respiratory function. Defibrillation is indicated in certain ventricular fibrillation, particularly ventricular fibrillation (VF) and pulseless ventricular tachycardia. Defibrillation does not appear when the heart is completely closed, such as Estelle or pulseless electrical activity (PEA). Defibrillation does not indicate that the patient is conscious or has a pulse. Improperly transmitted electric shock can lead to dangerous arrhythmias such as VF. Survival rates for outpatient cardiac arrest are often less than 10%. More than 20% welcome for hospital cardiac arrest. Within a group of people with cardiac arrest, certain heart rhythms can significantly impact survival. People with stunned rhythms (e.g., VF and pulseless ventricular tachycardia) may have a 21%–50% improvement in survival when there are no stunned rhythms (e.g., Astor and PEA).

Table 6.2 Defibrillator milestones.

Year	Achievement
1960s	Edward and Lown et al. produce fewer symptoms than AC defibrillators. The DC beat waveform was also improved.
1970s	Exploratory inner and external gadgets intended to naturally identify ventricular fibrillation
1980s	First programmed interior defibrillator was embedded in a human
Present-day	A great deal of progress acquainted with the defibrillator with the point of improving the endurance pace of the cardiac arrested patient

Types of defibrillators

- *Manual external defibrillator*

    These defibrillators require more understanding and preparation to successfully deal with them. Consequently, they are just basic in emergency clinics and ambulances where skilled hands are available. In conjunction with an ECG, the provider decides the cardiovascular rhythm and afterward physically decides the voltage and timing of the shock-through outer paddles to the patient's chest.

- *Implantable cardioverter-defibrillator*

    Another name for this is the automatic internal cardiac defibrillator (AICD). They continually screen the patient's heart, like a pacemaker, and distinguish VF, ventricular tachycardia, supraventricular tachycardia, and atrial fibrillation. When an unusual mode is identified, the gadget consequently decides the voltage of the shock to reestablish cardiovascular capacity.

- *Manual internal defibrillator*

    The manual interior defibrillators use an inner paddle to send the electric shock straightforwardly to the heart. They are used on open chests, so they are just regular in the working room. It was designed after 1959.

- *Automated external defibrillator*

    These defibrillators use PC innovation, subsequently making it simple to analyze the heart's beat and successfully decide whether the rhythm is shockable. They can be found in clinical offices, government workplaces, air terminals, inns, sports arenas, and schools.

- *Wearable cardiac defibrillator*

    Further examination was done on the AICD to deliver the wearable heart defibrillator, a versatile outer defibrillator, for the most part, demonstrated for patients who are not in an immediate requirement for an AICD. This gadget is equipped for observing the patient 24-hour-a-day. It is useful when worn and sends a shock to the heart at whatever point it is required.

- *Semiautomated external defibrillator*

    It has the carrier features of both full manual and automated defibrillator functions. It has an ECG display and manual override. Paramedics and emergency medical technicians mostly use these types.

Working principle

Energy is stored in the capacitor, which is relatively charged at a slow rate from the AC line. Energy stored in the capacitor is then delivered at a rapid rate to the chest of the patient. A simple arrangement involves the discharge of capacitor energy through the patient's resistance.

Fig. 6.2 shows the block diagram of complete defibrillator setup.

Table 6.3 summarizes the advantages and disadvantages of defibrillation threshold testing.

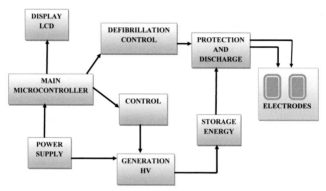

Figure 6.2 Block diagram of a defibrillator.

Table 6.3 Advantages and disadvantages of defibrillation threshold testing.

Advantages	Disadvantages
Extremely fine with protection gauge of 0.04% major antagonistic occasion rate	Major antagonistic occasions can happen including • pulmonary embolism • stroke • hypotension requiring intervention • cardiogenic shock
Patient and provider consultation in explicit clinical situations: • right-sided transverse implant • secondary prevention • generator change with hazard alert lead • concern regarding pacing, sensing, or impedance values	Low yield <3% with inadequate safety margin
Assurance of suitable detecting capacities	Expanded expense and procedural time

Human interface

The connections between the defibrillator and the patient were each equipped with a conductive gel to ensure a good connection and reduce the chest obstruction's electrical resistance (regardless of DC discharge) that burns the patient. It contains a pair of electrodes. Gels are either moist (related to surgical lubricants) or hard (like treated candies). Solid gels are more convenient because the gel used after defibrillation does not need to be removed from the skin. However, wet gel electrodes conduct electricity evenly throughout the body, so using a hard gel increases the risk of burns during defibrillation. The first type of paddle electrode has no gel to develop, and the gel has to be applied at another stage. Self-adhesive electrodes are presynthesized with gel. There is a general debate about which type of electrode is better in a hospital environment. The American Heart Association does not support all the latest manual defibrillators used in hospitals or may not quickly switch between sticky pads and conventional paddles. Each type of electrode has its characteristics and disadvantages.

Paddle electrodes

The most well-known of the electrodes (widely displayed in film and television) is a metal thimble metal paddle with a heated (usually plastic) handle. This type requires about 25 pounds (11.3 kg) of force or application to the patient's skin during a series of push-ups. Paddles have several advantages over sticky pads. Since these electrodes can be placed and used in most cases, they continue to be used in many hospitals in the United States in conjunction with disposable gel pads. This is important during cardiac arrest because every second of nonperfusion means tissue reduction. Modern paddles allow observation (electrocardiography), but the leads of various observations in hospital situations are often predetermined. The paddles are reusable, cleaned after use, and reserved for the next patient. Therefore, the gel is not disseminated, and these paddles should be attached to the patient before use. Paddles are usually found in manual outdoor units.

Self-adhesive electrodes

A new type of resurfacing electrode is designed as a sticky pad formed by hard or wet gel. Once supported, like any other sticker, peel it and stick it on the patient's chest as needed. The electrodes are then attached to the defibrillator much like a paddle. If defibrillation is required, charge and shake the machine without applying additional gel or regenerating and placing the paddle. Most adhesive electrodes can be used not only for defibrillation but also for TCP and synchronous cardioversion. These sticky pads are available in most automatic and semiautomatic units and replace the paddles in completely nonhospital settings. In hospitals, sticky pads can be professionally placed in cases of a high probability of cardiac arrest (but not yet). The pads offer benefits to trained users and healthcare professionals who work in the subatomic state of the field. No additional signals are required to monitor the pad, and no force is required during the injury. Therefore, the adhesive reduces the risk of entering physical (and thus electrical) contact with the patient by being affected by allowing the electrode operator to stay a few feet away. (Similar to electric shock due to operator misuse, the risk of electric shock to others does not change) Glue electronics are used only once. These can be used for multiple vibrations in the same course of treatment, but if the patient recovers (or in that case), the patient is removed when cardiac arrest occurs.

Placement

The regenerating electrodes are placed according to one of the two schemes. The previous plan placed the scheme electrodes above the left vestibule (lower chest, front of the heart) suitable for long-term electronic placement. The other electrode is located in the scapular center behind the heart. This measure is recommended as it is ideal for nonaggressive pacing. Front-back planning (front-back position) can be used when front-back planning is inconvenient or pointless. In this scheme, the anterior electrode is placed at the bottom right side of the hammer. The upper electrodes are applied to the patient's left side, just below, and to the left of the pectoral muscles. This scheme is suitable for defibrillation and electrical defibrillation, as well as ECG monitoring. Researchers have developed a software modeling system that can map a person's breasts and determine the optimal position of an external or internal cardiac defibrillator.

AC defibrillators

An AC defibrillator is the most established and least complex sort. The development of the AC defibrillator aims to make suitable qualities accessible for inside and outside defibrillation. In AC defibrillation, a shock of 50 Hz AC recurrence is applied to the chest for a period of 0.25−1 s through anodes. The system of applying electric stun to resynchronize the heart is known as countershock. Defibrillation proceeds until the patient reacts to the treatment. An AC defibrillator comprises a step–up transformer with essential and auxiliary winding and two switches. Switches and breakers provide AC flexibility to essential twisting of the transformer. The planning circuit is associated with a switch used to preset the ideal opportunity for the defibrillator to convey shock to the patient.

A resistive and a basic capacitor arranges or monostable multivibrator structures the planning circuit. It is activated with a foot switch or a press button switch. Different tapping is accessible along the optional winding. They are associated with terminals that convey electric shock to the core of the patient. Voltage value running between 250 and 750 V is applied for AC outer defibrillation. For security reasons, the auxiliary loop ought to be segregated from earth to avoid shock. For inward fibrillation, a voltage between 60 and 250 V is applied. To deliver uniform and concurrent compression of heart muscles, enormous flows are used for external defibrillation. In any case, this results in skin consumption under cathodes and severe withdrawal of heart muscles. It likewise brings about chamber fibrillation and stops VF.

Disadvantage of AC defibrillators
- They cannot be used to correct atrial fibrillation.
- Continuous attempts to correct the VF is required.
- Attempts to correct atrial fibrillation by this method often result more serious VF.
  Fig. 6.3 shows a circuit diagram of an AC defibrillator.

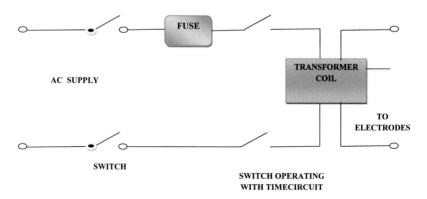

Figure 6.3 Circuit diagram of an AC defibrillator.

DC defibrillators

A DC defibrillator does not create symptoms and produces a regular heartbeat. VF is evaded when a high-voltage shock is sent through a releasing capacitor presented to the heart or chest of the patient. A DC defibrillator consists of an autotransformer T1 that is much like the high-voltage transformer T2. A diode rectifier redresses the yield voltage from T2 and is associated with a vacuum-type high-voltage changeover switch. At position A, the switch is associated with one discharge of the capacitor. At this position, the capacitor recharges to a given voltage. A foot switch on the electrode handle is used to administer the shock to the patient.

Presently, the high-voltage switch transforms its position to B, making the capacitor release electricity to the heart through anodes. To hinder the release from the capacitor, an inductor (L) is set in one of the anode leads. This L initiates a countervoltage that lessens the capacitor release esteem.

Dual peak DC defibrillator

If peak voltage is as high as 6000 V is used, there is a chance of harming the myocardium and the chest walls. It produces a double-pinnacle waveform of a longer duration at a lower voltage. Effective defibrillation is accomplished in grown-ups with a lower level of conveyed energy. Energy extends between 50 and 200 W-sec or joules. Effective defibrillation at attractive lower voltage levels is additionally conceivable with the shortened waveform. The abundance of the waveform is generally steady but is shifted to achieve the required vitality.

Large anodes are used for the proper conveyance of enormous current through the outside of the skin. These electrodes are called paddles.

External defibrillator

A unit dependent on PC innovation intended to examine the heart rhythm itself and afterward exhort whether a shock is required. It is intended to be used by laypeople who require limited preparation. It is typically restricted in its intercessions to conveying high-joule shocks for VF and VT rhythms. The programmed units likewise require some serious energy (for the most part, 10–20 s) to analyze the rhythm, where an expert could analyze and treat the condition far faster with a manual unit. Automated outer defibrillators are commonly either held via prepared faculty who will go to occurrences or are free units that can be found in places including corporate and government workplaces, malls, air terminals, cafés. AEDS requires self-adhesive terminals rather than handheld paddles, and the ECG signal gained

from self-adhesive electrodes contains less noise and has higher quality, with handheld paddles preferred for two reasons:
- quicker and more exact investigation of the ECG
- better shock decisions.

Hands-off defibrillation is a more secure strategy for the administrator, particularly if the administrator has practically zero preparation.

DC defibrillator with synchronizer
Synchronization implies synchronized working of the heart with the pacemaker. A synchronized DC defibrillator permits the electric shock at the correct point on the ECG of the patient. Electric shock is conveyed roughly 20—30 ms after the peak R wave of the patient's ECG.

Working principle
The patient's ECG waveform is followed. The R wave in the yield of the ECG enhancer triggers the time-defer circuit, which gives a delay of around 30 ms. From that point onward, the defibrillator circuit is turned ON for the capacitor to release the electric shock to the patient's heart. The second when the electric shock happens is noted by delivering the marker pulse on the monitoring display. This kind of circuit is favored in cardiovascular crises. Abrupt heart failure can be dealt with using a defibrillator, and 80% of the patients will be relieved from heart failure if this occurs within one minute of the attack.

Defibrillation electrodes
These paddles have metal circles of 8—10 cm in breadth for outside use:
- For internal use, smaller paddles are used on babies and youngsters.
- For external use, a pair of electrodes are immovably squeezed against the patient's chest.

Fig. 6.4 shows a block diagram of a DC defibrillator with a synchronizer.

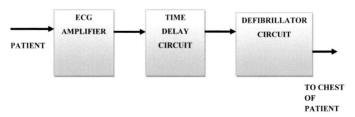

Figure 6.4 Block diagram of a defibrillator with synchronizer.

Defibrillation should be combined with intensive care known as advanced cardiac life support (ACLS). ACLS considers the underlying cause of malignant ventricular arrhythmias and prevents recurrence. ACLS includes

a. repeating CPR and defibrillation as needed

b. continuous monitoring of vital signs, cardiac rhythm, and other activities

c. drugs that suppress excessive arrhythmias, reduce cardiovascular load, and improve blood pressure, vital signs, and other functions

d. oxygen therapy, often intravenously (tracheal space), with the trachea connected to a ventilator to allow the patient to breathe

e. correcting arrhythmias using a temporary pacemaker

Other mechanisms depend on the underlying causes of life-threatening ventricular arrhythmias. The general procedure is as follows:

a. Coronary angioplasty is a narrowing of the coronary artery (heart) that hinders blood flow. Coronary artery stenosis is a risk factor for heart attack.

b. Permanent pacemaker implants provide continuous or necessary automatic pacing of cardiac rhythms.

Defibrillation immediately treats life-threatening ventricular arrhythmias:

a. VF occurs when the bottom of the heart or the ventricle pushes too fast and unevenly—running fast or vibrating. As a result, the heart pumps very little or no blood to the brain or body. Death occurs within 5—10 min without defibrillation.

b. Ventricular tachycardia can be pulseless and occurs when the ventricles beat too quickly, thus reducing the heart's efficiency. Thus, the amount of blood the heart can pump to the brain and body is reduced. If blood flow is insufficient to produce a pulse or causes unconsciousness, defibrillation is required. Pulseless ventricular tachycardia can rapidly lead to VF.

Pulseless VF and ventricular tachycardia can result from

a. overdose, addiction, or drug addiction, cocaine, methamphetamine, digoxin (Lanoxin), tricyclic antidepressants, some other antidepressants, some antipsychotics, and venlafaxine (effects). Exposure to certain chemicals, such as benzene and vinyl chloride, can also cause heart attacks.

b. abnormal amounts of potassium, calcium, or magnesium in the blood

c. other cardiac arrhythmias, some congenital heart diseases (congenital deficiencies), cardiomyopathy, heart failure, and cardiac arrest, including previous cardiac arrest.

Defibrillation is performed by a physician or specially trained nurse, emergency care provider, or another healthcare provider. Patients with leprosy can defibrillate in public places where automated external defibrillators (AEDs) are available. Some people with ICDs are at risk of developing severe ventricular arrhythmias. The ICD will automatically defibrillate as needed.

Treatment for those who perform defibrillation

a. Cardiologists specialize in the treatment of cardiovascular disease.
b. Critical care centers specialize in diagnosing and managing life-threatening situations.
c. Emergency physicians and pediatric emergency physicians specialize in rapid diagnosis and treatment of acute or sudden illness, trauma, and chronic illness
d. Cardiac surgeons specialize in treating heart and vascular conditions. Cardiac surgeons may also be known as cardiothoracic surgeons.
e. Thoracic surgeons specialize in treating chest disorders, such as those of the blood vessels, heart, lungs, and esophagus. A thoracic surgeon may also be known as a cardiothoracic surgeon.

The defibrillation process depends on the type of device. It usually involves the following steps:

a. The provider starts CPR until the cardioversion device is available.
b. The provider lubricates two viscous defibrillator electrodes or paddles in the upper-right chest and lower-left opposite region with a special jelly. The other positions are the lower-right and upper-left chest. In rare cases, the doctor makes an incision in the chest (thoracotomy) and places the electrode directly on the heart muscle.
c. The provider or external defibrillator (ED) analyzes the heart's rhythm and controls heartbeats as needed.
d. The provider or ED will resynthesize the cardiac rhythm of the result and provide more shock if needed. During this time, the team of healthcare providers will continue CPR and Advanced Life Support treatment as needed.

ICDs automatically perform defibrillation as needed. Defibrillation shock from ICD is kicking in the chest.

Risks

Defibrillation carries risks and potential complications that offset its life-saving characteristics:

a. burning skin
b. myocardial necrosis (death of myocardial tissue)
c. other cardiac arrhythmias, including asystole (no heartbeat, or "flat-line"), VF after vascular tachycardia, and other mild arrhythmias.

CHAPTER 7

Ventilators and anesthesia systems

Sudip Paul[1], Angana Saikia[1,2], Vinayak Majhi[1] and Vinay Kumar Pandey[1]

[1]Department of Biomedical Engineering, School of Technology, North-Eastern Hill University, Shillong, Meghalaya, India; [2]Mody University of Science and Technology, Laxmangarh, Rajasthan, India

Contents

Introduction to Biomedical Instrumentation and Its Applications
ISBN 978-0-12-821674-3
https://doi.org/10.1016/B978-0-12-821674-3.00003-6

Introduction

Treatment devices are an important part of health care, with increasing daily use. They play an important role in identifying, selecting, and purchasing low–cost medical equipment.

Technological and organizational sustainability promotes effective competition and reduces economic performance while ensuring staff training compliance and equipment maintenance. Change is driven by special techniques, many of which match equipment purchases with the national health infrastructure. Technical indicators are necessary for advising physicians, healthcare professionals, maintenance professionals, and industry representatives and for services/processes to prevent health defects. In consultation with health professionals, common standards have been developed that are adapted according to each professional category's unique requirements. Efforts have been made to reach an agreement on communications technology as an important part of technical training, with experts emphasizing the incorporation of carefully considered current methods:

(1) Health care (mainly electric/electronic) must be provided for hospital admission before installation.

(2) A property access grid is established. An adapter/upgrade is provided when used as a generator, uninterruptible power supply, or solar power source.

(3) Safety equipment is provided to prevent drug equipment downtime.

(4) The client/healthcare provider must ensure that repair work is prepared and planned. It is expected that parts age after some time. It is advisable to design a medical device. Caution/protection is required; the healthcare facility must be informed about training for staff/employee use as well as medical equipment.

(5) The health services department may play an important role in market research/management, as may health agencies and family welfare programs.

(6) The healthcare professional must agree on being a worker or an IPD/OPD manufacturer not by considering the number of beds at all levels but rather the number of medical devices sold.

Mechanical engineers have experimented with the noble idea of artificial organs to sustain human life. Five centuries ago, references to modern mechanical ventilation support appeared in the seminal work of Andreas Vesalius. Vesalius introduced mechanical ventilation by inserting a reed or cane into an animal trachea and then blowing air into the tube with his mouth. In the late 19th century, the first successful use of an endotracheal tube was reported. In 1908, George Poe demonstrated his mechanical respirator by asphyxiating dogs and seemingly bringing them back to life. The Roman physician Galen may have been the first to describe mechanical ventilation. Modern mechanical ventilation results from the popularity and technology of the iron lung developed by Drinker and Shaw in 1929. Mechanical ventilation has been used to support respiratory function in patients with various degrees of respiratory distress or failure. Mechanical ventilation is a substitute for the bellows action of the thoracic cage and diaphragm. Patients who have weak or absent spontaneous respiratory efforts usually require mechanical support to facilitate ventilation and oxygenation. The mechanical ventilation can maintain ventilation automatically for prolonged periods. In medicine, mechanical ventilation is a method to replace spontaneous breathing when patients cannot do so by themselves. In an emergency setting, it is typically administered after invasive intubations, a procedure of tracheostomy tube is inserted into the airway through which air is directly delivered. The iron lung, used primarily for polio victims, has provided ventilator support without intubation or tracheostomy. The mechanical ventilator categories are negative-pressure, positive-pressure, and modular concept. Current-generation ventilators incorporate computerized systems to deliver and monitor ventilator parameters. In the future, we can expect high-frequency oscillatory ventilators. Coachman and colleagues formulated "care of the mechanically ventilated patient as a fundamental component of a nurse's clinical practice in the intensive care unit (ICU). Published work relating to the numerous nursing issues of the care of the mechanically ventilated patient in the ICU is growing significantly, yet is fragmentary by nature." This chapter describes the components and technologies of ventilators and anesthesia delivery systems, with a particular focus on breathing circuits, circle breathing systems, generators, and ventilator adjuncts, as well as the technologies underlying these elements.

The word anesthetic refers to every component of respiratory anesthetics. After suffocation, gas is given for anesthesia. Under general anesthesia, drugs should be developed to ensure that the patient remains unconscious during infection for error response and prevention. Anesthesia is the accumulation of gas (e.g., nitrous oxide) or volatile water (e.g., ethyl ether, or desflurane). An active anesthesia study area is new equipment that performs functions in addition to general anesthesia. "Anesthesia" comprises several specific types. The drug delivery system guides gas cooling and heat to produce antibiotics.

Respiratory anesthesia includes inhalation and exhalation. Meanwhile, the perfect gas cleaning system is a chain system. An anesthesia workstation is an anesthesia delivery system that includes data analysis and control. In describing the specific purpose of a drug, the patient may be reminded that the pain cannot be removed surgically and that additional treatment may be required.

Achieving treatment goals can powerfully influence safety and interfere with normal operation.

As concerns respiratory function, the most important function of blood circulation in most cases Patient care is a powerful factor in patient development under anesthesia. The amount of anesthesia and gas is easy to control to reduce overheating risk. Encouraging a spirit of adaptation control is an important part of the system. The system is designed to supply oxygen and remove gas from the lungs.

When expired, gas escapes the patient's lungs from the airway circuit and returns to the anesthesia machine container. This dense gas flows through the trachea or supplies ventilation to the patient's lungs. This cycle describes the sum of each breath and is called tidal volume and trading volume. One minute is called low volume. Over time, patients inhale oxygen and anesthetics, emitting carbon dioxide and other absorbed gases into the environment. Respiratory gas without interruption circles reduces oxygen and anesthesia concentrations, so the anesthesiologist initiates a fresh gas respiratory circuit to replace exhaust gas illness. Using an anesthesia machine, the anesthesiologist directs flow and concentration gases other than fresh gas. These anesthesia machines can supply fresh ingredients in gas flow greater than the amount of gas absorbed from illness. When using higher fresh gas flow, the concentration of gas in the breath changes rapidly. Excess gas is injected into the appropriate fuel system for evacuation from the operating room. To conserve anesthesia, anesthesiologists provide a less fresh gas volume than the patient's minute volume. In this situation, the patient absorbs gas previously released into the respiratory tract (called revival). Carbon dioxide absorbed inside the respiratory circuit prevents the patient from breathing harmful CO_2. Gas residues (oxygen, laughing gas, nitrogen fumes, and anesthetics) are usually safe. While normal anesthesia takes place, anesthesia suppliers use more fresh gas. At the start and finish, the concentration of anesthesia changes rapidly, with anesthesia and low fresh gas flows resulting from small changes in concentration. This is related to the closed anesthesia technique, which matches the amount of fresh gas flow to prevent used patient gas from escaping into the gas circulation system.

Anesthesia performs a necessary life-insuring function, and anesthetic device failure and abuse can have devastating consequences. In 1974, the American National Standards Institute launched minimum anesthesia machine activity and safety standards for anesthetics.

Quality is important as the first systematic approach to safety standards for medical devices. Similar rules later written for medical devices and other anesthetic machine standards are updated regularly.

Breathing circuit

The system usually uses seminotification; this is the only type of anesthetic used in the gastrointestinal tract. The gas can be returned to the patient in cycles, including parts of the carbon dioxide layer, two-way valve inhalation, respiration, and anesthesia associated with the patient.

In collection equipment and systems, the level of gas flow during start-up movement is more than 25. The supply of natural gas in the front line is quite large. As a result, the patient can use gas stored simultaneously in a storage bag or vent pipe. Respiration gas travels in paths that carry carbon stop stimulating patience in the same way. When inhaled, a patient's gas escapes gas oil (or gasoline) depending on selected wind gust conditions and the direction of the gas produced by the return valve respiratory and systemic systems. If carbon dioxide exhaust gases are reproduced after consuming carbon dioxide, cancer recovers. Heat is generated by nearby anesthesia devices in the respiratory cycle. The gas mixes in the ventilator provocative procedures and guidance provide support to the patient. When breathing, fresh air enters the lungs through the carbon loop, and vice versa, carbon dioxide storage at home (apparently), and when breathed, the gas goes to the other garbage dump, and carbon can be absorbed from the feet and legs. Once more, the gallbladder fills and excretes gas. Blood flow (semilunar) is color-regulated for the waste collection system (if a great deal of gas is used in the controller). Therefore, gas flow regulates the amount of gas stop hot gas and gas wash yourself on the street corner inspiration. It reduces airflow and reduces gas emissions, disrupting the system and causing one to take in mostly gas on the next breath.

Circle system components

CO$_2$ absorbers

Hydroxides of sodium, potassium, calcium, and sodium barium are mainly used in various sludges as carbon dioxide. These are important hydroxides that reacts uncontrollably with carbon dioxide and eventually form carbonates that release water and heat. Strong oil has a size of 4 to 8 stitches (25−35 cm balls). It is found in chemical reactions, and detoxification repels the flow of gas through the container. Ethyl violin is steamed to the pH of the grocery store road sign; fresh white and purple bridges mean that the shock absorbers need to be replaced. The equipment produced, therefore, has clear sides insert the paint, which is easy to control when using. The instrument's normal volume is 900−1200 cm^3; a sharp object is used at home for 10–30 h according to working conditions. Many sharp objects have a chance to experience anesthesia that breaks down the anesthesia. One-way valves allow one-way flow and produce small amounts of air that can be toxic carbon monoxide. This is especially true for difficult catches because of the excessive flow of solid gas (for example, natural gas falls under anesthesia). Constant, strong correction is necessary, especially on weekends. New, more expensive equipment is designed to reduce or eliminate potential toxic gases by removing sodium, barium, and potassium hydroxides.

Unidirectional valves

One-way valves for the inhalation and breathing of air are easy, jewelry. Each has a closed lead for the delivery dish bowl. If the traction line is forced upwards, the opening valve is open for the gas to flow into the river. Small lower valve plates airflow response. Each valve has a clear dome that allows a visual display of valve performance. Often, valves become disabled because of improper opening or closing. If valves are present, carbon dioxide may be inhaled (in the sense that the valve does not close). If the valve is bent, it can open because of moisture but not in the right order.

Tank bags

The flexible tank bag has three functions for the respiratory tract. First, it is the same as a strong breathing circuit that allows for a changing air volume with no change in regional breathing pressure. Second, it can be used by hand to direct coercion control or in ventilation assistance. Third, it provides full access to coastal conservation within the respiratory tract. The shortage of gas supply in the spring exceeds the adjustable pressure-limiting (APL) output volume. Tank bags are formed so that the current gas flow is less than 15 L min 1 pe respiration remains <35 cm H_2O (3.4 kPa) until the bag reaches the maximum volume more than twice.

Thus, the respiratory pressure can reach 70 cm H_2O (6.9 kPa), which is achieved by quickly drying the tank bag.

Adjustable pressure-limiting valve

The APL valve (euphemistically called the opening valve) is a spring device for gas flow control from airways to waste-disposal systems. The valve opens or closes the gradient circuit—the force beyond the force acting on the spring (e.g., we later discuss force into the medical system corresponding to combined air pressure or removing a few centimeters of H_2O). If the patient is breathing, the anesthesia provided can be reduced immediately, and the electric motor containing the valve can be opened for better inhalation (usually <3 cm H_2O or 0.3 kPa). If the anesthetist pushes the tanks bag for operation or APL valve to aid ventilation open while breathing, some gasoline from the tank bag enters the waste disposal program for the patient. Pressing the rotation button, the anesthetist increases the pressure within a year so that the APL valve remains closed until the pressure is raised and the region reaches a full level of lung inflation; The APL valve then opens directly at the inhaled end, or very swollen lungs refuse it. Sometimes an APL valve design is required to regulate changes in the gas flow, current leakage, lung mechanics, and ventilation parameters.

Circular assets and restrictions

The first benefit from respiratory system maintenance to the upper respiratory tract and anomalies is improved airflow or stream with improved humidification and heating of inhaled gas. As mentioned, anesthetics are stored when the seal uses a very weak fresh respiratory system airflow. This is the true minimum flow; it simply replaces the gas the patient receives. "As a rule, adults can drink less than 0.5 L/day. As a rule, they more often use pure gas flows of $1-2$ L/min, but only that much. That flow level is still below the ventilation rate of $5-10$ min L min, which requires fresh airflow for ventilation systems. The spherical respiratory chain is useful because the same gas width can be used for various patient measurements. The weight of each adult can be between 100 and 40 kg. Anesthesia is performed with a new sealing system gas flow per minute. At the maximum, patients receive more anesthesia, and the more oxygen carbon provides, the more carbon dioxide there will be. Minimal flow is required for elderly patients, who are carbon dioxide absorbers. For greater convenience, the tank bag is smaller, with small breathing tubes selected for a young patient; otherwise, the system is the same for both patients' gas humidification and heating, another benefit of breathing again. Stir in fresh compressed air without water vapor inhalation, and this dry gas negatively affects the lungs. The ring function is caused by the respiratory system. The gas resin is moistened with a mixture of gases, etc. formed from water vapor as a by-product of carbon dioxide absorption. Both systems work with inhaled gas. With adequate heat using small currents, moisture is maintained to eliminate the need to find heat to moisten the inhaled air. Many disorders in the respiratory system take on the circumference of a large circle.

The absorbent cylinder is determined by the size of the tank bags for breathing; $3-6$ L is normal. If necessary, they also continue to change the gas concentration caused by a large container of pure gas constantly added to the previously extracted gas. After all, big chains are more important, making them weaker in ventilation efficiency and accuracy.

Anesthesia machine

Antibiotic anesthetics are used for proper distribution. The airways become normal, and the engine for the air conditioner has been upgraded to provide a variety of fuels. This provides everyone with oxygen, nitric acid, or air and helium or two carbon monoxides. They have one or more species the aqueous solution breathes fresh air.

The machine has several safety features that make it sound and protect from several serious prevention. Anesthesia is a mixture of gases (pictured). Dead gas enters the medical equipment, which retains the gas or offers it to the cylinder. The shut off gas is processed according to the operating description. These gases are silent, and the mixture can be used in airflow.

So it comes out of the fuel (it is also called clean gas). Oil is extracted from the medical tube equipped with large wind turbines.

Both types are properly installed according to the CGA's Diameter-Index Safety System (DISS) standard. It has a brown (O_2) color and a nitrous oxide (N_2O) color. Anesthetics are also equipped with airbags of compressed power supply; these steels are supplied with compressed gas for emergency storage for local use designed to protect against pipe risk and other electric cylinder malfunctions. Two yards are required for the game steam cylinder or refuse valve. By noting where the poles and holes continue every breath and watching the front of the anesthesia machine, the pressure can be seen for each strong gas pipeline inserted into the anesthesia tube.

Flow and frequency controller

Each pulse is controlled by a valve opened with a compressed gas button. The vaccine is distributed through valves and drains. The flow can be stopped by turning the valve to the right. The top speed of each item can be measured to determine which are fastest. Each flowmeter is specifically designed for gas. It has a strong influence on the amount of gas and its content. Using his size and strength. With the curve provided by the section tube, temperature and measuring device pressure lead to gas accumulation and cause significant changes that can be adjusted via the navigation buttons. The rest are also damaged between pipe powder or floating nanotech oil, and the tube-compatible technology tremble compatibility of some machines has been improved. A single shut-off valve indicates the value of the previous gas leak. It is housed in a 2-meter canal, and each linear fracture can be observed; instead of setting the speed, the speed limit is shown. Quickly returning to high speed is not possible. People run after each gasoline frequency connection to the junction controller. The velocity value should help calculate the gas flow rate and mixture.

Client evaporator

Gas converters are designed to increase zero value anesthesia disappearances after adding weight. This vapor is medicinal and therefore has limited effect, whereas volatile oils increase the percentage of production. Reducing the concentration of other compressed air oils makes users responsible for these effects because they are not visible. Fortunately, the front panel of the device includes habit care and relaxation. Although primarily used for anesthesia, the machine has several motors for one-time use to prevent evaporation. Another evaporator is used when it is turned on. He is a musician, do not use a locking mechanism with keys to fill the evaporator with anesthetic injection fluid. All anesthesia systems are now installed with automatic evaporation that adds some detail to the recorded absorption of the gas mixture change and normal production differences. Pressurized oil flow is divided into two flows, one created through the cross path and the other by evaporating the room water, an anesthetic active gas. In an evaporation room filled with anesthesia, the weight of the steam depends on the pressure of the steam anesthetic fluid—for example, sevoflurane.

It has a working pressure of 157 mm Hz (20.9 kPa) at 208°C, and the oil inside evaporates into the housing with 20% sevoflurane (at sea level). Anesthesia leaves the room (current flow) larger than the door of a house because of joint bloating and anesthesia associated with gas.

Ignoring changes, the rotating lens on the evaporator regulates the reliability of silent meditations, and administration stops the flow from all sides. To determine the pressure of sevoflurane vapor, divide the mixing speed with the oil into a 1% mixture. All 21 cause evaporation and a loss of room. Separate automatic reading of test variables is suitable for a specific agent because each anesthetized liquid has a large oxygen volume. Vapor pressure varies within a range for painting machines. The temperature is measured so that steam temperature is compensated; at high temperatures, the valve is subjected to at least the insertion temperature to ensure that the oil is liquid in the room. Cotton saturation is added with liquid indoor valve surfaces. Heat is necessary to dissolve the evaporated liquid (evaporation temperature). The solution must be wiped off to cool the device; it is made of pure gold and iron for a while at heat condition so that the heat can be easily transferred from that region. Changes in water vapor vary with atmospheric pressure; distribution further increases when barometric pressure decreases. Desflurane evaporators are designed differently because they use difluoride (664 mm Hg) with a high vapor pressure (88.5 kPa, 20°C) and boiling point (22.8°C). This is easy to manage without the many by-products generated when desflurane is administered with evaporation and transformation at room temperature. Desflurane oil is an anticorrosion agent, and the temperature in the control room is set to 39°C. This temperature is an increase of 1500 steam mmHg (200 kPa) and anesthesia. The gas evaporator button on the surface controls the flow system with the help of a computer; the pressure gauge is set to desflurane when the pressure is stressed. Finding a consistent product from a desflurane—oil mixture at exhaust room temperature does not affect evaporation rates and atmospheric pressure. Heating vessels need electricity, a desktop computer, and two electronic devices.

Safety methods

There are as many anesthesia machines as possible safety systems designed to prevent errors during use. As security systems, DISS and the Pin Index Safety System have improper forced-gas connections; a mechanistic plant does not use too many vaporizers at once by changing the filling system to prevent evaporation from filling properly. In safe operation with an oxygen alarm, the device has security systems that alert the user to stop the flow of certain gases in the oxygen supply. They have become obsolete (e.g., oxygen tank) voice/beep and message displays that seem to warn the user about oxygen supply under the limited pressure of ps0 psig. The valves (207 kPa) are error-resistant on each fuel supply to oxygen, and a regulating flowmeter is installed. Accident safety standards have been abolished.

Control valve or pressure drop

When oxygen is delivered, this mark is much lower than the initial level of the valve. Stop the flow or reduce the width accordingly. Pay attention to all other gases. This hinders the management of oxidizing gases in the blood (e.g., nitrogen oxides, helium, nitrogen, carbon dioxide) rapidly decomposes oxygen patient infection is also prohibited oxygen without air. Do not stop the safe practice from being successful; providing chemical gas mixtures at the proper oxygen supply pressure is a convenient gas system prevented by a gas balance system. An anesthesia machine equipped with a locking system provides high levels of commonly used nitrous oxide oxygen-free gases. Air contact between oxygen and nitrous oxide lines prevents the release of nitrogen oxides and adequate oxygen flow. One of these strategies is taxation Ohmeda Link 25 is a chain-linkage system.

In an oxygen flow needle, the valve is in contact with the nitrogen oxides. If the oxygen flow is adjusted three or more times or is less than one-third nitrate, the oxide stream limits the concentration of nitrous oxide to 75% maximum oxygen. The Drug Oxygen Price Index (OMC) involves a thin flow control valve in a nitrogen oxide line that is connected to the oxygen line by air pressure. This system primarily limits the circulation of nitrogen oxides to 72%–73% oxygen. The two systems just mentioned control nitrates and oxygen levels. Do not replace with other gases in the finished mixture. This can result in mixed oxidation (less than 21% oxygen) when a third gas is added in large quantities of density and oxygen flow. All anesthesia machines have an oxygen-washing system supplying 1.7 L per minute of oxygen in natural gas. The oxygen flow valve for a high-flow oxygen anesthetic must be closed if the oxygen supply increases rapidly or when replenishing large fluid volumes in the respiratory tract. Active pressure exerted on the patient during breathing is a challenging oxygen flow system because it is so large that it acts as a protection system.

Anesthesia pipeline

The assembly of control valves and flow control devices always works when the main circuit is closed. Writing quality decided to open all anesthetic security screens as a built-in user interface characteristic. Each is combined to prevent blood sugar oxygen analysis from monitoring oxygen storage in the respiratory tract during anesthesia. Oxygen regulation must be one alert call when oxygen is supplied not full yet adjust it to less than 18%. Dangerously high- or low-resistance respiratory pressure, stability of the respiratory cycle pay attention to the crash system atmospheric pressure is stable. High barometric pressure or high barometric pressure airway prevention: breathing cycle closure, breathing pressure regulation corresponding positive pressure, wait at least a few minutes for the air conditioner low-pressure wind signal (connect one by one) every time the breathing cycle is loaded. If the custom level is not reached within 15 s, the gap protects user interface functions from errors. In the gas flow setting, the oxygen regulator works all the time. The other gas flow is on the right side of the controller. Oxygen flow includes a special form of control from other gas control rods. Flow control is mine collision avoidance to prevent accidents change in gas flow. The total gas flow evaporation parameter increases the setting at the same time.

Prevention: Airflow can be heavy and strong, and variables are designed for single flow and use. A bad relationship or internal breakup usually causes the vertical position of the weight board. In addition, the contact method reduces the chance of air or drug abuse. The problem for the buyer is the lack of a guarantee in the food guide. The fog full of the right buyer flowmeters is sent directly through the air. Customs, in addition to oxygen, barometric pressure, and consumer rating properties are not monitored. In response, paramedics can also obtain results. In solving cold and water problems, common drugs affect the respiratory system in two ways. Reduce airway stimulation may cause airway obstruction. In addition, nerve-muscle blockers are widely used, meaning that when eyesight is poor, muscles are weak. For this reason, breathing is helped by relevant drugs during treatment for free good air pressure weaving storage bags. The fans have great support. This is especially true in the short term when storing machine time.

The fan has been replaced with a checkmark on the storage bags and APL valves in electrical circuits that replace the third pot.

Continuous breathing helps extend time, and anesthesiologists can pay attention to other things. For most patients undergoing surgery, the lungs have normal mechanical functioning and can be properly ventilated by sophisticated fans designed for convenience. However, the most effective anesthetic ventilation is safe, provides effective ventilation for different patients, and includes newborns and serious illnesses. Most patient anesthesia works through the lungs under electronically controlled deadlines. Anyone can determine direct current (DC) at a constant speed (voice control). Most are configured for permanent distribution respiratory pressure at a constant speed (pressure control). All anesthesia is performed automatically on patient breathing via positive end-expiratory pressure (PEEP), which offers some adult patients a positive-pressure warning. The systems are configured with a PEEP valve integrated into the PEEP valve respiratory system, an obsolete body actively managed by the fan patient. Usually, anesthesia ventilation does not make patients uncomfortable, so it does not affect the comfort of synchronous ventilation, pressure support, or continuous positive airway pressure (CPAP). As mentioned, the anesthesia system maintains anesthetic gas by restoring previously leaked gas to the patient. Unlike heavy ventilation, every breath brings fresh gas to the patient. Anesthesia is an integral part of artificial respiration that provides anesthesia systems and timely treatment. Mechanical air conditioning for most people who love anesthesia is achieved by combining the installation of tires with an adjustable call. Long inhaling injects solid gas into the solid gas room for ringing and adjusting the gas to the patient by a breathing line. The feed gas is usually oxygen or air; it never hits the airways if turned off again. When you breathe in, the auxiliary gas locked inside the patient chamber is breathed into the atmosphere through the last respiratory system. The priceless configuration also has an outlet valve the gas is released from the respiratory tract as a purifier. The exhaust valve system of this fan serves the same purpose APL valve for mechanical ventilation works manually or automatically. However, unlike the APL valve, this time remains closed inspiration ensures the success of many channels. The fan call is delivered to the patient.

Excess gas passes through the respiratory tract during respiration. The volume of seawater in the anesthetic spray is small for a properly delivered patient; the young man can copy gas as it passes through the anesthesia machine, which reduces respiratory compression. New gas is injected into the respiratory tract as the machine increases the number of channels. The fan is a vent outlet valve, i.e., the only way to avoid gasoline closure during inspiration. For example, for a pure gas flow rate of 3 L 1 min (50 mL 1), adjust the fan 10 breaths 1, I/E ratio 1:2 (inspiration) from time to time; the volume of seawater supplied has increased 100 mL inhaled (2—50 mL inhaled). The opposite volume of seawater supplied lessens respiratory damage. This is the amount of damage from respiratory schemes and high respiratory pressure. Mostly circular breathing chain compatibility is 79 mm H_2O (70—90 mL kPa), much larger than the usual 1—3 mm cm adaptation to an H_2O (10—30 mL kPa 1) emergency circuit because the size of the fan is large. For example, when ventilating a patient with high airway pressure using a 20 cm H_2O (2 kPa) anesthesia syringe, respiratory compatibility is 10 mm H_2O, and the distributed seawater volume decreases by 200 mL.

Limitations: Until recently, anesthesia fans were just devices designed to provide a volumetric breathing gas regulator. The pair of intelligent controls are on/off and select breathing rate, breathing/breathing (I/E) setting respiratory rate and volume. Although easy to use, it has several limitations. As discussed, if conditions change, the breathing volume changes with new gas pressure and flow. Increased breathing volume was very dangerous for young patients and premature births and growth. The gas flow in the anesthesia machine may be random; the risk of respiratory failure is high and causes airway pressure. It was especially dangerous in fall and became much lower than the set bottom wave value delivered to unknown patients who need high-pressure ventilation for heavy airway respiratory distress syndrome. These were the pneumatic capabilities of the fan that sometimes are not enough to compensate for the loss of airway compression; Ventilation does not provide adequate ventilation. Patients with high airway pressure (>45 cm H_2O) have large minutes (>10 L min 1). Another disadvantage is that petitioners of anesthesia are a right pneumatic movement with compressed gases. For the combined gas consumption coefficient, which approximately equals a specific normal minute volume (5—10 L per inch) size for adults, compression in the middle is not a problem. A gas supply is used, but oxygen decreases rapidly in an emergency for fans in the presence of compressed gas. Reserve cylinders connected to an anesthetic device (e.g., auxiliary cylinder) can supply more than 10 h of oxygen during respiration at low power but only work for 1 h. It is impossible to synchronize the patient's efforts with auxiliary ventilation methods. This limit is important for self-ventilation, such as CPAP, without compensating-pressure anesthesiological devices that burden breathing. Most limitations on preventing the formation of chain and endotracheal tubes or lung volume layers and atelectasis in general anesthesia have been resolved in the last decade, as new technologies are now part of anesthesia systems.

Scavenger system

Operating gas removal is not sufficient to avoid potential health problems, and efforts should focus on flexible anesthetics. The environment in the operating room causes problems such as headaches, dysphoria, and psychomotor disorders. Old monitoring activity rates have been reportedly involved in the development of cancer, emergency abortions, and neurological and genetic diseases. However, many studies do not confirm these findings. The National Institute for Occupational Safety and Health recommends as practice criteria that halogen anesthetics should be less than 2 parts per million (ppm), and nitrous oxide should be <2 ppm. Migration should be from the middle of the room outside the vacuum system or express tube. Exhaust gas can also be supplied to the vessel contains halogenated activated carbon anesthetics. Interaction between hack systems releases the ventilation valves in the just-described system in the breathing cycle and ventilation (e.g., APL and fan exhaust valves). As it works, the tank that comes out of the gas release system is important because as the gas flows and exhaust valves run at a steady speed, the flow of the release system increases. This system also provides low-pressure exhaust valves that are not too high or negative. This puts substantial pressure on the exhaust valve made high air pressure, then baratrov and heart failure are still a long way off. Exhaust valves cause constant negative airflow. There are two types of open and treatment systems closed open cleaning system can be used and vacuum removal system. In an open cleaning system, exhaust gas enters the bottom of an open solid container at the top of the atmosphere, and gas is constantly being released in the space below the container. As air blows in the room, the vacuum flow rate is higher than the exhaust gas flow rate, and gas enters the room from the well water tank when the speed of increased exhaust gas vacuum flow is located at the body parts. This route prevents gas from escaping the storage area if the average vacuum speed is below the medium flow from the chimney. Closed treatment includes a suitable tack for those bags breathing and excited valves. An active release system produces two or more valves for the internal pressure of a closed cleaning system. A vacuum drain valve opens to indoor air when the system is under extremely negative pressure, <1.8 cm H_2O (0.88 kPa). If the discharge flow is greater than the flow of the path, excessive pressure is reduced, leading to opening and releasing gas into the room.

Integrated monitors in all systems are electrically integrated and protective processes to prevent harm to the patient and include oxygen analysis, airway pressure measurement, and spectrometer measurement of the oxygen concentration within the oxygen analyzer. Effective system isolation against the breath oxygen concentration polygraphs, or a glycanic probe (fuel cell) separate from the movement, is commonly used by experts. The chemical reaction of the electron depends on oxygen; although cheap and reliable, the affordable oxygen concentration balance changes.

The order requires a daily order of hours an abrasive material for oxygen dissolution equipped with an alarm, and when it is large, an anesthesia system is used. The barometer applies pressure inside to take a breath in the bike and obtain smarter negative stress also helps the root machine ventilation is the most common sewer system. There are two pressure gauges: the parallel pressure tube. The meter shows direct pressure on mechanical monitoring and electric load meters. It shows the size of the pressure preview. Normal electric pressure in the monitor alarm system is measured by various parameters without negative pressure, positive pressure, and stress relief. An alarm feature allows fans to watch at any time to be steadfast in ensuring that positive pressure is felt within the heart periodically when breathing upon awakening. The pressure of the transmission system is felt in the ring operating system; This can be considered in other systems diseased part of the guide valve indicates the air pressure. Spectrum measures the gas flow of air respiratory and trauma protection when the breathing rate is very dangerous or high. Methods commonly used to measure flow include circulating trucks, two-spirited sprayers, cone pressure, and others. Breathing rate, saline volume, and minute volume are the sensor signals then transmitted to the customer. Some of them show recent waves. Further, cabbage usually has an alarm system today to change the other warning borders for smaller and larger marine areas.

Oxygen and carbon dioxide have preventive measures against oxygen oxides and bacteria; however, complete gas experts are likely to be present because anesthesia has increased in workplace gas balance control and management, and control is placed with the anesthesia system. Some patient medications, such as electromyography, heartbeat, and unstable blood pressure add a thermometer to an anemic workplace, but most are under control with clothes worn under anesthesia. Both cases should involve regular patient follow-up of this procedural test of patient success with oxygen supply, ventilation, blood circulation, and body temperature. An appropriate level of fee management advances in anesthesia safety began to be published in 1986 by the American Society of Anesthesiologists along with regular research and updates.

New technologies

The anesthetic delivery system described in 2002 had changed only gradually from the anesthesia, oxygen, and nitrous oxide pneumatic device developed by Henry Boyle in 1917. The machine is made up of steps, and 56 liquid-controlled safety devices were added. Special attention was paid to treatments in the 1970s and 1980s, with the addition of security features such as gas-sharing systems, security alarms, electrically controlled fans and tickets, and user interfaces to minimize errors. In the end, anesthetic screens and electronic recordings from the 1980s and 1990s were integrated into the workplace. Since 2000, the focus has been on improving fan activism through integration, automatic self-control, switching electrical control, and flowmeters. Water vapor has also had new technologies introduced in recent years. As mentioned, the amount of ebb under anesthesia determines that ventilation cannot be sent to the patient exactly. The patient may not be able to communicate with the respiratory system. Fresh air flows into the respiratory circuit; many techniques have been used to release this water to reduce these effects with new anesthetic systems. This method was used to reduce the two effects of chemistry and air pressure. First, breathe-slowly zones are used. Reduced breathing can be achieved by using interchangeable hoses between fans to slow the respiratory circuits.

One solution is to absorb and replace it with a smaller bottle. The new airways are designed for carbon dioxide absorbers may vary in use. Seconds many new devices measure respiration automatically chain identification in the field of automated adaptation procedure, then fill in the gaps corresponding to the respiratory contour with positive ventilation. The pressure in the respiratory tract provides additional pressure that is constantly felt volume to cover losses by pressing. Several methods were used to eliminate it; by increasing the waste clean air chain connection one way, the fan is set up for automatic volume compensation for the gas supplied in the respiratory tract. To ventilate it or make it satisfying, some inspiration is measured with a spirometer and retained. The respiratory tract is responsible for maintaining a certain volume of the recorded wave-inspired member or modified volume, and accordingly, the total clean air consumption measured by electronic instruments in an anesthetic system. None of the aforementioned methods except transformation of the respiratory chain turns on the flow sensor connected to the fan. A radically different approach called new gas uses respiratory chain changes to a new gas flow far from fan-supplied gas inspiration, and the extra effect removes the volume of the new airflow. It is a long chain during inhalation, and the air is supplied to the cylinder fan directing it to the patient's lungs, feed flow, and closed with a new neutral gas shut-off valve exhaust system. The flow is blocked by fan-controlled ventilation, the valve that is actively closed during inhalation. Pure air does not affect the volume of the transmitted wave. Instead, it flows back to the unpublished part respiratory cycle. The fan is inspected during removal/exit valve left cylinder opens removed from a mixture of fresh and clean air. This design of the tank bag depends on a number of other factor functional changes. The first is the correspondence of the respiratory chain at low positive pressure because this is only a small part. The tank bag is on a long mechanical chain as a result of ventilation; it fills with air and becomes empty throughout the fan cycle; it makes a significant difference with no movement of the bag during mechanical ventilation with normal breathing cycles.

Mechanical anesthesia machines produce internal gas tube replacement sensors. Benefits include strong faith, reduced trust services, and minor improvements in accuracy in automatic gas recording and running function use (for example, when customizing the fan). The device operates on the electrical principle of heat transfer and measurement.

Heat is needed to save heat exchanger energy in the direction of gas flow. Every gas has a pointer because every gas reacts with certain other gases. The specific temperature index varies. Gas water comes out with special light bulbs or flat mirrors for mechanical anesthesia. Most anesthesia cartel still controls gas mechanical needle valve, but some valves replace the electric control valve. For computer-type test equipment, the wireless computer's short circuit determines the amount of gas. They help car and internet service providers select a gas filter (such as an air or gas filter) to determine the total required oxygen concentration for total food consumption. New anesthesia machines are also in operation and are similar to steam but with hot tabs and computers. They work on these two principles to avoid changing computer control or the computer-controlled injection molding machine. The computer directs heater control valve anesthesia gas management. A roundhouse is attached to it. Computer syringes continue to apply the recommended anesthetic ingredients.

Gas is mixed with complex large water volume gas flows; in anesthetics, this is a computer steam device. The first benefit is an automatic equipment record and steam check settings. Second, various drugs can be evaluated as recommended to use a single control unit computer to detect chemical dehydration reaction symptoms.

Breathing systems

Ventilation anesthesia technology has been in full swing for every new car in the last 10 years; continuing development includes new respirators, as previously described, and has a compensatory chain effect. Fresh gases flow so that the number of specific waves is correct to give up some of the antiques known for their short price—respiratory rate in patients with hypertension. However, the new fan better overcomes this problem with a flow generator and compatibility fee. An anesthesia device and breathing circuit are used in the anesthesia machine fabius deder. It features three one-way active valves controlled by a fan during mechanical ventilation. Many air conditioner regulations provide multiple anesthetic ventilation modes (out of normal size control), pressure supports pressure control. Regular ventilation is mandatory at that time. The procedure uses electronic flow pressure to assess patient effort. The sensor needs many new breaths to do this; some anesthesia fans use an electronically controlled cylinder instead of conventional pneumatic pressure to prevent injury. The electrical pressure decreases rapidly use gas in the anesthesia delivery system. They release active gas from the respiratory system explosion cycle (compare blowers with a negative charge). As such, it does not apply to the regular district theme system—for example, a new cylindrical fan connection of breathing to gas separation. Fig. 7.1 is a photo of an early ventilator.

As discussed earlier, mechanical ventilation is indicated when the patient's spontaneous ventilation is inadequate to maintain life. It is also indicated as prophylaxis for imminent collapse or ineffective gas exchange in the lungs. Neurologic disease or trauma, drug overdose, postcardiac or respiratory arrest, and postoperative anesthesia are examples of patient conditions that may require ventilator support. Because ventilator care is an

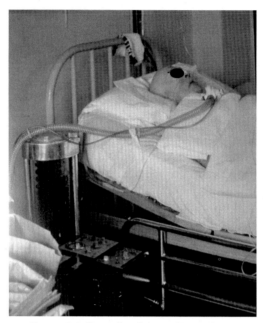

Figure 7.1 Example of an early ventilator.

integral and vital part of life support in the intensive care unit, it is a vital role of the nurse to provide Holistic care. When caring for a ventilated client, it must be concerned with the client first and the ventilator second. Nurses need to understand the clients' chronic health problems, chronic obstructive pulmonary disease, left-sided heart failure, anemia, and malnutrition. The three nursing goals in client care with mechanical ventilation are

1. to monitor and evaluate the response of the ventilator,
2. to manage the ventilator system safely, and
3. to prevent complications.

Another important aspect of ventilator care is weaning, which is moving from ventilator dependence to spontaneous breathing. The need for quality service increases daily, so nurses should enhance their knowledge and practice along all dimensions of patient care with mechanical ventilators. Long-term mechanical ventilation is associated with 47% in-hospital mortality, and 35% of clients survive for 1 year. Many survivors of long-term mechanical ventilation need continuous care in an extended care facility; many report a poor quality of life. Of the patients supported with mechanical ventilation in the intensive care unit, 5%—20% are not weaned within 2—4 days.

The start of home mechanical ventilation (HMV) resulted from the polio epidemic of the 1950s. In that situation, patients needed ventilator health support, and intensive care units were in utter need to set it up. As these patients needed this support for a longer continuous duration, alternative medical means were not available for staying in the hospital. This provides a new way to accommodate patients, which started the concept of HMV. It took years to get HMV in some countries only, and to date, it is limited to only a few countries. In the era of the Coronavirus pandemic, however, it needs to be distributed worldwide as we see enormous growth in the number of patients in need of ventilator support. Fig. 7.2 shows a home mechanical ventilator.

At the beginning of 2020, there were fewer patients with COVID-19, and the need for ventilators was at the usual level. But as the pandemic grew, demand increased in such a way that the global community was not able to fulfill current demand. The primary goal of HMV is to fulfill current ventilator demand and improve quality of life by COVID-19 recovery. It is of great concern that only a few countries are capable and efficient in ventilator production and can meet the demand from their populations, but economically disadvantaged countries more or less depend on developed countries or organizations like WHO. This dependency should be reduced, and those countries that are efficient in ventilator production should help in local ventilator production by providing local populations with technology for humanitarian reasons. Such actions could reduce global deaths significantly.

Figure 7.2 Home mechanical ventilator.

HMV could be a great help if it were available in the COVID pandemic where hospitals are overloaded and ventilator support is needed everywhere. This can be an effective treatment procedure in patients who need a support system during medication and patients with neuromuscular disease or respiratory problems. For example, neuromuscular diseases include diaphragm paralysis, muscular dystrophy, myotonic dystrophy, facioscapulohumeral dystrophy, limb–girdle dystrophy, amyotrophic lateral sclerosis, and others. HMVs result from rapid technological progress in ventilator support systems and wheelchairs technology, providing and improving patients' daily living, mobility, and participation in life, thus making HMV a more attractive option. Currently, HMV is judged to be an effective treatment that can improve both survival and quality of life.

Basic principles of ventilators

Mechanical ventilators are semiautomatic and automatic machines specially designed to perform all or part of the respiratory work of the body in moving oxygen and carbon dioxide gas into and out of the lungs. The action of moving gases into and out of the lungs is called respiration or breathing, or more formally, ventilation.

Syringe-type hand-driven pump devices are the simple mechanical devices that laid the foundation of modern-day mechanical ventilators. From this device, we could think to assist a patient's breathing that is fitted to the patient's mouth and nose using a mask. The masks are self-inflating and elastic resuscitation bags. This requires a one-way or two-way valve arrangement to move air from the system into the lungs during compression and an outward flow of carbon dioxide from the lungs to the outside as the device expands (see Fig. 7.3).

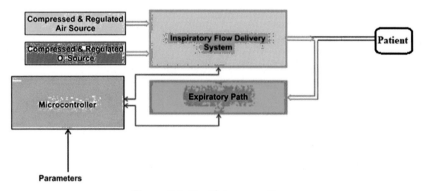

Figure 7.3 Ventilator operation.

This arrangement is not automatic and requires a person to push the gas into the lungs via mouth and nose. Thus, we cannot say such types of devices are in the category of mechanical ventilators.

Automatic operation of the ventilators is free from person/operator intervention for a safe and nonstop desired operation process. It requires three basic components:

1. a source of input energy to drive the device;
2. conversion of input energy to output energy in a meaningful way to form pressure and regulate the timing and volume of respiration; and
3. the monitoring of device performance and patient health.

Once upon a time, you were able to repair your vehicles at your home with the help of simple tools, but nowadays, things are entirely different. At that time, the average healthcare workers were completely disintegrated and assembled to do something new in mechanical ventilator to provide training, exercise, and repairs. In the late 1970s, ventilators received more attention in public and individual mechanical components and pressure pneumatic schematics. This philosophy came somewhat into being during that era. Modern-day cars and ventilators have increased manufacturing complexity for mechanical devices controlled by multiple microprocessors running sophisticated software. Fig. 7.4 shows a pneumatic schematic of an intensive care unit ventilator. The most rudimentary maintenance of ventilators is the responsibility of trained biomedical engineers. Ventilator design has been changed, focusing on individual system components for a more sophisticated and generalized ventilator

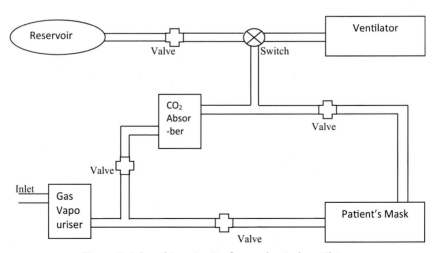

Figure 7.4 Breathing circuit of a mechanical ventilator.

model—a "black box." Ventilators are devices where we supply input and expect the desired output, whose internal processing is complex, input dependent, automatic and operator-independent. Such ventilators follow patient breathing patterns as a key feature of automatic mechanical ventilators. The most important design features of ventilators emphasize power requirements, pneumatic and electronic control systems, and pressure outputs in terms of pressure, volume, and flow waveforms. The interfacing between the various components of a mechanical ventilator and with the patient is crucial with a direct impact on the performance of a ventilator.

Generators

The various generators available in the market can be categorized as alternating current (AC) and DC. An AC generator is an electric generator that converts mechanical energy into electrical energy in alternative EMF or AC form. An AC generator works on the principle of "Electromagnetic Induction." A DC generator is a machine that converts mechanical energy into DC electricity. The energy alteration process uses the principle of energetically induced electromotive force.

Sl. No.	Differentiating property	AC generator	DC generator
1.	**Definition**	It is a mechanical device that converts mechanical energy into AC electrical power.	It is a mechanical device that converts mechanical energy into DC electrical power.
2.	**Direction of current**	In an AC generator, the electrical current reverses direction periodically.	In a DC generator, the electrical current flows only in one direction.
3.	**Basic design**	In an AC generator, the coil through which the current flows are fixed while the magnet moves. The construction is simple, and the costs are lower.	In a DC generator, the coil through which the current flows rotate in a fixed field. The overall design is very simple, but construction is complex because of commutators and slip rings.
4.	**Commutators**	AC generator does not have commutators.	DC generators have commutators to make the current flow in one direction only.

Sl. No.	Differentiating property	AC generator	DC generator
5.	**Rings**	AC generators have slip rings.	DC generators have commutators.
6.	**Efficiency of brushes**	Since slip rings have a smooth and uninterrupted surface, they do not wear quickly and are highly efficient.	Both brushes and commutators of a DC generator wear out quickly and thus are less efficient.
7.	**Short circuit possibility**	The brushes have high efficiency, so a short circuit is very unlikely.	Since the brushes and commutators wear out quickly, the possibility of sparking and short-circuiting is high.
8.	**Rotating parts**	The rotating part of an AC generator is a low-current high-resistivity rotor.	The rotating part of a DC generator is generally heavy.
9.	**Current induction**	The output current of an AC generator can be induced in either the stator or the rotor.	In a DC generator, the output current can only be induced in the rotor.
10.	**Output voltage**	AC generators produce high voltages that vary with amplitude and time. The output frequency varies (mostly 50–60 Hz).	DC generators produce a low voltage compared with AC generators, which have constant amplitude and time—i.e., the output frequency is zero.
11.	**Maintenance**	AC generators require very little maintenance and are highly reliable.	DC generators require frequent maintenance and are less reliable.
12.	**Types**	AC generators are of various types like three-phase, single-phase, synchronous, induction, and other generators.	DC generators are primarily of two types, separately excited DC generators and self-excited DC generators. According to the field and armature connection, they can be further classified as DC series, shunt, or compound generators.

Continued

Sl. No.	Differentiating property	AC generator	DC generator
13.	**Cost**	The initial cost of an AC generator is high.	The initial cost of a DC generator is less than that of an AC generator.
14.	**Distribution and transmission**	The output from AC generators is easy to distribute using a transformer.	DC generator output is difficult to distribute, as transformers cannot be used.
15.	**Efficiency**	AC generators are very efficient as the energy losses are less.	DC generators are less efficient because of sparking and copper, eddy current, mechanical, hysteresis, and other losses.
16.	**Applications**	It is used to power smaller motors and electrical appliances at homes (mixers, vacuum cleaners, etc.)	DC generators power very large electric motors like those of subway systems.

These were the main AC and DC generator differences. The differences shown for AC and DC generators include several higher-level concepts to provide in-depth insights about the two types of generators. This is extremely important for students wishing to pursue engineering (especially electrical).

AC generators are categorized in the following sections.

Rotating armature generators

Rotating armature AC generator consists of the stator provides a static electromagnetic field, whereas a rotor acts as the armature, rotates in the magnetic field, cutting magnetic lines of force to produce the desired voltage output. Output voltage comes out from the rotor through the slip rings and brushes. A slip ring is fixed to each end of the rotating loop. Brushes make sliding electrical contact with the slip rings. The generator's AC output voltage can be transferred from the slip rings through the brushes to an external circuit.

Rotating field generators

The rotating field AC generator is the most widely used generator. In such a generator, DC from a separate source is passed through windings on the rotor through slip rings and brushes. This maintains a rotating electromagnetic field of fixed polarity (similar to a rotating bar magnet). The rotating magnetic field of the rotor extends outward and cuts through the armature windings embedded in the surrounding stator. As the rotor turns, alternating voltages are induced in the windings because of magnetic fields of one polarity and the other cutting through them. Because the output power is taken from stationary windings, the output may be connected through fixed terminals. The advantage of this type of construction is that larger amounts of currents can be handled because there are no sliding contacts, and the whole output circuit is continuously insulated.

Polyphase generators

As described next, the electricity distribution system generally follows a three-phase rather than a single-phase distribution system for several reasons.

Transmission cost is less at the same voltage level compared with that of the single-phase state. A three-phase generator has nearly double (180%) the capacity of a single-phase generator of the same physical size.

A three-phase system can provide single-phase voltage and power easily by tapping any two of the power leads, whereas obtaining three-phase voltage and power from a single phase is a complex task.

Three-phase AC generator coils are placed 120 degrees apart around the inside of the stator to produce three-phase AC power. The armature coils are wired so that the generator has three different output voltages that differ in phase by 120 degrees. Each of the three coils generates an AC voltage sine wave. One voltage wave (phase) begins one-third of the way into the other wave cycle, and another wave begins two-thirds into the first wave cycle. This relationship is caused by the position of the coils in the stator. When the voltage in phase first wave has reached its peak positive value and is returning to zero, the voltage in the second phase reaches its peak negative value and begins to return to zero. The voltage in the third phase passes zero, and a negative voltage is induced. During a three-phase voltage cycle, the overall voltage induced is never zero.

Air-cooled generators

The power generator generally operates with a high magnetic field with a higher current density, producing heat. Therefore, a smart cooling module is required to remove this heat. This work is done by installing fans at each end of the rotor shaft and circulating cooling air (coolant) throughout the generator section. Cooler air reduces the generator temperature by reducing the heat produced by each unit of the generator. This arrangement of coolant, which is pressurized by the fan, cools the stator and rotor unit of the generator and returns to the fan section. However, only one cooler is required for each unit of closed-loop ventilation; coolant temperature itself rises by cooling heating elements downstream. The rise in the temperature of the coolant is due to the initial circulation of air through the fan.

Inspiratory phase and expiratory phase

The purpose of mechanical ventilation support is to provide medical gas to sustain life. Ventilator interaction is dependent on various factors like respiration, breathing pattern, disease states, patient's health, neural function, and clinical inputs. Patient-ventilator interaction provides comfort during illness in positive-pressure breaths. Improper patient-ventilator interaction may create difficulties for the patient when it is not synchronized properly. The mismatch in synchronization can cause increased muscle work, longer duration of ventilation, more oxygen gas consumption, and can lead to injurious pulmonary pressure. The proper placement of parameters, optimization, and synchronization is very important from the viewpoint of patient health.

Breathing patterns in mechanical ventilators can be divided into two parts as inspiratory phase and the expiratory phase. The ending phase of inhalation where oxygen flow ceases is known as the inspiratory phase, whereas the expiratory phase switches from the starting of exhalation. The transition point from inspiration to expiration is known as cycling, represented by the inspiratory phase to the expiratory phase when ventilator gas flow ceases and expiratory flow begins. Parametric mechanical ventilator settings are cycling and preset, time, volume, and flow. Modern-day ventilators provide graphs and other parameters for better assessment, understanding, and manipulation for clinicians (see Fig. 7.5).

Ventilator adjuncts

There are several adjuncts related to mechanical ventilators. This section focuses on adjuncts aimed to increase carbon dioxide elimination and optimization of lung requirements. A brief description and related issues are provided in the following section.

Adjuncts to increase CO_2 elimination

Inappropriate settings and operations may lead to mortality, causing ventilator-induced lung injury. Increasing the respiratory can enhance CO_2

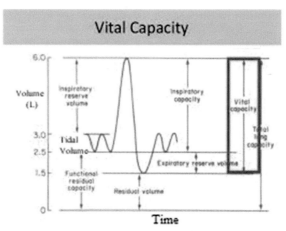

Figure 7.5 Respiratory flow waveform.

elimination. Other factors can do this with better administration of physiological dead space and expiratory washout. The three primary means of increasing CO_2 elimination are the following:

1. increased respiratory rate
2. decreased instrumental dead space
3. expiratory washout

Increase in respiratory rate

Patients with pulmonary dysfunction sometimes require ventilation support. Increasing the ventilator respiratory is one of the best ways to increase CO_2 exhalation due to decreased tidal capacity. Uncontrolled increases in respiratory rate can generate intrinsic PEEP that results in excessive intrathoracic pressure and lung overinflation. Inspiratory time may increase in proportion to the increase in respiratory rate; the resulting intrinsic PEEP can also cause the right ventricular function to deteriorate and result in inappropriate ventilator settings like a high respiratory rate with a high inspiratory to expiratory ratio. Along the same lines, airflow limitations may cause bronchial injury that promotes air trapping. In the reverse direction, external PEEP reduces intrinsic PEEP and provides a better homogeneous alveolar recruitment, whereas lung stiffness accelerates lung emptying.

Decrease in instrumental dead space

When our respiration system cannot remove CO_2 because of tidal volume reduction, the CO_2-laden gas is present in the physiological dead space at the end of expiration. It can be re-administered at the beginning of the following inspiration of the patient. There are three basic parts of physiological dead space:

1. Instrumental dead space is the ventilator volume tubing between the Y piece and distal tip of the endotracheal tube.
2. Anatomical dead space is the volume of the patient's tracheobronchial tree from the distal tip of the endotracheal tube.
3. Alveolar dead space is the volume of ventilated and nonperfused lung units.

Instrumental dead space can eliminate CO_2 in a better way by medical intervention. Replacing moisture and heat exchanger with conventionally heated humidifier positioned on the initial part of the inspiratory limb induces a decrease in CO_2 in rebreathing. Conventional humidifier, the temperature of the inspired gas can be increased to 40°C at the Y piece to reach 37°C at the distal tip of the endotracheal tube. In sedated patients, the tubing connecting the Y piece to the proximal tip of the endotracheal tube can also be removed to decrease instrumental dead space.

Expiratory washout

Expiratory washout replaces unwanted gas with fresh gas, i.e., oxygen and unwanted CO_2-laden gas present at exhalation end in the instrumental dead space. This aims to further reduce CO_2 rebreathing without increasing tidal volume. In contrast to tracheal gas insufflation, the administration of a constant gas flow is continuous over the entire respiratory cycle. The gas flow is limited to the expiratory phase during expiratory washout. A flow sensor connected to the inspiratory limb of the ventilator provides the signal to interrupt the expiratory washout flow when inspiration starts.

Currently, expiratory washout is still limited to experimental use. It is entering a phase in which overcoming obstacles to clinical implementation could lead to the development of commercial systems included in intensive care unit ventilators that aid in CO_2 elimination.

Neonatal ventilators

Mechanical ventilation for respiration is one of the well-known and common therapies in the neonatal care unit. This field has several technical complexities that overlap individual preferences caused by a lack of scientific evidence. The infant's management in receiving mechanical ventilation support remains largely dependent on individual preferences because of higher charges and the cost of treatment. Mechanical ventilation is a highly specialized area of neonatology, with lots of parameters made more complicated. It comes with different modes, techniques, and devices

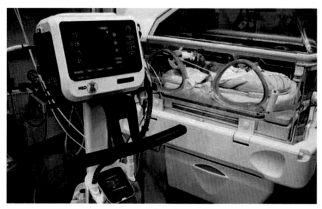

Figure 7.6 Neonatal ventilator.

depending on the cost of equipment it varies. As there is a lack of clear and strong scientific evidence for several aspects of mechanical ventilation support in preterm infants, achieving consensus is not easy. Design, development, and implementation of mechanical ventilation support protocols to not harm the newborn baby at any point of time (see Fig. 7.6).

In neonates, evaluation and impact analysis of a ventilation support protocol on the respiratory outcomes of preterm infants is crucial. Implementing a respiratory therapist-driven protocol is associated with earlier extubation, decreased rate of extubation failure, and shorter duration mechanical ventilation, without not even a single side effect.

Ventilator modes

Modes of ventilation can depend on the manufacturer, model, and series of different ventilators. Ventilator modes and their rationale for selection are based on the patient's pathophysiology and assessment.

Continuous mandatory ventilation refers to mandatory ventilation support with a continuous flow of medical gas that provides the neonate spontaneous breaths between ventilator breaths. In this, the ventilator delivers a breath regardless of the neonate's efforts. It leads to the potential of asynchronous ventilation between the neonate and the ventilator. This mode requires maximum support in the present regardless of breathing effort to avoid asynchronization. The ventilator delivers mechanical breathing for the neonate who attempts to breathe.

Synchronized intermittent mandatory ventilation

It delivers a preset number of breaths per minute whereby breaths are triggered by detection of the neonate's spontaneous breathing pattern and synchronizes the delivery of ventilator breaths to match the neonate's breaths. Ventilator-assisted breaths provide additional spontaneous breaths to the neonate. By setting preset rate and pressure over time, it weans the ventilator support to move toward extubation. In case of a higher respiration rate, it is challenging for neonates to adjust breathing to the preset values. This mode is used widely and is a choice of clinicians in neonatal practice.

Patient trigger ventilation or "assist control"

This mode triggers the ventilator in the breathing process and assists the neonate's breathing pattern within a set pressure level. Therefore, the delivery rate and recorded breaths are determined by the neonate. Under an apneic or nontriggering breath condition, the ventilator delivers the set backup rate with the predetermined pressure. It can be used to wean ventilation support by reducing pressure as the neonate controls the rate. Supporting a neonate's respiratory efforts is to be encouraged by using triggered ventilation with an optimum backup rate while allowing him to take control of his breathing in time.

Target tidal volume or volume guarantee

A desired tidal volume is set by a ventilator operator and is delivered using a ventilator at the lowest possible pressure that can achieve the set volume. It guarantees the neonate's optimal tidal volume with minimal pressures to avoid and minimize the risk of barotrauma and volutrauma to the lungs. Peak inspiratory pressure (PIP) varies with each breath with lung compliance changes or expands the lung. Setting the appropriate maximum pressure level is important, as delivering the set tidal volume in deteriorating lung conditions becomes difficult. In contrast, improved lung compliance makes it easy to obtain the desired volume delivery at lower pressures; therefore, the recorded PIP is lower and cannot reach the maximum limit. The ability to change the tidal volume for lung compliance is a major benefit of this mode.

High-frequency ventilation

It uses breath rates or "frequencies" much more than normal breath rates with a tidal volume near anatomical dead space. For example, high-frequency jet ventilation introduces small pulses of oxygen under pressure into the respiratory tract at a higher rate? It is thought earlier that it may reduce the lung injury associated with mechanical ventilation. This mode is used for short-term ventilation during airway surgery because of its capability to ventilate in the presence of air leaks. However, the real scenario of administration and necessity of two machines makes other forms of high-frequency ventilation are more suitable; hence, it is not as widely used.

Proportional assist ventilation

It provides assistance in proportion to the neonate's effort when the pressure applied increases in proportion to the tidal volume, and the neonate generates airflow with the frequency, timing, and rate of lung inflation controlled by the neonate. It is not common in the present scenario as compared with other operational modes. For effective use of proportional assist ventilation, there is protection for leakage and a mature respiratory system.

p-Based ventilator

Pressure support ventilation (PSV) assists the infant's breath by pressurizing the breath to the set pressure support limit. The flow termination sensitivity is set to terminate at a predetermined percentage of the peak flow. Full pressure support is a mode in its own right and is useful for neonates who are weaning from their support, allowing them more control in line with their breathing dynamics. PSV can also be used in conjunction with other modes by turning this on as an additional feature.

Ventilator testing

The minimum tests performed and documented for safe ventilator use are provided below. It should be remembered that measuring and standardization criteria may change from time to time. Users can replace a specific part, and equivalent and inappropriate tests can be deleted for which a ventilator does not have corresponding features or accessories.

1. **Battery test/power loss alarm:** When the unit is switched on, connect and disconnect the power plug when not used in patient-ventilator support. The ventilator's battery backup, lights, and disconnection alarm system should function appropriately.
2. **Lamp test:** Lamp test varies from manufacturer to manufacturer's procedures.
3. **Audible and visual alarms:** Disconnect the oxygen supply hose and air supply hose. An appropriate alarm should result and reconnect the hoses. Using a test, lung check for the appropriate activation of all audible and visual alarms. Specifically, momentarily disconnect the

circuit to test the low pressure, low exhaled volume, and apnea alarms. Occlude the circuit to test the high-pressure alarm. Or, as an alternative to these steps, change the alarm setting parameter to trigger the alarm under test.

4. **The inverse inspiratory:expiratory (I:E) ratio**: This can be verified momentarily by adjusting peak flow to create an inverse-ratio condition.

5. **Proximal airway pressure gauge and positive end-expiratory pressure control:** Set the PEEP level required for the patient. The manometer reading should cycle and return to the appropriate baseline (± 1 cm H_2O) at the end of each breath delivered to the test lung. Check the manometer zero (± 1 cm H_2O) by momentarily disconnecting the circuit's pressure line or inspiratory limb.

6. **Leak tests:** Perform either or both of these tests as the machine allows: Occlude the patient connection, set the pressure limit and tidal volume to their maximum levels and the peak flow and rate to their minimum levels, and initiate a breath. The manometer should reach the set maximum level, and the high-pressure alarm should activate. Set the inspiratory pause to 2 s, if possible, and set the PEEP level to 0. When the ventilator cycles, observe the pressure during the pause (i.e., the plateau pressure); the drift should not exceed $\pm 10\%$ of the plateau pressure.

7. **Modes:** Set the mode to be used for the patient. Using a test lung, verify proper operation for that mode as the ventilator cycles.

8. **Ventilator rate (and rate display):** Count the number of breaths delivered during a convenient interval timed using a clock or watch with a second hand. The measured rate should be within ± 1 breath per minute of the set rate (and the rate display, if so equipped).

9. **Volume (and volume display):** Show tidal volume, sigh volume, and minute volume. Use an external device such as a right spirometer or equivalent to independently measure exhaled volume. Connect a test lung to the circuit, cycle the machine, and compare the measured exhaled tidal volume and minute volume with their respective settings. Manually trigger a sigh of breath, if possible, and compare the measured value with the setting. All measurements should be within $\pm 5\%$ of the settings (and displays, if so equipped).

10. **Sensitivity:** Put the ventilator into assist mode. Squeeze and release the test lung; an inspiration should result when the airway pressure drops to the intended sensitivity level.

11. **Oxygen calibration:** Expose the oxygen monitor (or analyzer) used with the ventilator to room air and wall oxygen (100%) and calibrate it. Final readings should be within ±2%. Set the oxygen concentration to be delivered by the ventilator. Verify this concentration (±2% FiO_2) using the oxygen monitor (or analyzer).

12. **Nebulizer (if present):** Turn the ventilator's nebulizer on and verify that flow is produced at the nebulizer output port during each inspiratory cycle.

13. **Filters:** Ensure that a high-efficiency particulate-air (HEPA) filter is present on the main inspiratory line.

14. **Remote alarm operation (if present):** Set up a convenient alarm condition (e.g., power loss, high pressure), and verify that appropriate alarm notification occurs at the remote location.

Anesthesia

Anesthesia is an amalgamation of medications to provide patients with a sleeplike stage before surgery. General anesthesia does not cause pain because the patient is completely unconscious (see Fig. 7.7).

Anesthesia usually uses a combination of intravenous drugs and inhaled gasses (anesthetics). General anesthesia is not making the patient asleep, but the brain stops responding to pain signals or reflexes. During this time, an anesthesiologist monitors the body's vital functions and manages breathing. Typically, hospitals have an anesthesiologist and a certified registered nurse anesthetist to look after an anesthetic procedure.

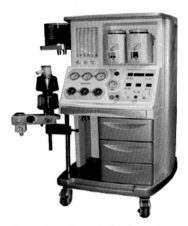

Figure 7.7 An anesthesia system.

Need for anesthesia

Anesthesia is an important part of surgery. It enables patients to undergo an operation safely without experiencing distress and pain. Anesthesia comes in three types:

(1) General anesthesia renders patients completely unconscious, preventing them from moving during the operation.

(2) Regional anesthesia causes a loss of feeling only in the body part to be operated on. This type of anesthesia is often performed in knee joint and hip joint surgeries.

(3) Local anesthesia is similar to regional anesthesia but influences a smaller area of the body.

General and regional anesthesia must be performed by anesthetists. Anesthetists evaluate patients before an operation to determine suitable anesthetic plans. An anesthetist is required to obtain patient medical histories, interview them, and order required blood for transfusion. An anesthetic plan considers the nature and duration of the operation, patient health conditions, and other technical support factors. Patients are required to fast for about 6 hours before surgery. During surgery, the anesthetist continues to administer anesthetic drugs or gases to keep the patient unconscious throughout the operation. An anesthetist also monitors the patient's heartbeat, blood pressure, blood glucose level, and oxygen level throughout the operation and should be ever alert to any unexpected and life-threatening development (such as blood loss and allergic reaction) occurring during the operation.

Gases and their sources

Surgical anesthetic inhalation was first used in the USA when diethyl ether was used in 1842. Since after, many chemical gases have been in use to anesthetize patients without pain during surgical procedures. Some anesthetic agents like diethyl ether, cyclopropane, divinyl ether, and ethylene are effective in their use but can cause fire and explosion with sufficient oxygen and a spark from static electricity or electrical equipment. Chlorofluorocarbon chemistry has produced nonflammable, halogenated, and volatile agents that can replace explosive agents. Some currently used anesthetic gaseous agents are described below.

Nitrous oxide

Nitrous oxide is one of the common gases used in anesthesia/analgesia in dentistry, surgical operations, and other medical uses. It is most commonly and generally known as "laughing gas." Inhalation and occupational exposure can result in neurotoxic effects by decreasing mental performance and audiovisual dexterity. It is often used with sevoflurane or desflurane to strengthen the anesthetic effect.

Halothane (Fluothane)

A nonflammable, hydrocarbon anesthetic that provides rapid induction with little or no excitement. It is less used because halothane may not produce sufficient muscle relaxation, supplemental neuromuscular blocking agents may be required. It reduces blood pressure, pulse rate, and respiration rate. It creates muscle relaxation and reduces sensitivity by changing tissue excitability. Its action starts with decrease the extent of gap junction mediated cell–cell coupling and altering the activity of the channels that underlie the action potential.

Enflurane (Ethrane)

It is a vasodilator acting directly on smooth muscles. It is nonflammable halogenated ether used as a general inhaled anesthetic primarily in veterinary procedures. It is a structural isomer of isoflurane. Extremely lipid (fat) soluble; may have prolonged action in obese individuals.

Isoflurane (Forane)

It is used primarily in veterinary procedures and is often combined with other anesthetics such as nitrous oxide. It is widely used and has a stabilizing effect on the cardiovascular system. It is less soluble in blood than halothane and therefore results in more rapid recovery. Common signs of exposure are dizziness, headache, and unconsciousness (in extreme cases). Pregnant women and operating personnel should minimize exposure.

Desflurane (Suprane)

It is a highly fluorinated methyl ethyl ether used for the maintenance of general anesthesia. It has lower blood and body tissue solubility; therefore, its uptake and elimination from the body are faster. It undergoes minimal metabolism and should have a low potential for toxic effects.

Sevoflurane (Ultane)

Sevoflurane is the preferred agent for mask induction because it irritates the mucous membranes less; along with desflurane, it replaces halothane and Forane. Like other halogenated ethers, it is administered in a mixture of nitrous oxide and oxygen.

Gas blending and vaporizers

The anesthesia machine is complex, consisting of a gas source, a gas mixing system known as a blender, an anesthetic gas evaporator, a breathing circuit, a ventilator, an exhaust system, a monitoring system, and an information management system. These elements result in higher user safety, accuracy, reliability, and ease of use. After more than a half-century of development, the function of the anesthesia machine has not noticeably changed. The gas supply system, mechanical gas mixing system, mechanical anesthetic gas evaporator, and exhaust emission system have become more mature, and the breathing circuit and respirator, electronic gas mixing systems, gas monitoring, respiratory monitoring, patient monitoring, electronic anesthetic gas evaporators, self-inspection functions before equipment use, and information management systems are constantly evolving and improving. As for technology, the worldwide anesthesia gas blender market has been further segmented into dual-tube and single-tube flowmeters. Dual-tube flowmeter fragments ruled worldwide because of their advantages over tube flowmeters, for example, accuracy, efficiency, and less energy use (see Fig. 7.8).

Vaporizers are interfaced with the anesthesia machine. It continually adjusts with other components of the machine throughout each case. Vaporizers are complex system comprises gas laws, specific heat, vapor pressure, and all. In **vapor pressure (VP)**, molecules escape from a volatile liquid to the vapor phase, creating a "saturated vapor pressure" equilibrium. VP increases with temperature and is independent of atmospheric pressure; it depends only on the physical characteristics of the liquid and its temperature. So, even although evaporation proceeds at a rate governed by liquid temperature and is independent of altitude (barometric pressure), individual vaporizer types may or may not function the same at altitude. **Latent heat of vaporization** is the number of calories needed to convert 1 g of liquid to vapor without temperature change in the remaining liquid. Thus, the temperature of the remaining liquid drops as vaporization proceeds, lowering VP unless this is prevented. **Specific heat** is the number of calories needed to increase the temperature of 1 g of a substance by 1°C.

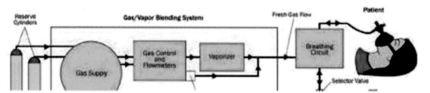

Figure 7.8 Gas vaporizer and blending system.

Manufacturers select materials for vaporizer construction with high specific heats to minimize temperature changes associated with vaporization. **Thermal conductivity** is a measure of how fast a substance transmits heat. High thermal conductivity is desirable in vaporizer construction.

Anesthesia delivery system

The first continuous flow anesthetic machine was named Boyle's machine in honor of anesthetist H E G Boyle. Modern anesthetic machines were developed from Boyle's machine. The anesthetic delivery system consists of four parts: (1) gas source (gas cylinder), (2) anesthetic machine, (3) breathing system, and (4) waste gas scavenging system. The functions of an anesthetic delivery system are to deliver oxygen and a precise amount of anesthetic gas to the patient, remove CO_2 from the breathing system, allow for ventilation (manual or mechanical), and scavenge waste gases.

Breathing circuits

The function of breathing systems is to deliver oxygen and anesthetic gases to patients and eliminate carbon dioxide. All breathing systems are composed of similar components but are configured differently. The common components include fresh gas flow, tubing to direct gas flow, an APL valve to control pressure within the system and allow waste gas scavenging, and a reservoir bag to store gas and assist with ventilation. Each breathing system receives three sources of gas: fresh gas, exhaled dead space gas, and exhaled gas from the alveoli. The proportions of each within the system are most greatly influenced by fresh gas delivery. Gas is delivered to spontaneously breathing patients at subatmospheric (negative) pressure during inspiration and atmospheric pressure during exhalation. Conversely, ventilated patients receive gas at positive pressure during inspiration and atmospheric pressure during exhalation. In this tutorial, we explore the different components and types of breathing systems used in common practice.

Components of breathing systems

A breathing system comprises components that connect the patient to the anesthetic machine and usually comprises the following components: 1. The APL valve allows a variable pressure within the anesthetic system using a one-way, spring-loaded valve. At a pressure above the opening pressure of the valve, a controlled leak of gas is allowed from the system, which enables control of the patient's airway pressure. The minimum pressure required to open the valve is 1 cm H_2O. A safety mechanism exists to prevent pressure exceeding 60 cm H_2O; however, pressure below this can lead to barotrauma. 2. The reservoir bag allows collection of fresh gas flow during expiration, which minimizes the amount of fresh gas required to prevent rebreathing. In addition, it allows anesthetists to monitor the breathing pattern of a spontaneously breathing patient. These are usually plastic or rubber and can come in sizes between 0.5 and 6 L. However, the most common size in the adult system is 2 L. Laplace Law states that pressure equals twice the radius divided by the radius of the bag. Therefore, as the bag increases, the pressure within it is reduced. This is an important safety measure, as the expansion of the bag to accommodate gas limits the system pressure. 3. The inspiratory limb allows the passage of fresh gas flow to the patient for inspiration. The expiratory limb allows the passage of expired gas from the patient. Although tubing length varies depending on the system in use, the diameter is of standard size: 22 mm for adults and 18 mm for pediatric systems.

CHAPTER 8

Physiotherapy equipment

Sudip Paul[1], Angana Saikia[1,2], Vinayak Majhi[1] and Vinay Kumar Pandey[1]
[1]Department of Biomedical Engineering, School of Technology, North-Eastern Hill University, Shillong, Meghalaya, India; [2]Mody University of Science and Technology, Laxmangarh, Rajasthan, India

Contents

Introduction to Biomedical Instrumentation and Its Applications
ISBN 978-0-12-821674-3
https://doi.org/10.1016/B978-0-12-821674-3.00006-1

What is physiotherapy?

Physiotherapy is a procedure to heal various injuries and maintain a patient's proper mobility and function. Physiotherapy is the application of various procedures in the diagnosis and treatment of numerous physical diseases, disabilities, and disorders. Regular physiotherapy can benefit patients in various ways:

Long-lasting relief from pain in various body parts can be achieved.

In some cases, surgery can be avoided with physiotherapy.

When everyday work becomes physically difficult, physiotherapy can help the body gain muscle strength.

Bone-, joint-, or muscle-related complications such as osteoporosis and arthritis can be reduced by regular physiotherapy.

A clinical physiotherapist may use multiple types of equipment as part of physiotherapeutic treatment. Sometimes equipment is also used for fitness, physical function, or pain relief. Equipment used in therapeutic treatment is popularly known as physiotherapy equipment. Commonly, this equipment helps in easily and painlessly executing daily tasks. But some equipment is used for relief from pain or disorders. For this therapeutic process, different kinds of machines and equipment are required to treat different diseases or disabilities.

Electrical stimulation

Electrical muscle stimulation (EMS), or neuromuscular electrical stimulation (NMES), is commonly used to evoke muscle contraction. EMS devices use electrical impulses to imitate the signal of a nerve muscle. In this process, electrodes are set on the skin of the critical area. This stimulator sends a signal to the muscle, and by that signal, the muscle starts contracting. This procedure helps strengthen muscles and develops the cycle of contraction and relaxation. This device is quite useful for patients who have survived from stroke and need to relearn primary motor function.

EMS has increasingly attracted attention for several reasons. It can be used by healthy subjects (e.g., athletes) as an energy training tool. It can be used as a rehabilitation and prevention tool in patients who are partially or completely stable. It can be used as an experimental tool to determine nerve or muscle function in vivo. It can be used as a postexercise recovery tool for athletes. The impulses are generated by the device and transferred to the skin near the stimulation of the muscles through the electrodes. The electrodes are usually pads that are attached to the skin. Emotions mimic the activities that come from the central nervous system and cause muscles to contract. Sports scientists have used EMS as a complementary method of sports training, and public research is available on the results obtained. Different types of Stimulation are used for recovery from different problems like reducing pain or increasing muscle strength.

EMS is used for rehabilitation purposes in medicine, such as physio-therapy for preventable muscle atrophy due to inactivity or neuromuscular imbalance that can occur after muscle damage (bone, joint, muscle, liga-ment, tenderness). It differs from transcutaneous electrical nerve stimulation (TENS), where electric current treats pain. For tennis, the currents are usually below the threshold, which means no muscle contraction is observed.

During EMS training, a set of complementary training groups (e.g., biceps and triceps) is targeted turn-based with specific training goals [12], often improving the ability to reach such objects.

Different forms of stimulation can be used for various purposes:

NMES

EMS

functional electrical stimulation—FES

TENS

EMG-triggered stimulation—ETS

reciprocal ETS—RETS

Fig. 8.1 is an example of an electrical stimulator.

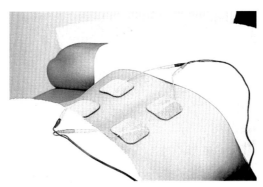

Figure 8.1 Electrical stimulator.

Strength–duration curve

A strength–duration curve is plotted between electrical stimuli with various intensities and pulse durations by each stimulus to start the response. It indicates the strength of impulses of various durations required to produce muscle contraction by joining the points that graphically represent the threshold value along the ordinate for the various durations. As the duration of a test stimulus increases, the strength of the current required to activate a single fiber's action potential decreases. Two important points exist on the curve—one is the rheobase, and the other is the chronaxie, which correlates to twice the rheobase. If the current requirement changes according to a change in pulse duration, the strength–duration curve is needed to study it (see Fig. 8.2).

When a stimulus is provided at the maximum pulse width available on the stimulator, the intensity of the current required to produce a twitch is called the rheobase of the muscle. At the double intensity of rheobase, the minimum pulse width required to produce the twitch is called the chronaxie of muscle.

Louis Lapicque proposed an exponential equation for the strength–duration curve. According to his equation,

$$I = b\left(1 + \frac{c}{d}\right)$$

where I = current, b = rheobase value, c = chronaxie value, and d = total duration.

Lapicque's hyperbolic formula combines the threshold amplitude of a stimulus with its duration.

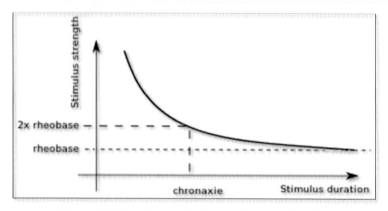

Figure 8.2 Stimulus graph of strength versus duration.

G. Weiss proposed another linear equation using a charge Q duration curve:

$$Q = b(d + c)$$
$$\text{or,} \, Q = Id$$

where   Q = electrical charge, I = current, d = duration, b = rheobase value, and c = chronaxie value.

Types of stimulators—two types
Therapeutic stimulators
These stimulators are usually used in therapeutic treatment. EMS is used to treat muscle spasms and pain. A therapeutic stimulator helps prevent atrophy and also increases patient strength. These stimulators can be quite useful for maintaining muscle function after a spinal cord injury or stroke. Therapeutic stimulators imitate the working procedure of body muscle with muscle contraction produced by the impulses of electrodes attached to the skin. Stimulators can improve a patient's range of motion and body circulation Sprains, arthritis, back pain, scoliosis, and sciatica can be treated with therapeutic stimulators. EMS stimulators use electric impulses to contract muscle. EMS is mainly used to stimulate skeletal muscle. Chronic pain can be relieved using a transcutaneous electrical nerve stimulator. The general therapeutic stimulator works to cure wounds and alleviate pain.

Neuromuscular electrical stimulators
NMES, or EMS, uses a device to help muscles initiate contraction through electrical impulses. This device can be used for recovery and prevention for persons who are partially or completely disabled. The device can be used as diagnostic equipment for evaluating neural or muscular function inside the body. This device is used in sports as a training machine. A neuromuscular electrical stimulator generates electrical impulses and delivers them through electrodes to the skin. These electrical impulses copy the action potential of the central nervous system, which helps muscles contract. NMES is required for rehabilitation in physical therapy to prevent the loss of muscle tissue caused by disease, lack of physical activity, or injury (see Fig. 8.3).

Peripheral nerve stimulation

Peripheral nerve stimulation (PNS) is a strong and useful procedure to treat chronic or refractory neurological pain. PNS works by sending tiny

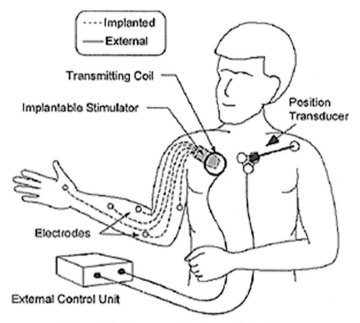

Figure 8.3 Neuromuscular electrical stimulator.

electrical transmissions from implants under the skin to pain-causing nerves. Its stimulator is powered by a small battery embedded under the skin. This stimulus helps block pain signals that travel to the brain. A tingling sensation can be felt in the painful area as pain is relieved. The amount of pain relieved varies from person to person.

In this procedure, a wirelike electrode is placed next to the peripheral nervous system with minor surgery. This electrode delivers rapid electrical pulses, but the procedure does not affect the central nervous system. Nerve stimulation can recognize intraneural or intrafascicular needle placement injection, prevent further intraneural needle advancement, and reduce the risk of nerve injury.

PNS employs a low-intensity (up to 5 mA) and short-duration (0.05—1 ms) electrical stimulus (1—2 Hz repetition rate). A certain response from muscle is obtained to locate a peripheral nerve or nerve plexus with a needle before injecting local anesthetic to close the area and block nerve conduction. This procedure is commonly done for surgical purposes or pain relief.

PNS is of two types. The first is direct PNS stimulation of a peripheral nerve, and the second is peripheral nerve field stimulation, wherein the target is

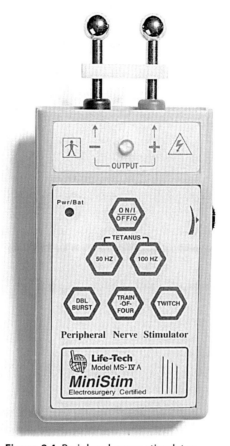

Figure 8.4 Peripheral nerve stimulator.

the terminal, sensory, cutaneous nerve fibers in a local area between the stimulus and nerve to determine the functional capability of a peripheral nerve stimulator to produce a motor response. This also determines the intensity and duration of the current used. A motor response from a current less than or equal to 0.5 mA is commonly recommended before injecting a local anesthetic. Block failure can be caused by stimulation at currents higher than 0.5 mA because the needle is too far from the nerve. Nerve damage can even be caused by injection after stimulation at a current lower than 0.1 mA (see Fig. 8.4).

Ultrasonic stimulation

Ultrasound or ultrasonic diathermy products used in physiotherapy equipment contain high-frequency sound waves that travel deep into the tissues and produce a mild therapeutic fever. Ultrasound creates deep heat in body tissues to treat selected conditions such as diathetic pain, muscle spasms, and joint contractions, but it is not intended to treat the disease. Sound waves are transmitted through a circular head rod applied to the skin with a gentle circular motion. The hypoallergenic gel helps transfer the ultrasonic energy and prevents the applicator surface from excessive heat. This procedure usually lasts 5—10 min.

An ultrasonic stimulator is a device that produces ultrasound to heal physical problems mainly related to nerves and muscles. Ultrasound stimulators are sometimes used to repair bones, treat cataracts, break kidney stones or gallstones, etc. Ultrasound is a type of mechanical energy (rather than a type of electricity), so strictly speaking, it falls within electrophysical therapy rather than electrotherapy. Mechanical vibrations at increasing frequencies are known as sound energy. The average human term ranges from 16 Hz to 15—20,000 Hz (for children and young adults). Above this limit, mechanical vibrations are known as ultrasound. The frequencies used in medicine are usually 1.0—3.0 MHz (1 MHz = 1 million/sec) and are not audible to human beings.

Ultrasound stimulation is a procedure by which neural activity of targeted brain regions can be excited or inhibited through pulsed ultrasonic waves. So we can say that ultrasound stimulation has a beneficial impact on specific neuronal pathways or nuclei in basic and clinical neurosciences. Ultrasound stimulation affects neurological disorders like Alzheimer's disease, epilepsy, and stroke.

Ultrasound therapy does not cause injury (but one may experience a fever sensation or constant fatigue). However, holding the stick for more than a few seconds can require high strength and become uncomfortable.

Ultrasound therapy can be used to treat the above conditions, but some situations and areas of the body do not allow for its safe use. If any of the following are true, the patient should notify the doctor performing the ultrasound:

- cardiac pacemaker
- malignancy in the area being treated
- healing fracture in the area being treated
- pregnancy

In addition to pacemakers, implantable medical devices like implantable deep brain stimulators may indicate that ultrasound use is inappropriate.

It should also be noted that off-the-shelf ultrasonic insulation devices have not been formally evaluated by the FDA. In general, these devices

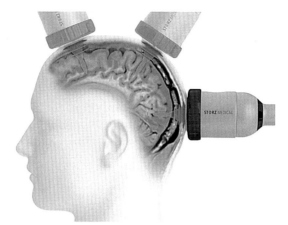

Figure 8.5 Application of ultrasound stimulator on the brain.

have been claimed to treat various ailments, disorders, and other uses not included in the descriptions above (for example, to reduce facial wrinkles) (see Fig. 8.5).

Stimulators for pain relief

Sometimes stimulators are very popular, and these are used to relieve physical pain. Spinal cord stimulators send a minimum level of electricity directly to the spinal cord to relieve pain.

The spinal cord stimulus consists of a thin wire (electronic) and a small pacemaker-like battery pack (generator). The electrodes are placed between the vertebral canal and the spinal cord (the epidural space), and the generators are placed under the skin, usually near the buttocks or abdomen. Spinal stimuli allow patients to use remote controls to transmit electrical transmissions when they feel pain. Both the remote control and its antenna are outside the body.

Experts still do not fully understand the process behind spinal stimulation, but now they know that they can target multiple muscle groups directly from the spinal cord and even feel pain in the brain. I will.

Traditional stimulators replace the sensation of pain with a light tingling called paraesthesia. But new devices provide "subperception" stimulation that patients cannot feel. Highly specialized and trained physicians use the latest devices under X-ray and/or ultrasound guidance in conventional pain management.

Spinal cord stimulation is often used because painless pain treatment options do not provide adequate relief. Spinal cord stimulants can be used to treat or manage a variety of chronic pain, including

- lower back pain, especially lower back pain that persists after surgery (surgical failure)
- postoperative pain
- arachnoid inflammation (painful inflammation of the arachnoid, the lining of the brain and spinal cord is a thin film)
- heart pain that cannot be used otherwise (angina)
- spinal cord injury
- neurological pain (e.g., severe diabetic neuropathy due to radiation, surgery or chemotherapy, and cancer-related neuropathy)
- peripheral vascular disease
- complex syndrome of regional pain
- pain following separation

Spinal cord stimulation can improve the overall quality of life and sleep and reduce the need for painkillers. It is commonly used with other pain management treatments such as medication, exercise, physiotherapy, and relaxation techniques.

As with all treatments, physicians will want to ensure that spinal stimulation is right for you and can significantly reduce chronic pain. To make this recommendation, the pain specialist will probably order an imaging test and a psychological study. Some insurance companies require psychological screening to confirm pain-acute disorders such as stress and anxiety.

There are differences between patients, but in general, the biggest beneficiaries of spinal cord injury treatment are the following:

- those who cannot take medication before surgery or minimally invasive therapy or cannot achieve adequate pain relief
- those with mental disorders, as no mental disorder reduces the effectiveness of the procedure

Spinal cord stimulators—three types

1. A conventional implantable pulse generator (IPG) is a battery-operated spinal code stimulator. In this procedure, a battery is placed in the spinal cord. But the problem is when the battery runs out, replacing it requires another surgery. This can be a good choice for patients with pain in one specific body part with their lower output.
2. A rechargeable IPG works similarly to the conventional IPG. But the only difference is that it has a rechargeable battery. This can be

recharged without surgery. For patients with lower back or leg pain, this can be a good choice. This stimulator puts out more electricity.

3. A radio-frequency stimulator has a rechargeable battery located outside the patient's body. This device is useful for patients who have lower back or leg pain because of its power.

The surgeon will adjust the device's operating power to support the three types of stimuli in the electrical signal. Different body positions can have interesting settings, such as a setting for sitting and one for walking. Most devices allow a doctor to store two or three predefined programs to access the most commonly used settings. Some newer devices include multiple waves for power delivery, such as high frequency, explosive, and high-density excitation.

Spinal cord stimulators are commonly used after nonsurgical pain relief treatment and can be used to treat various forms of chronic pain:

- back pain following surgery
- arachnoiditis—inflammation of the arachnoid, a thin membrane that covers the brain and spinal cord
- angina, an untreatable heart pain
- nerve-related pain caused by diabetes or cancer
- visceral abdominal and perineal pain
- complex regional pain syndrome
- peripheral vascular disease
- pain following amputation of a body part

Spinal cord stimulators are used for pain management along with medicine and therapies. These stimulators reduce the use of pain medicines.

Two steps, trial and implantation, are required for the implementation of a spinal cord stimulator.

Spinal cord stimulator test

The trial period is the first stage. The surgeon attaches a temporary device to examine the patient. Using a specific type of X-ray called fluoroscopy to conduct the work, the surgeon carefully inserts electrodes into the spine's epidural space. The position of these electrodes along the spine affects the position of the pain. Surgeons may ask for feedback during surgery to ensure that the electrodes are in the best position.

This test method usually requires only one incision in the buttocks to hold the electrodes. The generator/battery is on the outside of the body, usually in a belt and later around the waist.

Evaluate how much the device reduces pain for about a week. If the pain level is reduced by more than 50%, the test is considered successful.

The cable can be easily removed to the clinic without damaging the spinal cord or nerves in case of failure. If successful, surgery is scheduled for permanent replacement of the device.

Spinal cord stimulator replacement

The generator is placed subcutaneously during the permanent implantation process, and the test electrodes are replaced with sterile electrodes. Unlike trial electrodes, they are used to reduce movement.

The transplant can take about 1–2 h and is usually done externally.

After local anesthesia, the surgeon makes an incision (usually next to the lower abdomen or buttocks), places a generator, and makes a second incision (next to the spine) to insert a permanent electron. The groove is about the length of a driver's license. Similar to the test procedure, fluoroscopy is used to determine the location of the electrodes.

When the electrode and generator are connected and effective, the surgeon closes the incision.

Anesthesia increases when most patients are discharged on the same day of surgery. This incision can last for several days after surgery. Do not stretch, bend, or reach as it may pull the slit. The dressing can be kept in the grooved place and can be removed after about 3 days. In most cases, the incision heals within 2–4 weeks of surgical treatment.

The doctor discusses the recovery plan, which usually recommends light activity for about 2 weeks after surgery. Once the surgeon has approved regular activities, the patient can return to work and drive again (with the cessation of stimulants). This usually occurs 1–2 weeks after surgery.

Complications of spinal cord stimulation

Complications of spinal cord stimulation surgery are rare, but there is no risk-free procedure. A small number of patients may experience the following:
- Infection, which may occur within the first 2–8 weeks
- Bleeding
- Device movement (i.e., the electrodes move away from their original position, and the stimuli do not effectively prevent pain), often requiring follow-up surgery to bring the electrodes back to their proper positions.

- Device damage (e.g., a fall or intense physical activity destroys the stimulus)
- Dural puncture. The dura surrounds the spinal cord. Spinal cord stimuli are implanted in the epidural space just outside the dura. If the needle or electrode is too deep and stabbed, cerebrospinal fluid may leak. These punctures can cause severe headaches.
- Spinal cord trauma is rare, as traumatic injuries to the spine can cause nerve damage and paralysis.

Pain relief provided by spinal stimulants usually allows the patient to do as much as possible before surgery, but there are some limitations to keep in mind.

X-rays and CT scans are safe as long as the spinal cord stimulus is off. Before receiving a scan, a physician, nurse, or technician should always be aware that the patient's spinal cord stimulation is ongoing.

MRI is not always safe for spinal cord stimuli. Some new devices are compatible with various MRI device models and scan locations, but doctors should first evaluate the specifications. If the device is not MRI compatible, MRI can cause serious injuries.

Consult a pain specialist to determine if the procedure interferes or damages the stimulator model.

Some people find that airport security gates cause uneasiness (but harmless) interference with stimulators. If it is necessary to pass the patient through a safety driver, turn off the device before taking action. Patients need to stop agitation when operating or operating heavy equipment, as sudden changes in stimulus levels can be confusing (see Fig. 8.6).

Diathermy

Diathermy is a heat generation procedure through the human body using a high-frequency alternate-polarity radio-wave electrical current or induced heat produced by electricity. This heat is produced because of the resistive property of muscle tissue, through which the current goes.

Heat is useful for several processes, including increased blood flow, pain relief, and improved tissue mobility for healing.

Diathermy uses high-frequency currents to produce deep heat inside the target tissues. It can reach an area 2 inches below the surface of the skin.

Diathermy machines do not directly heat the body. Instead, machine-generated waves allow the body to generate heat from the target tissue.

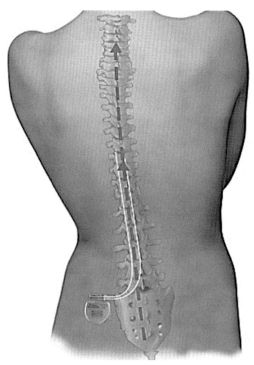

Figure 8.6 Spinal cord stimulator.

Diathermy is usually part of a complete physiotherapy or rehabilitation procedure. The frequency and length of treatment vary.

Diathermy is commonly used in therapeutic medicine for muscle relaxation and deep tissue warming. In physiotherapy, it applies moderate heat directly to pathological lesions in deep body tissues. The procedure is also used to stimulate blood circulation, provide pain relief, destroy damaged tissues, or clot bleeding vessels. German physician Carl Franz Nagelschmidt first used this procedure and gave it the name diathermy.

The same technique is used in a process called hyperthermia to create high tissue temperatures to destroy neoplasms (cancers and tumors), warts, and infected tissues. Diathermy is used in surgery to prevent excessive bleeding. This technique is especially valuable in neurosurgery and eye surgery.

The idea that high-frequency electromagnetic currents can have a therapeutic effect was simultaneously proposed by the French physician and biophysicist Jacques-Arsène d'Arsonval and Serbian American engineer Nikola Tesla (1890−91). d'Arsonval studied the applications of electrical therapy in the 1880s, conducted the first systematic study of alternative effects on the body in 1890, and observed electric shocks above 10 kg Hz. He observed that no physical response occurred at high frequency. They also developed three methods to carry high-frequency currents through the body: electronic communication, capacitive plates, and induction coils. Nikola Tesla first mentioned the ability of high-frequency currents to generate body temperature in 1891 and proposed its use in treatment.

In 1900, high-frequency was used in the body experimentally to treat various conditions in the new field of electronic therapy. In 1899, the Austrian chemist von Jenk determined the rate of heat production in tissues as a function of frequency and current concentration, first proposing high-frequency currents for deep heat treatment. [2] In 1906, the German physicist Karl Franz Nagelschmidt coined the term diathermy and used it for the first time in patients. [3] Nagelschmidt is considered the field's founder, revolutionizing it and writing the first textbook on diathermy in 1913.

In the 1920s, noisy spark discharge Tesla coils and Odin coil machines were used. These were limited to a frequency of 0.1−2 MHz known as "longwave" diathermy. Direct-contact electrodes were used to apply an electrical current to the body, which can cause skin irritation. The development of tube machines in the 1920s increased the frequency from 10 to 300 MHz, known as "shortwave" divergence. Body energy binds to the induction coil of the wire or capacitive plate, reducing the risk of burns. Microwaves were used experimentally until the 1940s.

Electrical current with a high frequency, around 0.5−3 MHz, is used in the diathermic process. This high-frequency current can avoid frequencies generated and used by the body.

Surgical diathermy is commonly referred to as "electrosurgery" and high-frequency AC in electrosurgery and diathermic surgery. Electrical current is used to run small blood vessels to stop bleeding. This technique causes local tissue to burn and damage, the field of which is controlled by the frequency and power of the device.

Some sources claim that electrosurgery applies to surgery performed by contributing radio-frequency alternating current and that "electrosurgery" uses only direct current-heated nichrome as in handheld battery-operated portable cautery tools.

Two types of diathermies are commonly used in surgery.

Monopolar—the single pole of an electric current passes through the electrodes to the tissue and processes the other stationary electrodes (neutral

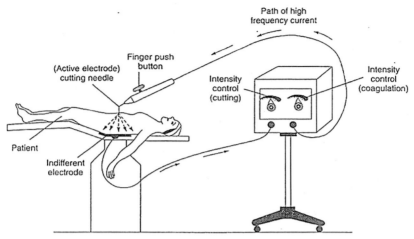

Figure 8.7 Diathermy procedure.

electrodes) in the body. Such electrodes are usually in contact with the buttocks or around the legs.

Bipolar—both electrodes are connected to the same penlike device, and current flows through the tissue. The advantage of bipolar electrosurgery is that it blocks the flow of electrical current through other body tissues and focuses only on the connective tissues. It is useful for microsurgery or cardiac pacemaker users (see Fig. 8.7).

Various diathermies are used in treatment.

Infrared diathermy

Infrared (IR) diathermy is used to heat a specific part of the body and works through direct body contact. The specific part of the patient's body where blood capillaries and neuron terminals exist can be heated properly. When an IR heat source comes into contact with the skin, heat transfers to the deeper layers of the body through conduction.

Infrared is part of the electromagnetic radiation spectrum. Wavelengths between 0.78 μm and 1 mm are considered IR spectra. This range is divided into three parts.

(1) IR-A, which ranges from 0.78 to 1.4 μm is the most penetrative part of the IR range. It goes a few millimeters under the skin. IR-A is used mostly in therapeutic treatment.

(2) IR-B, which ranges from 1.4 to 3 μm and goes under the skin about 1 mm.

(3) IR–C, which ranges from 3 μm to 1 mm and is mostly absorbed by the outer layer of the skin.

IR diathermy is the most beneficial process for heat transfer because the body absorbs these frequency ranges easily, so heat penetration occurs more inside the body.

Shortwave diathermy

Electromagnetic energy with high frequency is used in shortwave diathermy. This procedure is quite useful for generating pulsed or continuous energy waves. Two condenser plates are placed on the side of a body part being treated. Besides this process, flexible induction coils can be molded to fit a specific body part. Heat is produced when high-frequency waves pass through tissues between the condensers or coils. This high-frequency wave is converted into heat energy.

The absorption and resistance properties of tissues determine the total increase in heat and penetration depth. Shortwave diathermy uses the ISM band frequencies of 13.56, 27.12, and 40.68 MHz. A 27.12 MHz frequency and approximately 11 m wavelength are the maximums used by commercial machines. In general, shortwave diathermy is used to treat pain in muscles and joint soft tissues, sprains, strains, bursitis, tenosynovitis, and similar conditions.

Microwave diathermy

As the name suggests, microwave diathermy works through microwaves. Microwaves with a frequency above 300 MHz and a wavelength less than 1 m can produce heat without heating skin. This heat goes directly into the muscle. This diathermy is easy to use, but microwaves do not penetrate deep muscles, which can work in regions near the skin—for example, the shoulder. Over wet dressings or near metallic implants in the body, this diathermy cannot be used because of the possibility of burning. For a patient with an electronic cardiac pacemaker, this process also cannot be used. Microwave diathermy-induced hyperthermia has produced short-term pain relief in established supraspinatus tendinopathy. The majority of devices in clinical use have not provided satisfactory results in the therapeutic heating pattern to damaged tissue. New microwave devices working at 434 MHz are much improved over previous devices. Microwave diathermy is popular in radiotherapy and chemotherapy and now also used in physical medicine and sports traumatology (see Fig. 8.8).

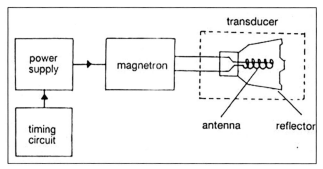

Figure 8.8 Microwave diathermy component.

Ultrasound diathermy

In ultrasound diathermy, ultrasound is produced and used for therapy. The ultrasound within tissues consists of high-frequency sound waves between 0.8 and 20 MHz, frequencies that humans cannot hear. High-frequency acoustic vibrations are used in ultrasonic diathermy. When the vibrations move through the tissues, they are converted into heat. This kind of diathermy is employed to deliver heat to specific muscles and structures. Some fibers are more absorptive, and some are more reflective, so there is a difference between the sensitivity of fibers to acoustic vibration.

Devices used for this procedure produce high-frequency AC that is converted into acoustic vibrations. The device is moved slowly across the surface of a specific part of the body. Ultrasound is a very effective agent for heat application, but its potential hazards and contraindications should be remembered when considering its use (see Fig. 8.9).

Surgical diathermy

Surgical diathermy is commonly known as electrosurgery. Surgical diathermy is when high-frequency electrical energy passes through the body, and patients obtain a desirable result from surgery. The diathermy process involves dielectric heating. Surgical diathermy involves multiple stapes.

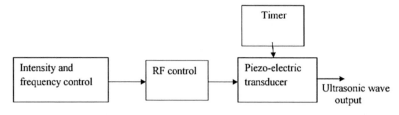

Figure 8.9 Block diagram of ultrasonic diathermy.

These steps are electrotomy (cutting), coagulation, fulguration, desiccation, hemostasis. A high-frequency current from 1 to 3 Mhz is used to execute the surgery. Machines used in this process depend on the heating effect of the current. Fine-wire voltage, high-radio-frequency voltage, and high cutting speed are required for this process. High-frequency AC flowing through tissue and heating it locally helps coagulation from the inside. For better coagulation, a high-frequency current can be used because this current does not burn the tissue from the outside.

Despite affecting deep-seated tissues, when superficial tissues are destructed, the process is called fulguration. A spark is passed from a needle or a tiny ball electrode to a specific portion of the tissue in this process. An electric arc is generated when the electrode comes very near to the tissue. In this stage, an electrode is inserted into the tissue and kept steady. Depending on the intensity and duration, the local current heat can be increased or decreased.

The concurrent and continuous use of radio-frequency current for cutting and burst–wave radio frequency for coagulation is called hemostasis (see Fig. 8.10).

This process has various advantages:

1. The separation of tissue by electric current takes place immediately in front of the cutting edge.

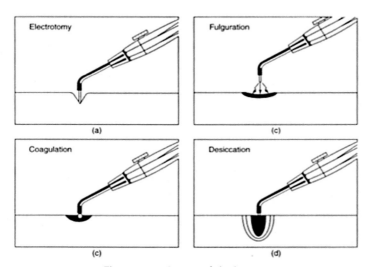

Figure 8.10 Stages of diathermy.

2. Cutting with an electric current is elegant and effortless.

3. The electrodes virtually melt through the tissues quickly and seal capillaries and other blood vessels.

4. A simplified coagulation process saves time, although bleeding can result from touching the spot immediately after the surgery.

Treating injuries with heat increases blood flow and makes connective tissues more flexible. It helps reduce inflammation and reduce tendencies for edema and fluid retention. By increasing blood flow to the injured area, the deep heat generated by the diathermy can accelerate healing.

Diathermy is used to treat the following symptoms:

- arthritis
- back pain
- fibromyalgia
- muscle obstruction
- myositis
- neuralgia
- sprains and strains
- tenosynovitis
- tendinitis
- bursitis

However, there is still very little evidence to prove that diathermy is the most effective treatment for these conditions.

Risks of diathermy

Burns from electrical disconnection usually come from faulty ground pads or outbreaks of fire. Monopolar electrocautery works because the radio-frequency energy is concentrated by the small surface area of the surgical instrument. The electrical circuit passes through the patient's body and enters the conductive pad connected to the radio-frequency generator. Since the pad's surface area is large compared with the tip of the device, the energy density of the entire pad must be low, and no tissue damage occurs in the pad space. However, if the circuit is interrupted or the energy is concentrated in any way, you may receive an electric shock or burnout. This can happen if the surface of the contact pad is small. An electrolytic gel is used if the pad is dry, if the pad is cut from a high-frequency generator, or with metal implants. Modern electric skin systems are equipped with sensors that detect high resistance in the circuit to prevent injury.

As with all types of heat applications, care should be taken to avoid irritation during diathermy treatment, especially in patients who are less sensitive to heat and cold. Flash fires have been reported in operating rooms equipped with an electronic device attached to a chemical flashpoint that produces heat, especially in the presence of an oxygen concentration associated with anesthesia.

People with embedded metal devices are at risk of injury with any kind of diathermy. These tools are pacemakers, prostheses, and intrauterine devices.

Treatment is not suitable for those with certain conditions:

- cancer
- skin sensation
- peripheral vascular disease
- tissue with limited blood supply (ischemia)
- infection
- fracture
- bleeding disease
- severe heart, liver, or kidney condition
- decreased skin sensation
- pregnancy
- sweat
- wound dressing

Diathermy is not considered safe for specific body parts, including the eyes, brain, ears, spine, heart, and reproductive organs.

Concerns have also been raised about the toxicity of surgical fumes produced by electrification. It contains chemicals that can harm patients, surgeons or operating room staff.

In patients using the surgically implanted spinal cord stimulator (SCS) system, diathermy implanted SCS components can damage tissues through the energy flowing through them, causing serious injury or death.

Electrotherapy

If pain and other symptoms cannot be properly controlled, electrotherapy (a treatment that leads to a mild electrical pulse in case of problems) may be an option. Electrotherapy includes treatments that use electricity to relieve pain, improve circulation, and repair tissues. It strengthens muscles, promotes bone growth, and improves physical performance.

Electrotherapy is a process of medical treatment that can be applied for various types of neurological diseases. It applies electrical current to the affected area and is primarily used by experienced physiotherapists to treat symptoms ranging from muscle pain to arthritis. This treatment option uses targeted and controlled electrical stimulation to help cure chronic pain, muscle wasting, muscular damage, and nerve pain. The term electrotherapy can be applied to many treatments, including electrical equipment used for deep brain stimulation for neurological disorders. The term is especially applied to the use of electric current to promote wound healing. In addition, the term "electrotherapy" or "electromagnetic therapy" also applies to the departments of alternative medical devices and therapies.

In a medical apparatus like a deep brain stimulator, this process is also used. But previously, electrotherapy was only used for physical therapy to relax muscle spasms, prevent and retard disuse atrophy, increase local blood circulation, rehabilitate and reeducate muscles with EMS, maintain and increase range of motion, and manage chronic and intractable pain, acute posttraumatic pain, acute postsurgical pain, and other conditions.

The electrotherapy unit is usually connected to a battery-powered device with an electronic pad attached to a cable attached to the skin. The electronic pads will stick to your skin so that when the electrodes are connected, and the unit is operated, a photocurrent is sent through the electrodes to the electric field.

The following are some effects of electrotherapy in treating the body:
- It transmits electrical impulses that suppress pain signals. It also reduces pain.
- Electrophysical therapy releases endorphins that reduce pain and discomfort.
- It improves blood circulation and has a healing effect on the body.
- The treatment stimulates the cells and reduces inflammation in the body.
- Electrotherapy devices also stimulate muscle tissue relaxation.
- Muscle atrophy can be prevented through EMS, including muscle stimulation.
- The treatment can relieve muscle cramps.
- Atrophy can be slowed or prevented.
- Local blood circulation can be increased.
- Muscles can be rehabilitated and reeducated.
- It can be used to manage chronic and persistent pain, including diabetic neuropathy.
- Acute posttraumatic and postoperative pain can be reduced.
- Surgical stimulation of muscles helps prevent venous thrombosis.
- Wound healing is enhanced.

The idea of using electricity for the body may seem painful, but many feel comfortable. For example, a person with tennis experiences shaking or trembling.

The exact mechanism of the beneficial effects of electrical stimulation remains controversial. Electrical stimulation can directly block the transmission of pain signals along the nerves. In addition, electrical stimulation has been shown to stimulate the release of endorphins, a natural analgesic produced by the body.

Many new electrotherapy devices bypass the wires, combine electrodes and battery power into one unit and have sex on the back, arms, legs, or anywhere else that cannot be worn regularly during work or other daily activities. Use the handheld controller to adjust the excitation level.

Many find electrotherapy useful, but others do not. There is a mixture of medical literature regarding the efficacy of electrotherapy, and not all electrotherapy treatments are supported by research.

Electrotherapy treatment can be performed in six different ways.

Types of electrotherapy

All electrotherapy devices have some similarities, such as using battery power to draw current through electrodes. However, the treatment has different frequencies, waveforms, and effects. The following are the most commonly used electrotherapies:

- TENS
- percutaneous electrical nerve stimulation
- EMS
- interferential electrotherapy
- electroacupuncture
- extracorporeal shockwave therapy (ESWT)
- galvanic stimulation

Ultrasound and laser treatments are often classified as a broad range of electrophoresis or electrophysical therapy, although they do not provide current. With ultrasound, sound waves speed up the healing process, directing it to the damaged area. Laser treatments can also be used to help heal tissues and provide more targeted and intensive care.

Transcutaneous electrical nerve stimulation

The most commonly used form of electrotherapy is transcutaneous electrical nerve stimulation or TNS. TENS therapy typically uses electrodes on a small viscous pad through the wires of a battery-powered device.

The electrodes are placed at the pain site, and an electric current is passed through electrodes to stimulate sensory nerves and create a tingling sensation that reduces pain.

The handheld controller allows the individual to choose from various options, including high- or low-frequency currents and complex excitation patterns. People often come in contact with TNS therapy during physiotherapy or in chiropractor offices. This gives an individual a chance to see if pain relief is sufficient to consider buying a tennis unit for home use.

In recent years, many TENS products have been sold as wearable devices. Some of these devices supply directly to the electronics of the battery-powered unit mounted on the body. The unit can be tied to one leg or attached to the back, shoulders, knees, or other body parts. These devices usually disappear under clothing.

Treatment can be 30 min or continuous. In some cases, treatment is possible overnight.

Tennessee's response is very different. Many consider TENS an important part of their treatment, but TENS does not relieve everyone's pain.

Dozens of treatment conditions.

Good response to TENS has been observed for the following:

Neck pain and stiffness—studies of people treated in several clinics have significantly reduced short-term pain through TENS therapy.

Back pain is often treated with TNS, but studies are inconsistent.

Diabetic neuralgia was significantly reduced in the three practice studies that were analyzed. The effect lasted to 6 weeks after the study but was not seen at the 12-week follow-up.

Fibromyalgia pain—one study found that patients' tolerance during physical activity improved significantly after a single TENS treatment.

TENS is usually applied to the painful area, but it can also be effective in other areas. Before buying the unit, it is advisable to test with different electrode configurations, and the tennis unit is used for a few days. Connecting electrodes to the upper part of the supply nerve or even to the other side of the body in the painful area may be effective.

Percutaneous electrical nerve stimulation

PENS and EMS are two treatments to be performed if TENS fails.

PENS refers to applying electrical stimulation through a small needle that enters the skin. Like TENS, the short wire connects to a battery-powered electric stirrer. The important difference is that the needle electrodes are now very close to the nerves or muscles under the skin, making the nerves less sensitive to pain. PENS therapy is used primarily in healthcare or physiotherapy settings but can also be used at home.

Some people will see immediate improvement but may require multiple treatments. The treatment lasts about 30 min. Reducing muscle spasms caused by pain increases the range of motion and improves overall physical function.

The period of publication from the pen is much longer. Treatment is long-lasting for many, but others require repeated convulsions.

Diabetic peripheral neuropathy is one of the conditions where pain is often observed.

Electrical muscle stimulation

EMS works on targeted nerves. This stimulates motor neurons and helps muscles contract. Generally, this process is used to treat and prevent muscle atrophy. EMS contracts muscles forcefully using electrical impulses in much the same way that our brains do. When used correctly, EMS can improve our health and well-being.

An electrical muscle stimulator looks like a regular TENS unit but is connected to a small battery-powered inductor. In EMS, the current is directed to the weaker muscles than the nerves, causing the muscles to contract and gradually regain strength.

EMS can be effective for rehabilitation after the muscles have become significantly weaker or for less severe conditions such as muscle spasms. EMS was tested in clinical studies to treat muscle atrophy after surgery and reconstruct the anterior ligaments. Half of the participants received EMS therapy and were not treated in a control group. People using EMS will develop more dense muscle and knee extension strength 3 months after surgery.

EMS helps build muscle slowly, but anyone considering EMS is advised about exaggerated marketing claims that EMS is a quick solution to build muscles (see Fig. 8.11).

Interferential electrotherapy

In this procedure, low-frequency electrical stimulation is used to contract and relax muscles, increase blood circulation, and relieve pain. Interferential electrotherapy has many clinical benefits, including reducing pain, stimulating healing, and reducing recovery time. Interferential electrotherapy cross-links the electro-leads with two currents produced by each channel;

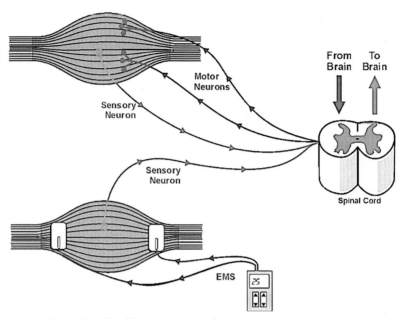

Figure 8.11 Working procedure of electrical muscle stimulation.

the inference of each other's pulses beats harder and deeper into muscle and tissue.

Nerve therapy reduces the flow of low-frequency currents to reduce neurotransmission. It is designed to relieve your pain and increase blood flow to the injured areas of your body.

Round soccer electrodes or flat pads are applied to the skin for treatment. When electrical signals stimulate the nerves, you will feel a tingling or "pain" effect on the muscles. This sensation depends on the settings that the doctor uses for your treatment. You should feel strong but not uncomfortable. If the patient feels uncomfortable during treatment, consult a doctor immediately.

Depending on the frequency of electrical current applied to the body, therapy interventions can relieve pain, improve blood flow, or improve muscle function. It is also believed to help reduce inflammation. Possible benefits are

- relief or elimination of pain;
- maximum reduction in weakness and swelling;
- recovery of lost movement and improved mobility and coordination;
- stimulation of natural hormones that help the body heal faster;
- effective treatment for chronic pain, according to many experts.

The above benefits of intervention therapy are desirable where they can be used in a variety of muscular workout situations.

The above benefits of intervention therapy are desirable where they can be used in a variety of muscular workout situations.

In very rare cases where intervention is not available, this treatment may not be appropriate:

- pregnancy
- epilepsy
- cancer
- electronic implants like pacemakers and stimulants; some pacemakers and motivational devices are relatively unaffected by electrical stimulation interventions, but others may exhibit deadly malignant behaviors; it is generally advisable to avoid electrical stimulation in this case
- problem with the heart or circulatory system
- on anticoagulant therapy or history of pulmonary embolism or deep vein thrombosis; do not treat with vacuum electrode applications because of risk of injury
- recently infected
- recently received radiation therapy
- not able to consent to treatment
- for children if treatment is required over active growth plates that are still growing (see Fig. 8.12)

Electroacupuncture

In the acupuncture process, a thin needle is inserted at a specific point of the human body. Similar to electromagnetic acupuncture, traditional herbal

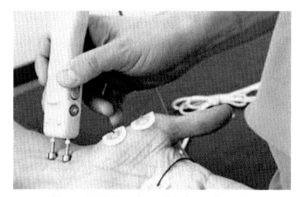

Figure 8.12 Interferential electrotherapy.

medicine (THM) is a widely used form. Acupuncture uses fine needles to stimulate specific acupuncture associated with unwanted symptoms.

In standard acupuncture, a needle is used at each treatment point. Electric acupuncture is a modified form that uses two needles.

A gentle flow flows between these needles during treatment. This current usually applies more stimulation to acupuncture using acupuncture than twisting needles or other hand manipulation techniques.

In electronic acupuncture (EA), the needle is inserted by hand but connected to a device that generates an electric current that stimulates the needle. It can be made using a simple tennis machine attached to the needle with wires and crocodile clips or commercially available units dedicated to EA. Regardless of the unit used, the physician should be able to change the frequency, intensity, and duration of the current. Two methods of EA are commonly used: low-frequency (1—4 Hz), high-intensity EA, and high-frequency (50—200 Hz), low-intensity EA.

Your health in THM depends on the flow of energy(qi) in your body. This energy travels along an invisible route known as the meridian. These are all over your body.

Qi is believed to help your body maintain balance and promote its natural healing skills. Blocking or interrupting the flow of qi can harm physical and mental health.

That is where electroacupuncture comes in. It stimulates the points related to your symptoms and helps resume the flow of Q. Two needles are placed around the point when the machine transmits electrical impulses to the needle.

The purpose of electroacupuncture is to maximize the potential healing effects of standard acupuncture.

Electroacupuncture is usually performed by an acupuncturist. The session looks like this:

- Your acupuncturist will evaluate your condition and select treatment points.
- They insert a needle into the treatment point and another needle nearby.
- Once the needle is inserted at the correct depth, the acupuncturist connects the electrode to a special electrical acupuncture device.
- After installing the electrode, they start the machine. The electroacupuncture machine has adjustable current and voltage settings. Acupuncture can adjust the current frequency and voltage during treatment, but a low voltage and frequency will be used first.

- The current is pulsed alternately between the two needles.

A typical session can last 10–20 min, which is less than the average acupuncture session.

Electrolysis is a fairly new treatment, so there is not much evidence to support its effectiveness in various applications.

However, some studies suggest that it may relieve the side effects of chemotherapy, arthritis, and acute (short-term) pain.

Arthritis

A 2005 review examined two studies examining the benefits of acupuncture for rheumatoid arthritis.

In one study, electronic connections were used. In this study, those who received an electronic connection reported a significant reduction in knee pain 24 h after treatment. This effect lasts for 4 months after treatment.

However, the review authors noted that the study had very few participants and was of low quality.

Another recent literary review in 2011 examined 11 randomized controlled trials of electroacupuncture for knee osteoarthritis. The results help electronic acupuncture reduce pain and improve mobility. The authors say the study suggested that 4 weeks of treatment was needed.

The study authors concluded by emphasizing the need for high-quality testing to support the healing effects of electronic systems.

Acute pain

Sources in the 2014 literature review examined several natural animal studies on electroacupuncture as a form of electronic synthesis. The results indicate that electronic connections help reduce a variety of pains.

The authors also found evidence that a combination of electroacupuncture and analgesics may be more effective than just medication. This is promising because it could mean that electroacupuncture to relieve pain may reduce the need for high doses of medication.

Note that these results were obtained from animal studies. Further research is needed to understand the effects of electroacupuncture on human pain.

Chemotherapy-related nausea

The 11 randomized trial sources reviewed in 2005 considered the use of acupuncture to reduce chemotherapy-related vomiting. The authors say

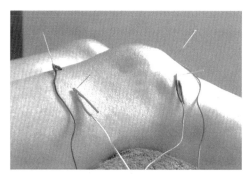

Figure 8.13 Electroacupuncture in knee arthritis.

that electroacupuncture helps reduce vomiting after chemotherapy rather than standard acupuncture (see Fig. 8.13).

Extracorporeal shockwave therapy

ESWT, also known as shockwave therapy, was first introduced in 1982 in clinical practice to control the location of urine. The success of this technique has quickly turned it into a front-line, noninvasive, and effective method for treating urinary stones. Later, ESWT was studied in orthopedics and confirmed to be able to loosen cement with total hip arthroplasty. In addition, animal studies conducted in the 1980s showed that ESWT can strengthen the bone-cement interface, enhance bone formation responses, and improve fracture healing. Although shockwave therapy has been shown to be beneficial in curing fractures, most orthopedic studies focus on tenderness disorders, facial disorders, and soft tissue conditions in the upper and lower extremities.

This is an effective, noninvasive treatment of injured soft tissues, specifically tendon and plantar fascia when an injury reaches a chronic nonhealing state. A safe alternative to surgery or steroid and other treatment injections, shockwave therapy is clinically proven to stimulate metabolic reactions. It activates the healing process by provoking an acute phase of healing and by stimulating enzymes involved in reducing pain, resulting in a high rate of success in treating these difficult to treat, often chronic and very painful injuries. The application ESWT is an evidence-based and effective treatment for plantar fascitis and most tendon conditions including achilles, hamstring, patellar, gluteal and tennis elbow tendinopathies.

In this process, acoustic waves with high energy are applied for treatment of the patient. stimulation of collagen production, release of painful trigger points, and relief from inflammation these things can be done by this process. Shockwaves are sound waves with specific physical properties such as opacity, low tensile amplitude after high final pressure, short erection time and short duration (10 ms). They have a single pulse, a wide frequency range (0–20 MHz) and a high-voltage amplitude (0–120 MPa).

These properties produce the positive and negative stages of the shockwave. The positive step produces a direct mechanical force, producing cavities and bubbles in the negative phase, then explodes at high speeds, producing a second wave of shockwaves.

The peak pressure of the shockwave is about 1000 times greater than the peak pressure of the ultrasonic wave.

Mechanism and principles of shockwave therapy

The effects of ESWT are unknown. However, ESWT's proposed measures include facilitating angiogenesis at the tender bone junction, tendon cell proliferation and bone marrow cell differentiation, and increased leukocyte penetration and growth factor enhancement. And protein synthesis which stimulates collagen synthesis and tissue regeneration.

A shockwave is a temporary pressure disruption that propagates rapidly in three-dimensional space. These are associated with a sudden increase in maximum pressure from atmospheric pressure. Significant tissue effects include cavities due to the negative phase of wave propagation.

The effects of direct shockwaves and indirect cavities cause hematoma formation and local cell death, facilitating new bones and tissues.

Shockwave therapy is primarily used for general muscular condition conditions. These include
- tender disorder of upper and lower limbs
- greater trochanter pain syndrome
- medial tibial stress syndrome
- patellar pain syndrome
- feeling of long bone fractures
- necrosis of the remnants of the femoral head
- osteoarthritis of the knee

There is no standardized ESWT protocol for the treatment of musculoskeletal conditions (see Fig. 8.14).

The therapy targets injured tissue with specially calibrated shockwaves which create tiny cavitation bubbles in the tissue, which then burst and stimulate blood flow, stem cell activity, and pain-reducing enzymes in the treated area. The application of shockwaves gives pain relief after treatment as well as stimulates long-term tissue normalization and regeneration.

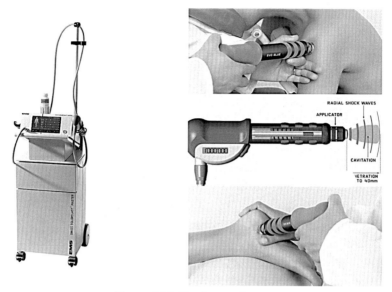

Figure 8.14 Shockwave therapy.

Patients receiving shockwave therapy can expect
- a reduction in pain felt by nerve fibers;
- an increase in blood circulation in surrounding soft tissues;
- stimulation of the healing process triggered by stem cell activation.
  Benefits of electrotherapy in physiotherapy include the following:

1. reduced nerve pain
2. promotion of healing of musculoskeletal injuries
3. noninvasive, drug-free pain control
4. prevention of muscle atrophy
5. increased circulation for wound repair
6. minimal to no side effects

Side effects of electrotherapy

The most common side effect of electrotherapy is skin irritation or rash caused by electrode glue or tape that holds the electrodes in place. Excessive use of electrotherapy can cause skin irritation. To avoid problems, you should strictly follow the guidelines for the duration of your treatment.

Electrical stimulation should not be applied to malignant tumors, skin damage, or infected areas. Rash, bleeding, or infection is possible through a variety of skin-penetrating electrotherapy.

Placing pads on a heart or pacemaker can cause cardiac arrhythmias, and fetal damage can occur from placing pads on a pregnant woman's abdomen. Indeed, persons with pacemakers and those who are pregnant are generally advised to avoid electrotherapy altogether.

Placing the pad over the neck can cause low blood pressure. It is not recommended to use electrotherapy while driving.

Repetitive transcranial magnetic stimulation

If a drug-based approach is ineffective in treating depression, doctors may prescribe other treatment options such as repetitive transcranial magnetic stimulation (rTMS).

Transcranial magnetic stimulation (TMS) is an invasive mechanism that uses magnetic fields to stimulate nerve cells in the brain to improve the symptoms of dementia. TMS is commonly used when other treatments for depression are ineffective. Treatment of this depression involves recurrent magnetic pulses and is therefore called recurrent TMS or rTMS.

This treatment uses magnetic pulses to target specific areas of the brain. People have been using it since 1985 to relieve the intense sadness and frustration that can be associated with depression.

Brain stimulation therapy helps individuals manage depression and anxiety-related issues. In this therapy, magnetic impulses are used to treat a target area of the brain.

Strong magnetic signals are generated to the brain by the rTMS device. The rTMS device consists of electromagnetic coils. When the device is placed on a patient's forehead, the coil passes the impulse to the brain nerves through the device. Initially, the area of the brain to be treated is identified. Then the specific nerve cells in the brain are targeted and induced. As this process goes on, the regulation of brainwaves improves, and the patient obtains relief by reduced symptoms.

An electromagnetic coil is placed on the scalp near the forehead during the rTMS session. Electromagnets inadvertently emit magnetic pulses, stimulating nerve cells in areas of the brain involved in mood control and depression. It is thought to activate areas of the brain where depression is less active.

The biology of why rTMS works is not fully understood, but stimulation affects brain function, which reduces the symptoms of depression and improves mood.

There are different ways to perform the procedure, and the strategy may change with more learning about the most effective way to treat the specialist.

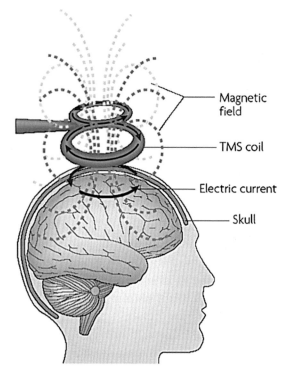

Magnetic field

TMS coil

Electric current

Skull

Figure 8.15 Repetitive transcranial magnetic stimulation mechanism.

This noninvasive process usually takes 30–60 mins (see Fig. 8.15).

Depression is a treatable condition, but for some people, the standard treatment is not effective. rTMS is commonly used when standard treatments such as medication and talk therapy do not work.

If someone is suffering from depression and anxiety-related issues, then rTMS can be a very effective treatment. rTMS is a noninvasive neuromodulation technique. This process has a greater effect on various mood-related issues and problems. In addition, other neurological and psychological disorders can be treated by rTMS, such as

- bipolar disorder
- Parkinson's disease
- obsessive-compulsive disorder
- panic disorders
- sleep-related issues
- chronic pain

The rTMS procedure has the advantages of being noninvasive and nonmedicinal, effective with mood-related issues, and a responsive treatment technique.

Risks of repetitive transcranial magnetic stimulation

rTMS is a noninvasive form of brain stimulation used in depression. Unlike vagus nerve stimulation and deep brain stimulation, rTMS does not require surgery or electrode implantation. In addition, unlike electroconvulsive therapy, rTMS does not cause seizures or sedation with anesthesia.

In general, rTMS is safe and well tolerated. However, it can cause some side effects.

Normal side effects of repetitive transcranial magnetic stimulation

Pain is not usually a side effect of rTMS, but some people report mild discomfort in the procedure. Electromagnetic pulses can tighten or stiffen the facial muscles.

This process is associated with mild to moderate side effects, including
- fatigue
- temporary hearing loss due to occasional loud magnetic noise
- mild headache
- aching face, chin, and scalp
  Although rare, there is a risk of seizures in rTMS.

Serious side effects of repetitive transcranial magnetic stimulation

Serious side effects are usually rare. They also include
- convulsions
- people with mania, especially bipolar disorder
- deafness if ear protection is inadequate during treatment
  Further research is needed to determine whether rTMS can cause long-term side effects.

  Before implementing rTMS, you need to do the following:
- physical examination and possibly clinical or other examination
- evaluate psychotherapy to discuss your depression
  This knowledge will help ensure that rTMS is safe and a good choice for you.

One should tell their doctor if they have one or more of the following conditions:

- pregnant or planning to become pregnant
- embedded metal implants or devices

  may be used for some people

  not recommended for those who use certain devices because of the strong magnetic field of rTMS: pacemakers and drug pumps, electrodes for monitoring brain activity, cochlear implants for hearing, embedded magnetic devices, or other metal devices in the body
- patient family history of seizures or epilepsy
- other mental health disorders such as substance abuse, bipolar disorder, and psychosis
- all brain injuries due to illness or injury, including brain tumors, strokes, and traumatic injuries
- uninterrupted or severe headache
- other medical conditions
- previous treatment with RTM and effectiveness in alleviating depression

If rTMS works, the symptoms of depression may improve or disappear completely. Weeks of treatment may be needed to relieve the symptoms.

The effectiveness of rTMS can be improved because researchers can learn more about the technique, the number of stimuli needed, and the best parts of the brain to use in treatment.

Following the rTMS treatment series, standard treatments for depression, such as medications and psychotherapy, can be recommended as continuous treatment.

It is not yet known whether this maintenance rTMS session will help your frustration. This includes continued treatment in the hope of preventing the withdrawal of symptoms in the absence of symptoms.

CHAPTER 9

Therapeutic equipment

Sudip Paul[1], Angana Saikia[1,2], Vinayak Majhi[1] and Vinay Kumar Pandey[1]

[1]Department of Biomedical Engineering, School of Technology, North-Eastern Hill University, Shillong, Meghalaya, India; [2]Mody University of Science and Technology, Laxmangarh, Rajasthan, India

Contents

Introduction to Biomedical Instrumentation and Its Applications
ISBN 978-0-12-821674-3
https://doi.org/10.1016/B978-0-12-821674-3.00001-2

Treatment devices are generally categorized into two groups, those that support the daily life of patients and those that support the provision of medical services by healthcare professionals that could not otherwise be performed because of physical obstructions or disabilities. Treatment devices used by healthcare professionals include oxygen supply systems, catheters, venous lines, and physiotherapy devices. For people with mild visual or hearing impairments, treatment devices such as hearing aids and spectacles can help improve vision and hearing. Even in severe cases, these devices enhance the ability of users to perform daily tasks such as reading, watching TV, and interacting. Accessories are also available for those with complete loss of hearing or sight. Such accessories may include text-to-speech programs for those with visual impairments and Braille books and closed-caption technology for those with hearing impairments. These devices do not recover lost impairements but allow users to participate in activities they would otherwise be forced to avoid.

People with disabilities that interfere with their ability to walk use many types of treatment devices. Crutches and canes can help people with temporary or mild disabilities, and walkers and wheelchairs are available for those whose restrictions are more stringent. Artificial joints can improve a patient's injured body parts to improve overall mobility. Two common examples are knee and hip replacement. Physiotherapy equipment such as massage tables, weight machines, and hot tubs can be used to help patients regain mobility after surgery or an accident.

Physicians and nurses use medical devices in treatment every day. Oxygen therapy systems, masks, tanks, and other equipment used in respiratory therapy are the most widely used devices in hospitals. A catheter can be used to deliver drugs to the body and remove certain fluids, such as urine. Wound care equipment such as suction devices and bandages are therapeutic because they help heal wounds and prevent infection.

Therapeutic devices may be modified in-house to provide the patient with greater mobility or safety. For example, a person in a wheelchair or who has difficulty climbing stairs may have a ramp installed on the home's exterior. For indoor stairs, a special lift device can be installed to safely lift the patient upstairs. Rails and seats can be placed in shower rooms or bathtubs. These devices make it easier and safer for physically challenged patients to stay at home without outside assistance.

Therapeutic purposes

Purposes for therapeutic equipment include the following:
a. prevent, diagnose, monitor, treat, cure, or compensate for disease, illness, defect, or injury
b. influence, interrupt, or modify physiological processes
c. examine susceptibility to illness
d. influence, control
e. test for pregnancy
f. examine, replicate, or modify human organs

Products with a therapeutic purpose

Devices used primarily to treat or prevent disease include
a. surgical instruments
b. pacemakers and defibrillators
c. contact or standard lenses for eyes
d. dental equipment
e. surgical laser system
f. assembling pump

Diagnosing or monitoring disease

Disease diagnosis and monitoring include a wide range of diagnostic medical devices:
a. in vitro diagnostic devices
b. heart rate monitors
c. high blood pressure monitoring devices
d. thermometers
e. stethoscopes
f. ultrasonic scanners and probes
g. medical imaging systems
h. ovulation kits

Preventing conception

For medical devices, this includes
a. condoms
b. diaphragms
c. intrauterine devices

Replacing or modifying parts of the human anatomy

Examples of devices that meet these criteria include
a. orthopedic implants
b. prosthetics and orthotics
c. orthodontic devices

Influencing, inhibiting, or modifying physiological processes

A wide range of products and sophisticated embedded devices are available to help people with disabilities:
a. gastric banding devices
b. transcutaneous electrical nerve stimulation machines
c. hearing aids
d. continuous positive airway pressure (CPAP) machines and breathing circuits

Preventive medicine

The rationale for preventive therapy is to identify individuals' risk factors and reduce or eliminate those risks to prevent disease. Early prevention is a natural action that attempts to eliminate risks to prevent disease before it can develop. An example is the vaccination of children. The goal of secondary prevention is to cure the disease before it is detected or its symptoms appear. Examples include regular uterine Papanicolaou test (Papanicolaou smear) screening and mammography. Clearly, early prevention is the least expensive way to control disease. The leading causes of death worldwide include heart disease, cancer, cerebrovascular disease, traumatic injury, and chronic lung disease. Smoking is the leading cause of preventable death and is associated with increased risk of heart disease (e.g., heart attack), cancer, stroke, and chronic lung diseases such as emphysema and chronic bronchitis. Many organizations around the world have established recommendations and guidelines for disease prevention. In the United States, for example, following the previous work of the Canadian Task Force on Preventive Health Care, the US Preventive Services Task Force conducted an important review to determine the effectiveness of various screening tests, vaccines, and preventive measures. This report, the *Guide to Clinical Preventive Services*, lists recommendations for various conditions assessed by the panel. Vaccination is the most effective way to prevent infection. Standard vaccines for children include diphtheria; tetanus; whooping cough; polio; measles, mumps, and rubella; influenza B; and hepatitis B. The 1-year vaccine against influenza should be given to children, adults over 65 years of age, those at risk of chronic cardiopulmonary disease, and those under long-term care. Adults over age 65 should be vaccinated against pneumococcal pneumonia, including 23 common strains of pneumococci. Acquired immunodeficiency syndrome, caused by the human immunodeficiency virus, is a major infectious disease problem. A vaccine is expected at some point, but the obstacles to its development are significant. Key precautions to prevent or reduce transmission include abstaining from sexual contact, using condoms, and not sharing needles among intravenous drug users.

The risk factors for coronary artery disease that can lead to a heart attack are smoking, high blood pressure, elevated serum cholesterol levels, a sedentary lifestyle, obesity, stress, and excessive alcohol consumption. In addition to improved total serum cholesterol levels, elevated low-density lipoprotein (LDL) levels and decreased high-density lipoprotein (HDL) levels cause significant risks. Improved (reduced) total cholesterol and LDL levels can be achieved through a proper diet, while lower HDL levels can be achieved by smoking cessation and increased physical activity. If these measures do not provide adequate control, various medications are available to lower cholesterol levels.

The main risk factor for stroke is high blood pressure, and smoking and type 2 diabetes significantly increase this risk. Transient ischemic attack occurs before a stroke in 20% of patients and is accompanied by the sudden onset of one or more of the following symptoms: temporary loss of vision in one eye, numbness on one side, temporary loss of speech, and local weakness lasting less than 24 h that is resolved without permanent damage until the stroke occurs.

The most important preventative measure against cancer is eliminating cigarette smoke. As 30% of all cancer deaths are due to smoking, an essential element of prevention is raising awareness among environmental nonsmokers and others about the risks of secondhand smoke. Early prevention of skin cancer includes reducing exposure to UV light with sunscreen and protective clothing. For other cancers, secondary precautions include mammography, clinical breast examination, and breast self-examination for breast cancer; pelvic and pancreatic examinations for uterine and ovarian cancer; and sigmoidoscopy, rectal finger examination, and blood stool testing for colorectal cancer.

Bone democratization and bone loss (osteoporosis) are common in men and women over age 70 and can cause fractures, back pain, and short stature. Osteoporosis is the most common symptom in postmenopausal women resulting from estrogen deficiency. The most effective way to prevent postmenopausal bone loss is through estrogen replacement therapy and increased calcium intake. Early precautions include increased physical activity and avoidance of tobacco and heavy alcohol intake.

Accidents are the leading cause of deaths in the United States that are not attributed to disease, and the majority of those deaths are due to alcohol abuse. Other factors include failure to fasten seat belts and motorcycle helmets, lack of sleep, and guns in the home.

Insomnia is the feeling of not being able to sleep. Transient insomnia can occur for those working certain shifts or traveling within multiple time zones, and stressful life events or reassignments may follow. Sleep disorders are associated with stimulant- or treatment-related conditions such as anxiety, depression, or pain. Older people spend less time sleeping, and wake up more often. Treatment of insomnia involves establishing good sleep hygiene, which consists of maintaining a regular schedule for retiring and waking up, setting a comfortable room temperature, and reducing destructive stimuli such as noise and light. Daily exercise is beneficial but should be avoided a little before bedtime. Stimulants like nicotine and caffeine should be avoided. Alcohol interferes with normal sleep patterns and should be avoided. Those who abuse alcohol sleep lightly and often wake up unconsiouly, so they feel out of control the next day. When medication is needed, physicians often prescribe one of the sleep-inducing benzodiazepines. These can have long-term, moderate, or ultra-short-acting effects. None of these drugs should be used regularly over a long period. Nonbenzodiazepines are also available in various lubricants and absorbers, and their benefits depend on individual preferences.

Design of a therapeutic regimen

A completely new condition is determined when the doctor diagnoses or identifies the most probable cause of the condition and decides on appropriate treatment. One of the conditions for consideration is the patient's treatment advice and expectations. The patient may have mild symptoms resulting from convulsive sensations. If testing can rule out this possibility, reassurance can have a therapeutic effect. If possible, physicians work to cure the disease and thereby relieve the symptoms, but the goal is to relieve or improve the symptoms or restore normal functioning, whether the disease is unknown or chronic. If completely curing the condition is not possible, knowledge about the disease, its causes, and expectations of significant relief can be provided. Patients often want to know the name of the illness, its cause, its duration, additional symptoms that may occur, and what the doctor can do to improve the speed of treatment. Providing information about the illness helps to alleviate anxiety and fears that otherwise hinder the patient's progress.

The key element of a successful treatment approach is the patient's positive attitude toward the physician. A relationship of trust and respect for physicians in reputation-based or long-term supportive care is one of the physician's most powerful treatment tools. When choosing a management plan, the physician has many options, each with a different outcome or outcome. Often, the best options are related to the patient. Oral medications provide faster relief than injections. Potential side effects of medications and treatments can affect treatment options—for example, people prefer dizziness over nausea. A new decision tree is made when a treatment course is selected, leading to new options based on feedback. Further testing, higher drug doses, or drug replacement may be required. Almost all treatments carry some risk of unwanted side effects and unexpected complications. Physicians explain these risks in terms of possibilities and encourage patients to accept or reject treatment based on these barriers and suffer complications to achieve their side effects or get remedies.

Another factor that affects successful treatment is patient consent, which is the degree to which the patient follows the regimen advised by the physician. Treatment often requires significant lifestyle changes such as following a specific diet, starting an exercise program, or stopping harmful habits such as tobacco use. In addition, the number and complexity of prescribed medications affect patient follow-through—patients are more likely to continue taking a medication once a day than to be prescribed four times at once. It may be more difficult to obtain consent from patients with chronic but asymptomatic disorders, such as hypertension. Patients who have no symptoms may need reassurance that the medication is necessary to prevent unpleasant events (such as high blood pressure, stroke, and other heart problems) from occurring. Similarly, patients with depression and anxiety may stop taking medication after symptoms have subsided. They do not accept the need to continue taking the drug until instructed to stop.

In deciding which treatment method is most likely to be effective, physicians must rely on scientific research that has been proven it more effective than another treatment or other drugs. The most reliable studies are truly objective and can influence their interpretation in anticipation of specific results, eliminating the possibility of patient drug prejudice. Such studies are "double-blind" and control both potential trends by comparing the active ingredient with an inactive placebo (an inactive drug). Neither bias can affect the outcome, as neither the patient nor the physician knows which drug the patient is taking. This is the best way to show drug effectiveness, but it is quite difficult to control all the variables that may affect the result, such as a factor present in varying degrees among the groups. The doctor will determine whether a specific drug is suitable to use for a particular patient. In doing so, the doctor will probably rely on previous experience with drugs or technologies that have worked in the same situation. Experience and patient understanding are the keys to maximum success in treatment.

Lithotripsy

Lithotripsy is the use of high-powered shock waves to pulverize and destroy kidney stones. The shock wave, made by using a high-voltage flash or an electromagnetic motivation, is centered on the stone. This shock wave breaks the stone, and this permits the parts to go through the urinary system. Since the shock wave is produced outside the body, the system is named extracorporeal shock wave lithotripsy (ESWL). ESWL is used when a kidney stone is too large to naturally pass through the urinary system on its own or when a stone becomes stuck in a ureter (a cylinder that carries urine from the kidney to the bladder) and will not pass. Kidney stones are incredibly agonizing and can cause genuine clinical complications if not evacuated.

ESWL should not be considered for patients with serious skeletal distortions, weighing more than 300 lbs (136 kg), with stomach aortic aneurysms, or with uncontrollable bleeding disorders. Patients who are pregnant should not be treated with ESWL. Patients with cardiovascular pacemakers should be assessed by a cardiologist acquainted with ESWL. The cardiologist should be available during the ESWL in case the pacemaker should be overridden. Lithotripsy uses the strategy of centered shock waves to pulverize a stone in the kidney or ureter. The patient is put in a tub of water or in contact with a water-filled pad, and a shock wave is made and centered on the stone. The wave breaks and sections the stone. The subsequent debris, called gravel, then goes through the rest of the ureter, the bladder, and the urethra during urination. There is an insignificant possibility of harming the skin or interior organs because biological tissues are strong, not fragile, and the shock waves are not focused on them.

Before the lithotripsy system, a total physical assessment is performed, trailed by tests to decide the number, area, and size of the stone or stones. A test called an intravenous pyelogram (IVP) is used to find the stones. An IVP includes infusing color into a vein in the arm. This color, which appears on the X beam, goes through the circulation system and is discharged by the kidneys—the color at that point streams down the ureter into the bladder. The color encompasses the stones, and X beams are used to assess the stones and life systems of the urinary framework. A few people are adversely affected by the color material, in which case it cannot be used. For these individuals, centered sound waves, called ultrasound, can be used to find the stones. Blood tests are performed to decide whether any potential draining issues exist. A pregnancy test is performed for women of child-bearing age to ensure that the patient is not pregnant, and older patients have an EKG performed to ensure no potential heart issues exist. A few patients may have a stent put before the lithotripsy methodology. A stent is a plastic cylinder put in the ureter that permits gravel and urine entry after the ESWL system is finished.

Lithotripsy is the treatment of kidney stones. Skin that is not used for skin stones (not punctured in the skin) cannot pass through the urethra. Treats kidney stones by sending waves. The shock wave breaks the large stone into smaller stones and passes through the urethra. Lithotripsy allows a person with certain types of stones in the urethra to survive the invasive surgical procedure to remove the stones. The doctor should be able to see the stones with the help of X-rays or ultrasound.

The introduction of lithotripsy in the early 1970s revolutionized the treatment of patients with kidney stones. Patients who once needed major surgery to remove the stone can be treated with lithotripsy without any incision. Lithotripsy, for example, is the only noninvasive treatment of kidney stones and does not require an incision or an internal binocular device. Lithotripsy involves the administration of several target rock waves. The shock wave produced by a machine called a crusher focuses X-rays on kidney stones. The shock wave proceeds through the body's skin and tissues to reach the stone and break it into tiny pieces. In the weeks following treatment, these pieces are excreted in the urine.

In the 20-plus years since lithotripsy was first performed in the United States, we have learned a great deal about how different patients respond to this technique. While it is possible to identify patients who are less likely to be successful with lithotripsy, other unidentified patients are likely to clear stones. Many of these parameters are uncontained, such as the size and location of kidney stones, but other operations can be performed during lithotripsy treatment with a positive effect on the procedure's outcome. At the Brady Urological Institute, our surgeons are researching important groups to make lithotripsy safer and more effective and discovering strategies to provide perfect, sophisticated treatment.

Other steps are also available for treating kidney stones:

➢ Urethral examination is an endoscopy using an instrument called an endoscope to remove stones from the urethra through a small, flexible, lighted tube.

➢ Percutaneous nephrolithotomy (tunnel surgery) is used for stones that cannot be treated with lithotripsy or endoscopy. It involves removing stones with a thin tube through a small incision in the kidney.

➢ Open surgery is more invasive and uses a large incision to penetrate the stone directly.

➢ Stents are synthetic cylindrical devices that can be used in other processes. Stents can be inserted into the urethra with a special scope to make stones pass more easily.

Kidney stones

The substances usually excreted by the kidneys are in the urethra, but they can crystallize and harden the kidneys with stones. Once the stones are removed from the kidneys, they can enter by removing the narrow passages in the urethra. Some kidney stones are small, smooth, and pass easily through the urethra without discomfort. Other stones have rough edges or grow as large as a pea and can cause extreme pain when caught while moving in the urethra. Bladder, ureter, and ureter kidney stones are the most common. Most kidney stones are small enough without intervention. However, at about 20%, the stone is more than 2 cm (about 1 inch) thick and may require treatment. Most kidney stones are made with calcium. However, other types of kidney stones can occur. The types of kidney stones are

a. Calcium stones contain calcium, a common component of a healthy diet used for bones and muscles, and are usually excreted in the urine. However, excess calcium in the body can form stones in combination with other waste products.

b. Struvite stones, comprising magnesium, phosphate, and ammonia, may develop after urinary tract infections.

c. Uric acid stones can occur when urine is highly acidic in cases such as gout or malignant tumors.

d. Cystine stones include cysteine and are one of the major blockages occurring in muscles, nerves, and other parts of the body.

The body absorbs nutrients from food and converts them into energy. After the body needs food, waste products are released into the intestines and blood. The urethra maintains chemical balance like potassium, sodium, and water and removes a waste product called urea from the blood. Urea is formed when protein-rich substances like meat, poultry, and some vegetables are broken down in the body. Urea is transferred to the kidneys in the bloodstream.

Urinary organs and their functions

Two kidneys toward the center of the back below the chest level can be identified as a pair of purple-brown limbs. Their functions are to

a. remove liquid waste products from the blood in the form of urine;

b. stabilize the balance of salt and other substances in the blood;

c. produce erythropoietin, a hormone that helps build blood cells;

d. control blood pressure.

The kidneys extract urea from the blood through a small filtration unit called a nephron. Each nephron has a small capillary called glomerulus and a small tube called a kidney tube that passes through and reaches the renal tubules of the kidneys.

a. Two ureters: A thin tube that carries urine from the kidneys to the bladder. The muscles of the urethral wall regularly tighten the urethra and push it away from the kidneys. Kidney infections develop when urine returns or stops. Within about 10–15 s, a small amount of urine is excreted from the kidneys into the bladder.

b. The bladder is a hollow triangular organ of the lower abdomen. It is held in place by ligaments that connect the other organs and bones of the pelvis. Relaxes and urinates the bladder wall, stores urine, contracts and relax through the urethra, and empties the urine. Two glasses of urine can be kept in a normal healthy adult bladder for 2–5 h.

c. Two sphincter muscles: A round muscle that holds the bladder tightly like a rubber band around the opening and helps hold urine

d. Urethra: When it is time to urinate or empty the bladder, they warn the person. The urethra is a tube that allows urine to pass out of the body. The brain directs the bladder muscles to tighten, and urine is suppressed from the bladder. At the same time, the brain signals the urethra to relax the resting muscles and exit the bladder.

Risks during procedures

One can ask the doctor about the amount of radiation during the procedure and the risks associated with a particular condition. It's a good idea to keep a history of past radiation, such as previous scans and other types of X-rays, to share with your doctor. The risks associated with radiation exposure may be related to the number of X-rays or long-term treatments.

Complications of lithotripsy include

a. bleeding around the kidneys

b. infection

c. obstruction of the urethra due to stones

d. remaining stone debris that may require further lithotripsy

Contraindications to lithotripsy include

a. pregnancy

b. patients with "anticoagulants" or bleeding disorders; aspirin or other anticoagulants should be discontinued at least 1 week before lithotripsy

c. patients with chronic kidney disease cannot completely remove bacteria from the kidneys

d. patients with urethral obstruction or scar tissue that may interfere with the passage of stones

e. patients requiring immediate or complete approval of the stone

f. patients with cysteine and certain types of calcium, who cannot break down these stones well through lithotripsy

Patients using cardiac pacemakers should inform their doctors. Lithotripsy can be performed in pacemaker patients with cardiologist approval and some precautions. Motion-responsive pacemakers implanted in the abdomen may be damaged during lithotripsy. There may be other risks depending on our particular treatment condition. Obesity and intestinal gas can interfere with the lithotripsy process.

Electrohydraulic lithotripsy

It is a strategy used for huge upper urinary lot calculi: a high-limit condenser makes a high-voltage flash between two terminals at the tip of a test; in a liquid-filled organ, this makes a pressure-driven shock wave that can be coordinated toward a calculus, making it cavitate and piece.

Extracorporeal shock wave lithotripsy

ESWL is the noninvasive fragmentation of kidney stones or gallstones with shock waves created outside the body. It requires no cuts, catheters, or nephroscopes. The method depends on the rule that shock waves are not ruinous until they arrive at a surface with a change in acoustic impedance, a type of protection from the section of sound waves. The impedance of calculi is not the same as that of water, bone, and delicate tissue; in this way, tissue through which the wave goes just as tissues encompassing the stone are not hurt.

Hemodialysis machine

A machine used in dialysis that channels a patient's blood to expel over-abundance of water and waste items when the kidneys are damaged, dysfunctional, or missing. The dialysis machine itself can be thought of as an artificial kidney. Inside, it comprises plastic tubing that conveys the evacuated blood to the dialyzer, a heap of empty strands that shapes a semipermeable layer for sifting through contaminations. In the dialyzer, blood is diffused with a saline arrangement called dialysate, and the dialysate is in turn diffused with blood. At the point when the filtration procedure is finished, the scrubbed blood comes back to the patient. Most patients who must use dialysis because of kidney impairment use a dialysis machine at a dialysis center.

Likewise, a peritoneal dialysis machine can be used for dialysis continuously at home, which eliminates the requirement for ordinary hemodialysis center medicines. Using this machine during the day and often during rest, patients can control their dialysis.

The blood gets separated in the dialyzer of a hemodialysis machine. The dialyzer has thousands of tubes inside it. The blood streams inside each tube, and the dialysate remains outside the cylinders. Small pores in the cylinders carry out the waste and abundance of liquids from the blood into the dialysate. The filtered blood leaves the dialyzer and flows back to the body.

Working principle

Dialysis chips away at the standards of the dispersion of solutes and ultra-filtration of liquid over a semipenetrable layer. Dissemination is a property of substances in water, which will generally move from a zone of high concentration to a territory of low concentration. One side of the blood-stream consists of a semiporous film, and the other consists of dialysate, or uncommon dialysis liquid. A semipermeable film is a slim layer of material that contains gaps of different sizes or pores. Smaller amounts of solute and liquid go through the film; however, the layer hinders the section of bigger substances (for instance, red platelets and large proteins). This duplicates the sifting procedure in the kidneys when the blood enters the kidneys, and the bigger substances are isolated from the little ones in the glomerulus. The two fundamental sorts of dialysis, hemodialysis and peritoneal dialysis, expel waste and excess water from the blood in various ways. Hemodialysis expels waste and water by coursing blood outside the body through an outer channel, called a dialyzer, that contains a semipermeable layer. Blood streams one way, and the dialysate streams the opposite way. The countercurrent progression of the blood and dialysate augments the fixation slope of solutes between the blood and dialysate, which assists with expelling more urea and creatinine from the blood. The centralizations of solutes typically found in the pee (for instance, potassium, phosphorus, and urea) are unfortunately high in the blood, yet low or missing in the dialysis arrangement, and consistent substitution of the dialysate guarantees that the convergence of undesired solutes is kept low on this side of the layer. The dialysis arrangement has minerals like potassium and calcium that are like their regular concentrations in solid blood. For another solute, bicarbonate, the dialysis arrangement level is set at a marginally more elevated level than in normal blood to support the dissemination of bicarbonate into the blood, to go about as a pH cushion to kill the metabolic acidosis that is regularly present in these patients. The degrees of the segments of dialysate are regularly recommended by a nephrologist as per the necessities of the individual patient. In peritoneal dialysis, squanders and water are expelled from the blood inside the body using the peritoneum as a characteristic semipermeable film. In the stomach cavity, waste and excess water move from the blood, over the peritoneal film, and into a unique dialysis arrangement called dialysate.

The instrumentation used in the hemodialysis machine is described below:

The blood is involved from the corridor of the patient and blended in with an anticoagulant—for example, heparin—constrained into equipment known as an artificial kidney or a hemodialysis machine.

> Dialysate pail: dialysate pail is the storage container of the dialysate. Through the dialysate, the blood from the patient is coordinated to

course through channels or cylinders limited by a cellophane layer. This film is porous to little solutes and impermeable to macromolecules.

> Dialyzer: the dialyzer is powerful hardware that channels the blood. Dialyzer layers have divergent pore sizes. Those with lesser pore size are called low-flux membranes, and those with greater pore size are called high-flux membranes. Dialyzer films are made of cellulose. Additional film is produced using manufactured articles using polymers. Nanotechnology is used in some high transition films to create uniform pore sizes. Dialyzers come in various sizes. A greater dialyzer with a greater layer can eliminate more solutes.

> Dialysate heater: the dialysate ought to be kept at a proper temperature. The temperature is constrained by the dialysate warmer.

> Check valve: the dialysate check valve controls the progression of the dialysate to the dialysate pail. On the off chance that dialysate flooding occurs, the check valve will guide dialysate overflow and the channel blender where the dialysate will be blended with depleted dialysate. The blended portion is tossed out as fumes.

> Dialysate recirculating loop: this loop takes depleted dialysate and reconverts it into new dialysate that is appropriately warmed and diverted to the dialysate pail. The dialysate may be disposed of after each use or reused. Patients do not share reused dialyzers. On the off chance that dialyzers are reused, if done appropriately, the results will be comparable to those from a single use of dialyzers.

> Dialysate holding tank: the dialysate holding tank holds the new dialysate, and whenever the dialysate of the pail must be changed, the new dialysate from the holding tank is taken to the pail.

> Flowmeter: a stream meter is used in machine setup to control the progression of new dialysate from the dialysate holding tank to the dialysate pail.

Heart-lung machine

The heart-lung machine is likewise called a cardiopulmonary detour machine. It takes over the heart's pumping activity and adds oxygen to the blood. This means the heart is inactive from the activity, which is vital when the heart must be opened (open-heart medical procedure). When the heart is associated with the heart-lung machine, it does the same work as the heart and lungs. The heart-lung machine conveys blood from the upper-right atrium of the heart (the correct chamber) to an uncommon store called an oxygenator. Inside the oxygenator, oxygen rises through the blood and enters the red platelets. This results in blood that is dark (oxygen-poor) to splendid red (oxygen-rich). At that point, a channel expels the air rises from the oxygen-rich blood, and the blood goes through a plastic cylinder to the body's fundamental blood course (the aorta). From the aorta, the blood moves all through the remainder of the body. The heart-lung machine can assume control over crafted by the heart and lungs for quite a long time. Prepared professionals called perfusion technologists (bloodstream experts, additionally called the "pump team") work the heart-lung machine.

Cardiopulmonary bypass (CPB) is a technique in which a machine temporarily processes heart and lung function during surgery to maintain a patient's blood circulation and oxygen levels. CPB pumps are often referred to as cardiopulmonary equipment or "pumps." The CPB pump is operated by a sprayer. CPB is a form of extracorporeal circulation. Extracorporeal membrane oxygenation (ECMO) is commonly used for long-term treatment. CPB bypasses the heart and lungs, mechanically circulating blood throughout the body and supplying oxygen. Heart-lung machines are used to maintain perfusion of body organs and tissues and by surgeons working in anemic surgery. The surgeon places a cannula in the right atrium, vena cava, or femoral vein to remove blood from the body. Venous blood is removed from the body through the cannula, then filtered, cooled or heated, oxygenated, and then returned to the body by mechanical pumps. The brachiocephalic artery (especially) can be sorted. Patients are given heparin to prevent clotting and proteome sulfate after the adverse effects of heparin. You can maintain hypothermia during the procedure. Body temperature typically ranges between 28°C and 32°C (82.4°F–89.6°F). During CPB, the blood cools and returns to the body. Cold blood lowers the body's basal metabolic rate and reduces the need for oxygen. Cold blood is usually highly viscous, but the bypass tube dilutes the crystalline liquid blood for priming.

CPB is commonly used in cardiac surgery. This technique helps the surgeon to oxygenate and circulate the patient's blood, allowing the surgeon to work on the heart. Many surgeries, such as coronary artery bypass grafting, make it difficult for the heart to beat, causing the heart to stop (or stop). Regular use of the CPB is required to avoid air and to increase the surgeon's visibility to provide bloodless areas for the surgery required to open the chamber of the heart, such as repair or replacement of the metal valve. The machine pumps blood and uses artificial lungs to take in oxygen from red blood cells and reduce carbon dioxide levels. It mimics the function of the heart and lungs, respectively. CPB can be used to induce systemic hypothermia. This is a condition where the body can be maintained for 45 min without perfusion (blood flow). When blood flow stops at body temperature, permanent brain damage occurs within 3—4 min and can soon be fatal. Similarly, CPB can be used to rehabilitate people with hypothermia. This procedure using CPB is again successful when the patient's core temperature exceeds 16°C. ECMO is a simplified version of the cardiopulmonary machine that includes a centrifugal pump and a heart-pulmonary machine that temporarily processes the function of the heart or lungs. ECMO is effective for patients with cardiac or pulmonary dysfunction, acute pulmonary insufficiency, large pulmonary embolism, lung trauma from infection, and many other problematic cardiac surgeries that affect cardiopulmonary function. ECMO provides the heart and lungs time to repair or heal, but this is only a temporary solution. Patients with a terminal illness, cancer, severe nervous system damage, uncontrolled sepsis, and other symptoms may not be eligible for ECMO.

The Austrian-German physiologist Maximilian von Frey built the first prototype of the heart-lung machine in 1866 at the Faculty of Physics at the Carl Ludwig University in Leipzig. However, before the discovery of heparin in 1916, blood clotting machines could not be used. In 1926, Soviet scientist Sergei Brukhonenko developed the heart—lung device for the perfection used in Kane's experiments. A team of scientists from the University of Birmingham, including chemical engineer Eric Charles, was one of the pioneers of this technology. Dr. Clarence Dennis, led by a team from the University of Minnesota Medical Center, performed the first human surgery, including open-heart surgery, on April 5, 1951, with the temporary mechanical acquisition of heart and lung function. The patient did not survive because of unexpected and complicated congenital heart disease. Nelson later became president of the Church of Jesus Christ of Latter-day Saints and was the first to perform open-heart surgery in Utah. The first successful mechanical support for left ventricular function was used by David Drill on July 3, 1953, with a Dodrill—GMR machine, a device codeveloped by General Motors. This machine was later used to support correct ventricular function. John Gibbon and Frank F. performed the first successful open-heart surgery in humans using a lung-lung machine on May 6, 1953, at Thomas Jefferson University Hospital in Philadelphia. Albrighton, Jr. repaired an atrial septal defect in an 18-year-old woman. John W. at the Mayo Clinic in Rochester, Minnesota. A surgical team led by Kirklin developed the Gibbon machine as a reliable device in the mid-1950s. CPB devices were used at the University of Michigan in the 1960s. The artificial lung was first conceived by Robert Hooke in the 17th century. Bubble oxygen suppliers do not interfere with blood and oxygen. These are called "direct contact" oxygen suppliers. Membrane oxygen donors introduce a gas-permeable membrane between blood and oxygen and communicate directly with the oxygen feeder to reduce blood stroke. Since the 1960s, much work has focused on overcoming the barriers to gas exchange in membrane barriers, leading to the development of high-performance micro-4 forward aids and eventually direct contact to artificial lung. In 1983, Ken Lizzie patented a closed emergency cardiac bypass system that reduced the complexity of circuits and components. The device can be quickly deployed to a nonsurgical setting to improve the patient's survival after cardiac arrest.

Fig. 9.1 shows a block diagram of the heart lung machine.

Some of the components used in hurt lung machine are as follows:

➢ In cardioplegia, a separate circuit for the injection of a fluid into the heart to produce cardioplegia to stop the heart from beating.

➢ A venous cannula removes oxygen-deprived blood from a body while an arterial cannula inserts oxygen-rich blood.

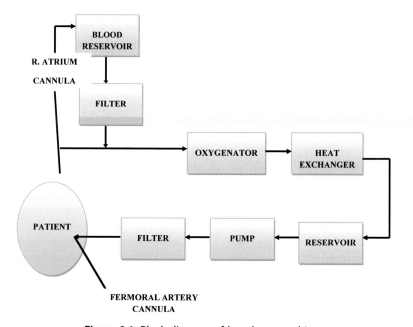

Figure 9.1 Block diagram of hurt lung machine.

> A blood reservoir serves as a receiving chamber for venous return.
> A heparin pump is given to a patient to reduce the blood's ability to clot.
> The roller pump compresses a segment of the blood-filled tubing, and then blood is pushed ahead of the moving roller.
> Oxygenators supply oxygen for the blood and also transport carbon dioxide, anesthetic, and other gases into and out of the circulation.
> The heat exchanger controls the body temperature by heating and cooling the blood passing the perfusion circuit.

Surgical procedures in which CPB is used include the following:

a. coronary artery bypass surgery
b. cardiac valve repair or replacement (aortic valve, mitral valve, tricuspid valve, pulmonic valve)
c. repair of large septal defects (atrial septal defect, ventricular septal defect, atrioventricular septal defect)
d. repair or palliation of congenital heart defects (Tetralogy of Fallot, transposition of the great vessels)
e. transplantation (heart transplantation, lung transplantation, heart–lung transplantation, liver transplantation)
f. repair of some large aneurysms (aortic aneurysms, cerebral aneurysms)
g. pulmonary thromboendarterectomy
h. pulmonary thrombectomy
i. isolated limb perfusion

CPB is not benign, and there are many problems associated with it. As a result, CPB is only used for a few hours when heart surgery can occur. CPB activates the clotted cascade and stimulates inflammatory mediators, causing hemolysis and coagulopathy causing this problem to be exacerbated by supplementing the membrane oxygen supply with proteins. For this reason, most oxygen manufacturers recommend using them for only 6 h, but care has been taken to allow them to be used for up to 10 h, allowing them to stop or stop working. It is not designed. For longer periods of time, ECMO is used. It works for up to 31 days, like in Taiwan for 16 days, after which the patient receives a heart transplant. The most common complication associated with CPB is the protamine response at the opposite time to anticoagulant therapy. There are three types of proteinuria reactions, each of which can be fatal: hypotension (type I), anaphylaxis (type II), or pulmonary hypertension (type III). Patients with preexposure to the protein, such as prevascular (sperm proteome) or diabetes (insulin preparations containing protamine-neutralized protamine Hagedorn), have a type II proteome response. The first step in managing the protamine response is to stop the protamine injection immediately. Corticosteroids are used in all types of protamine reactions. And the patient must return to the bypass.

There is no perfect contraindication to CPB. However, the care team should consider a few factors when planning surgical treatment. Heparin-induced thrombocytopenia (HET) and heparin-induced thrombocytopenia and thrombosis (HIT) are potentially life-threatening conditions associated with heparin administration. In HET or HIT, antibodies against heparin are formed, resulting in platelet activation and thrombus formation. Since heparin is commonly used in CPB, patients known to have antibodies to HET or experience HIT require another form of anticoagulant therapy. Bivalirudin is the most studied heparin alternative that requires CPB in patients with HIT. Antithrombin is a third deficiency in very few patients who may develop heparin resistance. These patients may need to receive appropriate anticoagulants from recombinant antithrombin III, such as excess heparin, fresh frozen plasma, or other blood products.

The persistent left superior vena cava (PLSVC) is a variant of the thoracic level system, and the left superior vena cava cannot be applied during normal development. It is the most common form of the thoracic venous system, and abnormalities occurring in about 0.3% of the population are often detected by preoperative imaging tests but can also be detected interstitially. PLSVC can achieve proper venous drainage or cause back pain. PLSVC management during CPB depends on factors such as PSLVC size and sewage site.

CPB can contribute to instant cognitive decline. Cardiopulmonary circulatory systems and connective surgery leave a variety of debris in the bloodstream, including debris from the blood cells, tubes, and blades. For example, when a surgeon clamps the aorta and attaches it to a tube, the resulting ambulance can block blood flow and cause brain stroke. Other cardiovascular causes associated with trauma can include hypoxia, high and low temperatures, abnormal blood pressure, arrhythmias, and postoperative fever.

Components

CPB consists of two main functional units, the pump and oxygenator, that remove relatively oxygen-depleted blood from a patient's body and replaces it with oxygen-rich blood through a series of tubes (hoses). A heat exchanger is used to control body temperature by heating or cooling the blood in the circuit. It is important that all components of the circuit are coated internally by heparin or another anticoagulant to prevent clotting within the circuit.

Tubing

The components of the CPB circuit are interconnected by a series of tubes made of silicone rubber or PVC.

Pumps
Centrifugal pump

Many CPB circuits now employ a centrifugal pump for the maintenance and control of blood flow during CPB. By altering the speed of pump head revolutions, blood flow is produced by centrifugal force. This type of pumping action is considered to be superior to the action of the roller pump by many because it is thought to prevent overpressurization, clamping, or kinking of lines and produce less damage to blood products (hemolysis, etc).

Roller pump

The pump console usually comprises several rotating motor-driven pumps that peristaltically "massage" tubing. This action gently propels the blood through the tubing. This is commonly referred to as a roller pump or peristaltic pump. The pumps are more affordable than their centrifugal counterparts but are susceptible to overpressurization if the lines become clamped or kinked. They are also more likely to cause a massive air embolism and require constant, close supervision by the perfusionist.

Oxygenator

The oxygenator is designed to add oxygen to infused blood and remove carbon dioxide from the venous blood. Cardiac surgery was made possible by CPB using bubble oxygenators, but membrane oxygenators have supplanted bubble oxygenators since the 1980s. The main reasons for this are that membrane oxygenators tend to generate many fewer microbubbles, referred to as gaseous microemboli and generally considered harmful to the patient, and reduce damage to blood cells compared with bubble oxygenators. More recently, hollow-fiber oxygenator use has become more widespread. These derivatives of membrane oxygenators further reduce the occurrence of microemboli by reducing the direct air-blood interface while simultaneously providing adequate gas exchange.

Another type of oxygenator recently gaining favor is the heparin-coated blood oxygenator which is believed to produce less systemic inflammation and decrease the propensity for blood to clot in the CPB circuit.

Heat exchangers

Because hypothermia is frequently used in CPB to reduce metabolic demands (including that of the heart), heat exchangers are implemented to warm and cool blood within the circuit. Heating and cooling are accomplished by passing the line through a warm or ice water bath. A separate heat exchanger is required for the cardioplegia line.

Cannulae

Multiple cannulae are sewn into the patient's body in a variety of locations, depending on the type of surgery. A venous cannula removes oxygen-depleted venous blood from a patient's body. An arterial cannula infuses oxygen-rich blood into the arterial system. The main determinants of cannula size selection are determined by the patient's size and weight, anticipated flow rate, and the size of the vessel being cannulated. A cardioplegia cannula delivers a cardioplegia solution to cause the heart to stop beating.

Techniques

Preoperative planning

The CPB needs important predictions before surgery. In particular, cannulation, cooling, and cardiac protection techniques need to be coordinated between surgeons, anesthesiologists, perfusion specialists, and nursing staff.

Cannula technique

Many cannula insertion techniques rely on surgical and patient-specific details. Simple arterial cannulation involves placing a single cannulation in a distant ascending aorta. The simplest form involves the position of a single cannula (known as a biphasic cannula) that passes through the right atrium and the inferior vena cava. Two cannulas are used in several surgical treatments, including tricuspid and mitral valves. One goes through the inferior vena cava and the other through the superior vena cava. This is known as a single-stage cannula.

Intertechnology

The CPB circuit should discharge fluid and all air from the arterial line/canal before connecting to the patient. The circuit is made with a crystal solution, and blood products can also be added. Before cannulation (usually used in central Peru after opening the pericardium), heparin or another anticoagulant was administered until the active freezing time did not exceed 480 s.

Arterial cannula insertion sites are examined for catheterization or other ailments. Preparatory images or ultrasound probes can be used to help identify aortic calculations that could potentially be disrupted and cause obstruction or stroke. When the cannula insertion site is considered safe, string stitches of two cubic diamond-shaped pearls are placed on the remote climbing banker. A scalpel puncture wound is made in a purse, and an artery passes through the cannula intersection. It is important to pass the cannula vertically to prevent cosmic separation. Purse-string sutures are secured to the cannula with a tourniquet. At this point, the perforator moves the arterial line of the CPB circuit, and the surgeon connects the arterial line from the CPB machine to the arterial line from the patient. When the two are connected, make sure that they are never in the circuit. Otherwise, the patient may suffer from air embolism. Other sites of arterial disorder include the axillary artery, the brachiocephalic artery, or the femoral artery. In addition to the location differences, the venous cannula performs in the same way as the arterial cannula. Calculations in the venous system are less common and do not require ultrasound examination or use for calculations in place of cannulation. Also, the venous system is under a lot less pressure than the arterial system, so you should only place one suture. When using only one cannula (dual stage cannula), it enters through the tricuspid valve, through the right atrium condyle, and into the inferior vena cava. If two cannulas are required (single-stage cannula), the first cannula usually passes through the superior vena cava and the second cannula passes through the inferior vena cava. In some patients, femoral vein rupture may also occur.

If we need to have a cardiac arrest for surgery, we also need a cardio-pulmonary cannula. Reflected heart pain (anterior to the flow of arteries in the heart), lost pain (flowing between the arteries of the heart), or both can be used as the surgeon chooses. A small incision is made in the aorta, and it is placed in the coronary artery to supply cardiac perigia. For retrograde cardioplegia, an incision is made on the posterior (anterior) surface of the heart through the right ventricle. The girth is placed at this intersection and coronary sinus, which passes through the tricuspid valve. The heart attack line is connected to the CPB machine.

At this point, the patient is ready to go bypass. Blood from the venous cannula enters the CPB machine by gravity, where it is oxygenated and cooled (if needed) and returns to the body through the cannula. Cardiac perdia can be operated to close the heart, and a cross-clamp is placed between the arterial cannula and the cardiac plaudia cannula to prevent blood flow in the arterial blood flow. When the patient is ready to leave the bypass support, administer protein sulfate to counteract the anticoagulant effect of heparin by removing the cross-clamp and cannula.

Automated drug delivery system

Drug delivery systems (DDSs) are technologies designed for the targeted or controlled release of treatment agents. The drug has long been used to improve health and prolong life. Drug supply practices have changed dramatically over the past few decades, and even greater changes are expected in the near future. Biomedical engineers have made significant contributions to understanding the physical barriers to efficient drug delivery, such as the transport of blood through the circulatory system through cells and tissues and the transfer of drugs. He has also contributed to the development of several new drug delivery systems that have entered clinical practice. Yet, with this advancement, many drugs discovered using sophisticated molecular biology techniques have unacceptable side effects due to their interaction with healthy tissues that are not targeted. Side effects limit the ability to design optimal drugs for many diseases, including cancer, neurodegenerative diseases, and infectious diseases. The drug delivery system determines the drug rate and controls where the red drug is in the body. Some systems can control both.

Physicians have historically tried to intervene in various parts of the body affected by risk and illness. Side effects can occur depending on the medication, the method of administration, and the body's response. These side effects may vary from person to person, depending on the type and severity. For example, oral medications for seasonal allergies can cause unwanted sleep and stomach upset. Organizing medications locally (rather than affecting the whole body) is a common way to reduce side effects and drug toxicity while maximizing the effectiveness of treatment. Topical (used on the skin) antibacterial ointment for temporary infections or painful joint cortisone injections may avoid some systemic side effects of these drugs. There are other ways to provide targeted medications, but some medications can only be administered regularly.

An "automated drug delivery system" (ADDS) implies a mechanical framework that performs tasks or exercises other than exacerbating or organization, comparative with the capacity, apportioning, or circulation of medications. An ADDS will gather, control, and keep up all exchange data to precisely follow the development of medications into and out of the framework for security, exactness, and responsibility.

➢ An automated unit portion framework is an ADDS for capacity and recovery of unit dosages of medications for an organization to patients by people approved to perform these activities.

➢ An automated patient apportioning framework is an ADDS for capacity and administering of recommended tranquilizes legitimately to patients as per earlier approval by a drug specialist.

DDSs are created to convey the necessary measure of medications adequately to fitting objective destinations and to keep up the ideal medication levels. Examination of more up-to-date DDSs is being completed in liposomes, nanoparticles, niosomes, transdermal medication conveyance, inserts, microencapsulation, and polymers.

Advantages of DDSs are the following:

➢ builds bioavailability
➢ can be used in long-term medicines for chronic disease
➢ supports upkeep of plasma drug levels
➢ diminishes unfriendly medication impacts
➢ a decline in the aggregate sum of medications acquired this way diminishes symptoms
➢ improved patient consistence because of decrease in number and recurrence of dosages required
➢ less harm is supported by ordinary tissue because of focused medication conveyance
➢ decreased costs by creating more up-to-date conveyance frameworks for existing atoms

Areas for future research in drug delivery systems

Scientists study how our bodies respond to disease and how the disease develops and progresses as they learn about the effects of specific environmental or genetic signals. This advanced understanding with technological advances suggests a new approach to drug distribution research. The key areas for future research are

a. Overcoming the blood—brain barrier (BBB) because of a brain disease or disorder: The various cells that make up the BBB, when functioning properly, constantly control the flow of blood and the movement of essential substances in the central nervous system, identify and attack substances that can cause brain damage. Prescribed drugs to the brain are important for the successful treatment of certain diseases such as brain tumors, Alzheimer's disease, and Parkinson's disease, but better ways to pass or bypass BBB are needed. One method under study used advanced ultrasound technology that quickly and safely destroyed the BBB, allowing the drug to directly target brain tumors without the need for surgery.

b. Intracellular distribution increased: Just as immunity protects the body from disease, so too every cell has internal processes for detecting and removing potentially harmful substances and foreign bodies. These foreign agents may contain drugs linked to the target delivery medium. Thus, researchers are working to develop reliable ways to provide treatment aimed at such cells, so further treatment engineering is needed to ensure access to the right structure within the cell. Ideally, future healthcare will include a smart delivery system that can bypass cell protection, transport the drug to interstitial targets, and release the drug in response to specific molecular signals.

c. A combination of diagnosis and treatment: The possibility of DDS is out of treatment. Using advanced imaging techniques, including targeted delivery, physicians may one day be able to diagnose and treat the disease with a new technique known as theranostics.

Physicians have historically tried to intervene in various parts of the body affected by risk and illness. Side effects can occur depending on the medication, the method of administration, and the body's response. These side effects may vary from person to person, depending on the type and severity. For example, oral contraceptives for seasonal allergies can cause unwanted sleep and an upset stomach. Organizing medications locally (instead of affecting the whole body) while maximizing treatment efficacy is a common way to reduce side effects and drug toxicity. Topical (used on the skin) antibacterial ointment for temporary infections or painful joint cortisone injections may avoid some systemic side effects of these drugs. There are other ways to provide targeted medications, but some medications can only be administered regularly.

Fig. 9.2 shows a newer drug delivery system in anesthesia.

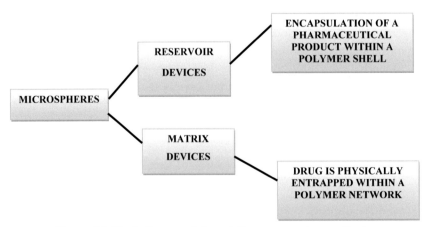

Figure 9.2 Block diagram of drug delivery system in anesthesia.

Baby incubator

A neonatal incubator is an average gadget used to keep up an ideal situation for the care of a newborn. For the most part, it is used for untimely children, which means those born preterm, or within 37 weeks of amenorrhea (a proportion of time used in pregnancy), as well as those with certain intrinsic ailments. Sterile and warmed to a perfect temperature, the incubator let the newborn children proceed with their typical turn of events. Incubators are outfitted with little openings, permitting the overseer and guardians to contact the infant. Exacting principles of cleanliness must be regarded before contact is made.

An incubator is an independent unit generally the size of a standard lodging outfitted with a reasonable plastic arch. The incubator guarantees the perfect ecological conditions by either permitting the temperature to be balanced physically or giving auto-alterations dependent on changes in the infant's temperature. An incubator likewise shields the preemie from disease, allergens, or unreasonable clamor or light levels that can cause hurt. It can control air stickiness to keep up the respectability of the skin and can be outfitted with exceptional lights to regularly treat neonatal jaundice in babies.

Types of incubators
➢ Closed box incubators have a natural air filtration framework that limits the danger of contamination and keeps the loss of dampness from the air. A closed incubator is a place where children are completely surrounded. There is a portal hole on the side for human hands, but it is designed to keep out bacteria, light, and other elements. Like living in a closed incubator temperature-controlled bubble. The biggest difference between a closed incubator and an open incubator is how the heat is transferred and how the temperature is controlled. The closed incubator allows the baby to carry hot air around the canopy. The temperature and humidity can be controlled manually using the external mass of the incubator or adjusted automatically based on the skin sensor attached to the child. (An incubator thus automatically adjusted is called a servo-controlled incubator). A closed incubator is actually a unique microenvironment. This means it is ideal for children who need extra bacterial protection, low light/noise, and humidity control. Some closed incubators have two walls to prevent heat and air loss. These are commonly referred to as double-walled incubators.

➢ Double walled incubators have two dividers that can additionally fore-stall warmth and air dampness misfortune.

➢ Servo-control incubators are naturally customized to alter temperature and humidity levels based on skin sensors attached to the baby. The levels depend on skin sensors connected to the infant.

➢ Open-box incubators, otherwise called Armstrong incubators, provide brilliant warmth beneath the infant yet are in any case open to the air, taking into consideration simple access. Sometimes called a bright warm, the baby's skin temperature automatically controls the generation of heat. You can see many monitors, but the incubator is open on top of the kids. Because of this open air space, the open incubator holds less control of the incubator than the closed incubator. However, they can still monitor the child's vital activities and warm them up. Since you can touch the baby directly from above, it is easy to match the skin with the baby in an open incubator. Open incubators are primarily suitable for babies who need to warm up temporarily and measure vital figures. Uncontrolled humidity and the inability to protect against bacteria in the air means that open incubators are not ideal for children who need more regulated environments and bacterial protection.

➢ Portable incubators, otherwise called transport incubators, are used to move the infant starting with one piece of the emergency clinic then onto the next. As the name implies, these types of incubators are usually used to bring babies between two different locations. A child can be used when moving to access services currently transmitted to another hospital that is not currently available or where a specialist is needed. The transport incubator usually includes a mini ventilator, cardiopulmo-nary monitor, ivy pump, pulse oximeter, and oxygen supply. Transport incubators are usually small, so they fit snugly in places that are not suitable for regular retractable incubators.

Incubator temperatures can shift depending on the gestational age, the useful condition of the infant's lungs, and other well-being intricacies. As a rule, the NICU is kept to a temperature of 82°F—86°F, while the Incubator is ordinarily set with the goal that the infant can keep up an internal heat level of somewhere in the range of 95°F to 98.6°F.

Table 9.1 shows some historical infant incubators that have been used. Fig. 9.3 shows the principle of operation of an infant incubator.

Advantages of an infant incubator:

➢ It provides the nearest condition to that of the mother's uterus.

➢ The temperature is at a uniform airflow, and the humidity can be controlled to the ideal level.

Table 9.1 Historical infant incubators.

Year	Invention
1891	First modern incubator invented by Dr. Alexander Lyon
1898	The first American incubator hospital was set up at the Trans-Mississippi Exposition in Omaha, Nebraska
1907	Pierre-Constant Budin released a study on the influence of body temperature on infant mortality
1932	Julius Hess, in his patents for incubators, proposed a mechanism for supplemental oxygen in the incubator
1933	Blackfan and Yaglaw released a report on improved survival for newborn infants nurtured in humidity-enriched environments.

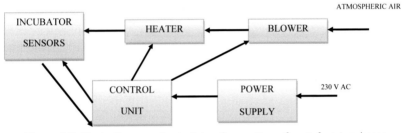

Figure 9.3 Block diagram of principle of operation of an infant incubator.

➤ It shields the infant from all the unsettling influences and disease in the NICU.

Disadvantage of infant incubators:

➤ The disadvantage is that since the infant lies in a shut hood, it is hard to get to the child for clinical methodology or care.

Additional equipment is used to assess and treat sick newborns. These include:

a. Sphygmomanometer: A sphygmomanometer is a machine attached to a small cuff that wraps around the patient's arm or leg. This caffeine automatically measures your blood pressure and displays data for review by our supplier.

b. Oxygen Hood: This is a transparent box that fits with the baby's head and delivers oxygen. It is used for babies who can still breathe but need some breathing assistance.

c. Ventilator: A shortness of breath that sends air into the lungs. Severely ill children accept this intervention. Ventilators usually play the role of the lungs but are treated to improve the efficiency of the lungs and blood circulation.

Functions of newborn incubators

a. Oxygenation with the hood or nasal cannula, or oxidation with CPAP or mechanical ventilation. Infantile respiratory distress syndrome is the leading cause of death in preterm infants, and the main CPAP treatment is to administer pulmonary surfactant and stabilize blood glucose, blood glucose, and blood pressure.

b. Observation: Intensive care of modern neonates includes advanced measurements of body temperature, respiration, cardiovascular function, and oxidation, and brain activity.

c. Protect from low temperature, pollution, noise, ventilation, and excessive handling. Incubators can be described as plastic-enclosed washbasins and temperature controllers designed to keep them warm.

d. Nutritional supply by an intrauterine catheter or NG tube

e. Drug administration

f. Provides moisture and maintains moisture balance by maintaining high air humidity to prevent excess damage due to evaporation in the vapor tract of the skin

Transport incubators are transportable forms of incubators used when transporting sick or premature babies, for example, from one hospital to another and from a regional hospital to a large medical facility as a suitable neonatal intensive care unit. It usually provides a tiny ventilator, a cardiopulmonary monitor, an ivy pump, a pulse oximeter, and oxygen built into the frame.

Incubators for control of temperature and environment

Temperature regulation in newborns: One of the most important reasons for the survival of the newborn is the thermoregulation of the baby. Mammals have the advantage of masturbating. This is because mammals can generate heat and maintain a stable body temperature. However, in winter weather and extreme summer conditions, homothermous can have an effect. The newborn has all the skills of a mature household, but the range of environmental temperatures under which a child can normally work is severely limited. There are many defects in the thermoregulation of children. Children have a relatively large surface area, low thermal insulation, and a small mass that works in heat sinks. Newborns have little ability to change posture and save heat, nor do they have the ability to adjust their clothing in response to heat. Reactions can also be prevented in adverse conditions such as illness and hypoxia. The heat exchange between the environment and the baby is equal to any other object or its environment. Heat is driven by conduction, evaporation, and radiation. Heat exchange by vehicle is relatively low. Conductivity depends on the thermal conductivity of the substance in contact with the body. Babies are usually placed in a mattress that has relatively low thermal conductivity, so the amount of heat the baby has from the mattress is relatively low. Heat loss from children due to condensation depends on airspeed and temperature. Evaporation loss depends on wind speed and the absolute humidity of the air. When a baby is cared for in a moderate or regular heated air incubator with moderate humidity, the heat loss from evaporation is only a small fraction of the total heat lost by the baby. However, in a nursery environment, evaporation is a major source of heat loss when a thin, immature baby is placed under a radioactive overhead heater. Bright heat loss is somewhat more complex than others. It depends on the region and shape of the surface, the body surface temperature, and the surface area temperature obtained. The baby's body reacts differently to high and low temperatures. At warm ambient temperatures, the baby's body sweats through the sweat glands—basal metabolic rate increases, and body temperature increases. The risk of hyperthermia is very high and will appear soon. Intense overheating can cause heatstroke and death, and low pressure can result in brain damage due to hypertensive dehydration. Babies born more than 8 weeks before maturity cannot sweat. Babies born just 3 weeks early are very sweaty, confined to the head and mouth. Sweating during childbirth becomes relatively early after delivery and allows the baby to be kept on a regular bend. At cooler ambient temperatures, babies can generate heat from vibrations and other muscle activity. Cool pressure is subtle in its consequences, but care should be taken. The newborn can also be kept in the newborn incubator.

Design and performance
The initial incubator and transport incubator design and performance features provide the basis for understanding the use and functionality of the device. The design features address the intended use of the device to meet the needs of users and patients and demonstrate features that ensure the device is safe and effective during use as instructed. The features listed in this section are important in determining whether the initial incubator or the initial transport incubator is equivalent to any marketing device.

Thermo-neutrality
Thermo-neutrality is one of the major environmental factors affecting premature or low birth weight babies in incubators. The sharp difference in the temperature of the incubator causes heat loss, hypothermia, and apnea

in the newborn. These are closely related to wind flow and wind velocity. (A physiologically correct initial model is built using a 3D laser scanner system and a rapid prototyping machine Flow) Flow visualization generates many small rotations in the airflow chamber, giving it many small static indications that it is seen in the area between the air inlet and the newborn. Heat ray measurement suggests that the velocity of the wind is not uniform at long inlets. Qualitative fluid dynamics shows a relatively equal temperature of about 34°C in front of the newborn and 36.1°C in the right armpit and crotch. Flow fields from airflow view, heat-ray measurement, and fluid calculation dynamics are very qualitative and quantitatively similar. Small varicose veins formed between the newborn and the mattress can interfere with fertilization and heat transfer from the newborn. Therefore, it is important to remove AD around the newborn when designing the newborn incubator.

Biocompatibility of materials

Biocompatibility testing is necessary for all parts of the device that come into direct or indirect contact with the patient to determine possible toxicity as a result of contact between the components of the device and the patient's body. The materials used in the manufacture of equipment certainly do not create an unreasonable risk of local or systemic adverse effects directly or indirectly through the expression of their components. Carcinogenic; Or adversely affects reproduction and development.

The evaluation of new devices for human use requires systematic testing data to ensure that the benefits offered by the final product outweigh the potential risks posed by the device component. If not, but the question of the safety or effectiveness of the formation of results, when it appears. These materials or chemical components include plastics, metals, coolants, plasticizers, fungicides, and other treatments for chemicals, devices, or device components. The biocompatibility test should be performed on the final product, and the test conditions should be used to mimic the actual patient's use as closely as possible. The biological structure of these substances and chemicals needs to be thoroughly discussed to support the lack of testing.

A baby may be kept in an incubator for the following reasons:

a. Premature birth: Early infants may need extra time to develop their lungs and other vital organs. (Their eye and eardrums are so sensitive that normal light and sound cause permanent damage to these organs). In addition, babies born long ago did not have time to develop fat just under the skin and needed help to keep their bodies warm and toasted.

b. Breathing issue: Babies may have fluid or meconium in their lungs. It can lead to infections and shortness of breath. Newborns may also be immature rather than fully developed lungs that require observation and extra oxygen.

c. Infection: Incubators can reduce the chances of bacteria and additional infections when the disease is slightly cured. The incubator also provides a protected area, allowing vitals to be monitored 24 h a day, 7 days a week when a child needs multiple IVs because of medications, fluids, etc.

d. Effects of gestational diabetes: If the mother has gestational diabetes, when it comes time to monitor blood sugar levels, many doctors easily give birth to a baby to keep the baby warm.

e. Jaundice: Some incubators contain special light that helps reduce jaundice and yellowing of the skin and eyes in children. Neonatal jaundice is common and can occur if the baby has high levels of yellow pigment bilirubin during the normal division of red blood cells.

f. Long or traumatic surgery: If the newborn is experiencing trauma, constant monitoring and additional medical assistance may be needed. The incubator can provide a safe uterine-like environment where the baby can recover from trauma.

g. Low birth weight: Even if the baby is not immature, if it is very young, it cannot be kept warm without the help of an additional incubator. In addition, very young children may struggle with the same important tasks that premature babies perform (e.g., breathing and eating) and may provide extra oxygen and incubators from a controlled environment.

h. Recovery from surgery: If the baby needs surgical treatment because of complications after birth, the baby needs to be monitored later and stay in a controlled and safe environment. The incubator is suitable for this.

Laparoscopy

Laparoscopy, otherwise called diagnostic laparoscopy, is a careful demonstrative strategy used to look at the organs inside the midsection. It's a generally safe, insignificantly intrusive technique that requires just little entry points. Laparoscopy is used to take a gander at the stomach organs. A laparoscope is a long cylinder with a light and a camera at the tip. The instrument is embedded with a cut in the stomach divider. As it moves along, the camera sends pictures to a video screen.

Laparoscopy is frequently used to analyze the pelvic or stomach. It's generally performed when noninvasive strategies cannot aid with analysis. As a rule, various stomach defects can be determined like,

➢ ultrasound
➢ CT filter
➢ X-ray filter

It is difficult to trust the person who is leading the laparoscopic procedure. George Killing of Dresden, Germany, performed the first laparoscopic surgery on a dog in 1901, and Hans Christian Jacobaeus of Sweden performed the first laparoscopic surgery on a human body in 1910. In the decades that followed, many individuals refined and improved their approach to laparoscopy. The invention of the computer chip–based television camera was a major event in the field of laparoscopy. This invention provides a way to project a broad view of the surgical field on the monitor while freeing both the surgeon's hands for surgeries and making it easier to perform complex laparoscopic surgery.

Patrick Steptoe, one of the pioneers of IVF, helped popularize laparoscopy in the United Kingdom. She published a gynecological textbook, Laparoscopy, in 1967. In 1972, Clark Buffalo invented, published, patented, presented, and recorded film laparoscopic surgery using instruments sold by Wayne Instruments Company, NY. In 1975, Pesofand began the experience of laparoscopy (salivary gland detection) for Tarasconi organs in the Obgyn Department of the Medical College (Brazil, RS, Pasofand). It was first published at the Hyatt Regency Atlanta at the third AAGL conference. This laparoscopic surgery was the first laparoscopic organ rhythm published in the medical literature. In 1981, the first laparoscopic appendectomy was performed at the gynecological clinic of Kiel University in Germany. After a lecture on laparoscopic appendectomy, the president of the German Surgical Society wrote a letter to the board of directors of the German Gynecological Society, recommending that beans be discontinued from medical practice. Sem then submitted a paper on laparoscopic appendectomy to the American Journal of Prosthetics and Gynecology. It was previously unpublished because the technique was "unethical," but it was eventually published in the journal Endoscopy. Regular wrapping of ovarian cysts in semis, myomectomy, treatment of ectopic pregnancy, and finally laparoscopic-assisted vaginal hysterectomy (also known as intrafascial hysterectomy of the uterus) are some of the standard steps we have taken. He also made a medical device, Visap, in Munich, Germany, and still makes a variety of endoscopic instruments.

In 1985, he created Pelvi-Trainer = Laparo-Trainer. It is a practical surgical model that allows colleagues to practice laparoscopic techniques. More than 1000 articles have been published in various journals. He has created more than 30 endoscopic films and more than 20,000 color slides to teach and inform interested colleagues about his techniques. Details of his first atlas, pelvic endoscopy, hysteroscopy were published in 1976, slide atlas pelvic endoscopy, hysteroscopy, embryology in 1979, Germany in 1984, gynecological endoscopy book, published in English and several other languages. 2002. In 1985, German surgeon Erich Muhe underwent the first laparoscopic cholecystectomy. Later, laparoscopy was quickly adopted for nongynecological applications. Prior to Muhe's cholecystectomy, the only area in which laparoscopy was performed extensively was gynecology, using a relatively simple and easy procedure of mostly laparoscopy and tubal ligation. In 1990, 20 laparoscopic clip applicants were introduced, instead of automatically upgrading clips (instead of having to reload and restart the loaded clips for each clip application). Made it more comfortable (removing the gallbladder). On the other hand, some surgeons continue to use a single clip applicator because they save 200 per patient, deducting nothing from the quality of the clip binding and adding only a few seconds. Laparoscopy can be performed using both tubular ligation and cholecystectomy stem and ligation, reducing the cost of single-clip and multi-clip (compared with a stutter). Again, this increases the length of the case but significantly reduces the cost (ideal for developing countries) and eliminates large accidents of loose clips. Laparoscopic gallbladder removal was the first transatlantic surgery in 2001. Remote surgery and robotic surgery have become commonplace since then.

Laparoscopy is performed when these tests do not provide enough data or understanding to a conclusion. The method may likewise be used to take a biopsy, or test of tissue, from a specific organ in the midsection.

Anomalous outcomes from laparoscopy show certain conditions, including

➢ attachments or careful scars
➢ hernias
➢ a ruptured appendix, an irritation of the digestion tracts
➢ fibroids, or anomalous developments in the uterus
➢ blisters or tumors
➢ malignant growth
➢ cholecystitis, an irritation of the nerve bladder

➢ endometriosis, a turmoil where the tissue that shapes the coating of the uterus becomes outside the uterus

➢ injury or injury to a specific organ

➢ pelvic fiery malady, a disease of the conceptive organs

Laparoscopic is a particular method for performing a medical procedure. Before, this strategy was usually used for gynecologic and nerve bladder medical procedures. Throughout the most recent 10 years, this strategy has ventured into intestinal medical procedures. In a customary "open" medical procedure, the specialist uses a solitary cut to go into the midregion. The laparoscopic medical procedure uses a few 0.5—1 cm entry points.

Every entry point is known as a "port." At each port, a rounded instrument known as a trochar is embedded. Specific instruments and an extraordinary camera known as a laparoscope goes through the trocars during the technique. Toward the start of the methodology, the midsection is swelled with carbon dioxide gas to provide a working and review space for the specialist. The laparoscope communicates pictures from the stomach cavity to high-goal video screens in the working room. During the activity, the specialist watches point-by-point pictures of the midsection on the screen. This framework permits the specialist to play out similar activities as customary medical procedures but with smaller incisions.

In specific circumstances, a specialist may decide to use an extraordinary sort of port that is sufficiently huge to insert a hand. At the point when a hand port is used, the careful strategy is classified "hand-assisted" laparoscopy. The cut required for the hand port is bigger than the other laparoscopic cuts yet is generally littler than the entry point required for customary medical procedures.

A laparoscopic medical procedure is as sheltered as a conventional open medical procedure. Toward the start of a laparoscopic activity, the laparoscope is embedded through a little entry point close to the midsection button (umbilicus). The specialist at first assesses the midregion to decide whether a laparoscopic medical procedure might be safely performed. On the off chance that there is a lot of irritation or if the specialist experiences different elements that forestall an away from of the structures, the specialist may need to make a bigger entry point so as to finish the activity securely.

Any intestinal medical procedure is related to certain dangers, for example, difficulties identified with sedation and draining or irresistible confusions. The danger of any activity is resolved partially by the idea of the particular activity. A person's overall health and ailments are likewise factors that influence the danger of any activity. You ought to talk about with your specialist your individual danger for any activity. Tubal sanitization is one case of a medical procedure that should be possible using laparoscopy. Laparoscopy additionally is one of the manners in which that hysterectomy can be performed. In a laparoscopic hysterectomy, the uterus is isolated from inside the body. It very well may be eliminated in pieces through little entry points in the midregion or eliminated in one piece through the vagina.

Issues that can happen with laparoscopy include the following:

➢ draining or hernia (a lump brought about by helpless recuperating) at the cut locales
➢ internal bleeding
➢ infection
➢ harm to a vein or other organ, for example, the stomach, gut, bladder, or ureters

Laparoscopy might be used to search for the reason for chronic pelvic pain, infertility, or a pelvic mass. On the off chance that an issue is discovered, it regularly can be treated during a similar medical procedure. Laparoscopy likewise is used to analyze and treat the accompanying ailments:

➢ Endometriosis: in the event that signs and manifestations of endometriosis and medications have not helped, a laparoscopy might be suggested. The laparoscope is used to see inside the pelvis. In the event that endometriosis tissue is discovered, it can frequently be taken out during a similar methodology.
➢ Fibroids are developments that structure inside the mass of the uterus or outside the uterus. Most fibroids are benign (not cancer), but a modest number are harmful (malignant growth). Fibroids can cause pain or heavy bleeding. Laparoscopy now and again can be used to eliminate them.
➢ Ovarian cysts: some women experience growths on the ovaries. The growths frequently disappear without treatment. In any case, on the off chance that they do not, the ob-gyn may recommend that they be eliminated with laparoscopy.
➢ Ectopic pregnancy: laparoscopy might be performed to eliminate an ectopic pregnancy.
➢ Pelvic floor disorders: laparoscopic medical procedures can be used to treat urinary incontinence and pelvic organ prolapse.
➢ Cancer: some malignancies can be eliminated using laparoscopy.

Benefits

a. There are small marks.

b. One will be released soon.

c. The pain heals less when the scar heals, and the scar heals faster.

d. One will soon be back to your normal activities.

e. One's internal scratches may be less.

In some cases, the surgeon may insert a camera or surgical instrument into the skin by opening the unit. This means less pollution. However, this is difficult for the surgeon because the instruments are very close together. In other cases, the surgeon may decide to use a device accessible with one hand. This is called "hand-assisted" laparoscopy. The skin cut should be no longer than 0.5 inches long, but it can still be shorter than traditional surgical surgery. This has made it possible to use laparoscopic surgery on the liver and other organs.

This technology helps the team of physicians to be accurate. In the robotic version of laparoscopic surgery, the surgeon first cuts the skin and serves the camera as usual. Instead of capturing surgical equipment, he installed robotic mechanical weapons. Then move to a nearby computer. Many surgeons find robotic surgery especially useful for the work of overweight people and for gynecological and urological surgeries. Most prostate removal surgery uses a robot. In robotic surgery, the monitor provides the surgeon with a 3D, high-resolution, magnetic image of the body. When viewing the screen, use the hand controls to operate the robot and surgical instruments. This will make the surgeon more accurate, will have less impact on the body, and can reduce bleeding. It also reduces postoperative discomfort.

Procedure

Laparoscopy is performed when you are lying with your head under your feet and in a slightly bent position. General anesthesia is given during surgery to relax the muscles and prevent pain. Then make a small incision near the navel. A laparoscope is inserted through this intersection. Swell the abdomen to make the limbs easier to see. The laparoscope can also be equipped with surgical instruments for tissue sampling and removal of scar tissue. After surgery, they usually stay in the recovery room for an hour. They were then taken to the outpatient surgical unit for continuous observation. It will be discharged from the hospital after being instructed to restore the house. In most cases, one may be relieved about 4 h after the laparoscopy. Patients rarely have to be hospitalized overnight after this procedure. Within 2–8 weeks of the laparoscopy, we will be asked to return to the healthcare provider's office to book a follow-up. Contact the healthcare provider for a follow-up appointment before leaving the hospital.

Advantages

a. Laparoscopy is performed with a slightly curved head under the foot. General anesthesia is given during surgery to relax the muscles and prevent pain.

b. The laparoscope can also be equipped with a surgical instrument to remove tissue samples and scar tissue.

c. This incision provides additional openings for the equipment needed to complete a minor surgical procedure.

d. After surgery, they usually stay in the recovery room for an hour. He was then taken to the outpatient surgical unit for continuous observation.

e. Patients are rarely admitted to the hospital overnight after this procedure.

Disadvantages

a. Laparoscopic surgery requires proper visual and pneumoperitoneum for surgical manipulation.

b. Surgeons lose skills because of limited speed at the surgical site.

c. Realization of deep depth.

d. Patients need to use instruments to manipulate tissues without manipulating them directly by hand. This results in the inability to accurately determine the force applied to the tissue and the risk of tissue damage by applying more force than necessary. This limitation also reduces the sensation of touch, making it more difficult for the surgeon to feel the tissue; sometimes, the surgeon feels the diagnostic tools like the palate of the tumor and performs subtle surgeries like the breast.

e. The spindle points cause the endpoints of the instrument to move in the opposite direction of the surgeon's hand, which makes it an unethical irregular motor skill to become proficient in laparoscopic surgery. This is called the fulcrum effect.

f. Some surgeries (such as carpal tunnel) are usually better for the patient if the area can be opened, and the surgeon sees the "big picture" around the anatomy and solves the problem better.

Risks

a. The main problems during laparoscopic surgery are related to the cardiopulmonary effects of pneumoperitoneum, systemic carbon dioxide absorption, venous leukemia embolism, inadvertent damage to the intraabdominal structure, and the patient's condition.

b. Trocars are usually placed visually, so there is a significant risk of damage to the trocar during insertion into the abdominal cavity. Injuries include hematomas of the abdominal wall, umbilical hernias, umbilical infections, and infiltration of blood vessels or small or large intestines. Patients with a lower classification of obesity indicators or history before abdominal surgery have an increased risk of such injuries. Although these injuries are rare, serious complications can occur and are primarily associated with the site of navel insertion. Vascular injury can lead to potentially fatal bleeding. Intestinal damage can delay peritonitis. It is very important to identify these injuries as soon as possible.

c. Oncological laparoscopy poses a risk of port-site metastasis, especially in patients with peritoneal carcinomatosis. This incidence of isogenic cancer can be reduced by special measures such as protection of the trocar site and placement of the trocar midline.

d. Some patients experience electric burns by surgeons who use electrodes that leak current into nearby tissues. Injuries can lead to perforation of the limbs and peritonitis.

e. About 20% of patients suffer from peritoneal trauma due to hypothermia during surgery and increased exposure to cold, dry gas. Surgical humidification therapy that uses hot and humid CO_2 when inadequate has been shown to reduce this risk.

f. Not all CO_2 that enters the abdominal cavity is removed during surgery. When the gas rises and the CO_2 pocket rises in the abdomen, it pushes the diaphragm (the muscle that separates the abdomen from the chest cavity and stimulates breathing) and acts on the excitatory nerves. Pressure can be applied. It causes a sensation of pain that can extend up to the patient's shoulder, for example, in about 80% of women. However, in all cases, the pain is transient because the body's tissues absorb CO_2 and remove it by inhalation.

g. Difficulty clotting and tight adhesions (scar tissue) from previous abdominal surgery increases the risk of laparoscopic surgery and is considered a relative contraceptive indication for this procedure.

h. The formation of intraabdominal adhesions is a risk associated with both laparoscopic and laparoscopic, and it remains an important, open-ended problem. These usually occur in 50%−100% of all abdominal surgeries and are at risk of adhesion in both methods. Adhesive complications include chronic pelvic pain, bowel obstruction, and female infertility. In particular, obstruction of the small intestine is the most important problem. The use of surgical humidification during laparoscopic surgery can reduce the incidence of adhesive formation. Other techniques to reduce adhesive formation include the use of physical barriers such as films or gels or the use of a wide range of fluids in individual tissues during the subsequent surgical treatment.

i. Gas used to create space and smoke generated during surgery can leak into the operating room through access devices and devices. Gas plums can contaminate airfields with particles of operation teams and patients, as well as viral particles of potential pathogens.

Robotic laparoscopic surgery

In recent years, electronic devices have been developed to assist surgeons. Several features include:

a. Visual Enhancement - Use a larger screen for better visibility

b. Stability-electric dampness of vibrations caused by mechanical or unstable human hands

c. Use of specialized virtual reality training tools to improve physician skills in simulator-surgery

d. Decreased number of pregnancies

Robotic surgery has been suspended as a solution in developing countries, allowing a single central hospital to operate multiple remote machines in remote areas. There was a strong military interest in the ability to perform robotic surgery to provide mobile therapy while keeping trained physicians safe from war.

CHAPTER 10

Patient safety

Sudip Paul[1], Angana Saikia[1,2], Vinayak Majhi[1] and Vinay Kumar Pandey[1]

[1]Department of Biomedical Engineering, School of Technology, North-Eastern Hill University, Shillong, Meghalaya, India; [2]Mody University of Science and Technology, Laxmangarh, Rajasthan, India

Contents

Background and introduction

Occupational health and safety workers are experiencing additional, lesser-known events: healthcare workers suffer from rates of nonfatal occupational diseases and injuries even higher than those in construction industries. What's more, a recent report based on health claims data shows that hospital staff are at greater health risk and more likely to be diagnosed with chronic illnesses and hospitalized. What do these statistics tell us about the safety of

Introduction to Biomedical Instrumentation and Its Applications
ISBN 978-0-12-821674-3
https://doi.org/10.1016/B978-0-12-821674-3.00002-4

399

patients and healthcare workers? Is there a link between labor safety and patient safety? Is there a synergy to improve patient and work safety? Synergy, according to Merriam-Webster, is the mutually beneficial combination or compatibility of different participants or business components. What is the best way to coordinate reform efforts for the benefit of all?

Medical errors and patient harm have occurred for more than a century. Other than some isolated pioneers, many medical and nursing professions do not seem to recognize the scope and severity of the problem, or if they do, they refuse to recognize it. The majority of medical staff have always been aware of their personal safety. However, patient safety is an intense effort that requires the patient to think beyond simply the public healthcare system.

One of the great achievements of the last decade has been medical officials now identifying medical errors in patient injuries and discussing the matter publicly with professionals, politicians, and the general public. Prior to this, medical errors had rarely been detected by patients and had never been mentioned in medical journals or even by the government. Millions of people are being unnecessarily harmed, and huge sums of money are being spent. It is like spreading a pandemic in a country to see someone get into trouble.

Protecting patients is sometimes tantamount to avoiding mistakes, which seems innocent enough, but such a view can result in limited estimates even though recognizing and understanding mistakes is fundamental to patient safety. This is special attention may also need to be paid to research and patient safety. However, when we consider the general purpose of patient safety, it tends to be seen as a mistake or a risk.

Harm is what patients care about most. We all bear mistakes and are careful until we are hurts and realize that not all damage is caused by an error. Think of the damage that arises in health care: surgical complications, dangerous infections and injections, infections from overcrowded hospitals, adverse drug reactions, and excessive volume due to poorly designed infusion pumps. Many errors are harmless and may be necessary to gain knowledge. For example, many minor surgical errors may not compromise patient safety or the final surgical outcome. We always have many reasons to think about danger.

Twenty years ago, medical errors were rarely mentioned in the medical literature except in a few books. One of these was published in 1994 by Harvard surgeon Lucian Leape presented a seminar paper on errors in medical treatment. The note started by citing numerous studies showing that drug abuse is particularly high. This topic is emotionally charged and has not yet been taken seriously. Leape went on to argue that there were address errors in other major security industries. He particularly insisted on the prevention of treatment errors, the "perfection model". Doctors and nurses—if encouraged and well trained—should not make such errors. Punishing them with denial or discipline if they make a mistake was the most effective means of protection from future wrongs. The argument extended to physicians, nurses, pharmacists, and administrative staff. They are essential for the successful reduction of hospital care errors by changing how they think about mistakes. Leape said the solution to medical malpractice is not primarily in medical science but in providing instruction to reduce mental and human factors and mistakes, recognize human limitations and mistakes, and depend more on terms of service better than training.

Drug-related side effects are a common risk in patients injured as a result of drug-related medical interventions. However, some medications may not have side effects, or they may be limited. For example, unexpected allergic reactions were observed in 11 of 4031 treatments of adults in two US state hospitals, of which 6.5 involved drug interactions, and 228 of 100 interactions were negative. Drug use was also more common in health care, from 13% to 55%. Adverse drug-related incidents have been reported.

Ensuring patient health from any side effect that may harm other parts of the body by creating immediate illness risk should be the highest priority for all healthcare services. In several cases, treatment for one disease led to suffering from allergy, constipation, headache, dermatitis, etc. This brings our attention to applying a treatment method to achieve several treatment goals with minimal or no side effects to create a trustworthy healthcare delivery system. This effort will minimize the adverse effect of malpractice incidents and improve patient recovery rates. As we know, the global population is rising, and the field of health care is expanding in accord with global demand. In this context, healthcare systems need more accurate and sophisticated strategies to fulfill the vast field of medicine. We cannot apply a random strategy to human beings for the treatment of any disease.

Health care is one of the fields where every minute mistake matters greatly and can create great damage and implications. Compared with other industries, this field needs high accuracy and competency where the scope of error is negligible. Medical error can be defined as a wrongly planned action to achieve the desired goal. Such events can occur because of lack of professional competency, healthcare products, treatment strategies, procedures and systems, consultation and prescription, communication gaps, and understanding of patient orders, product labeling and nomenclature, packaging, compounding, dispensing, administration, education, use, and consumption.

Medication errors can also be seen in well-developed countries, so we should think about the condition of developing and underdeveloped countries with healthcare systems that are not as developed as those in developed nations. Maintaining patient safety as the highest priority is the fundamental need of the hour. In the 21st century, around 10% of patients admitted to general hospitals have been harmed during treatment, and deaths related to medical error are always on the list of the top 10 causes of death. Errors in dispensing, prescription, and administration account for around 25% of total deaths caused by medication errors, whereas other subcategories account for much less than this.

Generally, people think that hospitals are safe places, but this is not true in all cases as we can see cases where patients are handled casually, leading to patient harm. Such evidence has turned patient safety into a discipline that provides a set of guidelines to integrate knowledge and expertise with the capability to revolutionize our current healthcare system. Several countries have already realized this and have begun to make changes in the field. Norway, Sweden, the UK, and other countries are doing well by tackling present-day challenges effectively, with the World Health Organization (WHO) playing a prominent role in increasing global awareness about patient safety. WHO is also providing financial assistance and guidance to economically undeveloped and disadvantaged countries where it cannot be implemented without external assistance and guidance—countries like Sudan, Syria, Libya, etc. The World Health Alliance is also concerned about implementing patient safety in a phased manner worldwide. Fig. 10.1 shows some patient safety parameters.

There are several incidents where patient safety is swept aside while practitioners are engaged in treatment. Traditional thinkers believe a well-trained practitioner hardly faces any error in practicing compared with the incompetent practitioner who is new to the field. This statement is quite correct, as we know a newcomer will not recognize the exact locations of internal organs and the exact cause of disease. In most cases where the practitioner makes a mistake, that information is suppressed within the four walls of hospitals. In this way, patients and their families are kept in the dark regarding maltreatment. Lack of knowledge and awareness plays a vital role in low reporting of such events. The lack of legal counseling and

Figure 10.1 Patient safety parameters.

encouragement to report practitioner errors favors clinics and hospitals. It supports low reporting that reduces factual information for several reports and nongovernmental organizations (NGOs). One will hardly find a single record of malpractice in primary care despite the many malpractice cases that have been exercised for day-to-day treatment. Primary care needs the best-quality professionals where average quality professionals are deployed. It is not a case of a single country but rather the real situation around the globe. If one goes through a random survey on patient safety and health care, it can be seen that general medical units and intensive care units performed, respectively, around 30% and 10% of adverse events where patient health has been harmed. That draws the attention of healthcare providers and policymakers to the extent of harm and safety. In the last few decades, the public has been more alert to patient safety. This can be seen not only in India but throughout the world.

Internet revolution brought a change in our thinking and reporting toward malpractice. Many things can be done from mobile phones nowadays to reduce the extra travel for physical reporting. We can obtain vast information about injuries and diseases and the probable treatment procedure, including appropriate medication. The 21st century provided revolutionary thinking, and the field of health care responded accordingly by bringing forth various types of new information related to patient harm. This information-enabled technology helped curtail patient harm in an advanced manner. Medical professionals and think tanks debated about manual and systemic errors that could be reduced by upgrading the outdated treatment system by adopting newly developed sophisticated technologies in health care. As we know, modern technology can reduce errors because of atmospheric factors, validation, and standardization. Fig. 10.2 shows the elements involved in the quality validation and improvement process.

Applying engineering and design concepts to upgrading healthcare systems has brought changes like intravenous pumps, computerization, and physician medication. The efficiency of doctors and nurses reached a new height with the implementation of high-end healthcare technology like ultrasound, brain mapping, MRI, and CT scanning. The effective change in this field became an eye-opener for people regarding healthcare delivery. The earlier record in the field of medicine placed more emphasis on biological data than on technological intervention. Technology-enabled medication can decrease the chances of human errors.

However, it is easy to help build a global system of improvement to improve patient outcomes. But the application of reality is not as easy as it seems above. Good things designed to help improve health will face serious problems from people who believe "how can it be better than ..." or "the

Figure 10.2 Quality validation and improvement.

rule is . . ." or their "patients have never seen such a situation ..." or the best excuse, "As a leader, the organization does not consider the work useful or relevant at this time." If there is a general problem that occurs in many of the clinics mentioned in the article on patient safety, the adverse events of the patient are broad and far-reaching. From physical abuse and harassment of patients to better health, medical and legal cases, and subsequent fines from the third salary. Patient safety issues. Adverse events often affect patients, doctors, organizations. In addition to the "direct" effects on patient safety, many "indirect" effects of the product are often difficult to measure or quantify. By helping to increase patient safety reports, directly or indirectly, hospitals gain knowledge and improve their ability to attract new clients. In general, the risk of exposure to health risks may also be directly linked to the organizational commitment to safety tolerance.

Increasingly, both public and private insurance companies refuse to pay their doctor's bills and services provided to care for patients affected by work (or pain). The financial responsibility for solving outcomes is often more important and outweighs the need to pay for the patient's initial solution. Because the modeled cost of employee loss and higher cost of money was higher than in the past, these additional costs were passed on to hospitals and healthcare providers. There are also many impulses from both the public and third parties to provide benefits based on performance and quality and not just "performance", resulting in thriving enterprises and carefully considered results.

Consecration and delivery to dentists and physician conditions are also associated with safe benefits for patients' general benefits and records.

Perceptions of patient safety

Hospitals should be organized to construct a conceptual safety framework through a comprehensive survey by considering all safety measures. Traditional beliefs and behaviors surrounding hospital safety should be identified throughout the qualitative analysis to develop a conceptual framework and typology of patient safety. Primary care safety begins with patient feedback inclusion with teamwork effort, evidence-based practice, communication, learning, and improvement that is patient-centered. Several studies and reports have concluded that because of misunderstanding, ambiguity, and the complex nature of safety norms, they are challenging to operationalize. The basic key to organizational and patient safety is found to be senior leadership accountability in most cases. Safety policy and regulations in integration with patient expectations in health care create pressures for hospital leaders to provide evidence of organizational safety that ensures patient safety. Reports suggest that this type of approach may improve a hospital administration's ability to answer questions about patient safety. Structural empowerment and patient safety among adult critical care unit nurses are interlinked. Conditions of workplace effectiveness and the hospital's role in patient safety are crucial in conducting territorial hospital surveys. Fig. 10.3 shows the internal and external factors involved in patient safety culture.

A significant positive correlation has been seen, in which an increment in structural empowerment is linked with an increment in staff-patient safety perception. Based on this, it is recommended that nursing leaders consider structurally empowered staff work environments to promote patient safety. Additionally, a research report suggests that improved structural empowerment can provide an indirect influence on patient safety as a method to decrease and eliminate medical errors. It has certain limitations, as it assumes the scenario of whatever is available in a geographically small area, and the response rate is fairly low. In addition, the limited geographical and hospital setting, along with the inclusion of only one type of healthcare professional, reduces its generalizability to countries without the same financial and infrastructural status. Although it may suit well to correlate an important link between structural empowerment and patient

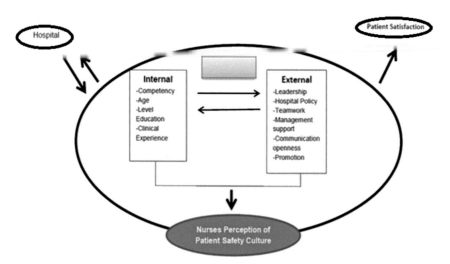

Figure 10.3 Patient safety culture.

safety in most cases, its limitations suggest the need for further research. Differences in the perception of patient safety among nurses and staff nurses are studied in a descriptive, correlation, and cross-sectional point of view at a high level. There are several categorical variables of safety, demography, education level, and length of time. Using some subscales from the hospital and surveys on patient safety, the dependent variables include overall perception of safety, number of events reported, teamwork within units, and safety grade. More positive responses on overall safety perceptions and teamwork can be achieved among hospital staff compared with overall. Significant differences have been seen based on the number of practical and observational experiences among nurses, in which those with 1 to 5 or greater than 5 years of experience in hospital duty are less positive in perceptions of teamwork within units, overall safety, safety grade for work area, and the number of events reported. This chapter provides insights into perceptions of patient safety among hospital administration and others and emphasizes assessment of the health professional's role as an important factor that may improve the effective use of medical staff. This section highlights important differences among casual and permanent health professionals. Where limitations exist, the generalization of results may be limited as well. Additionally, casual medical practitioners are not in designated positions and intermittently take on the lead role, making it difficult for researchers to determine true leadership experience. The relationship between collective safety behaviors and patient safety perceptions among nurses is examined in a cross-sectional study of variables that can be linked to patient safety perceptions, including time period. The Safety Organizing Scale is used to measure safety organizing behavior at the unit level, which includes measurement of five subconcepts: preoccupation with failure, sensitivity to operations, deference to expertise, reluctance to simplify operations, and commitment to resilience. Scales were used to measure perceptions of patient safety at the unit level, as well as patient safety grade and the number of events reported in previous years. Several studies found a relationship between increased safety organizing behaviors and positive nurse perceptions about teamwork, manager actions promoting safety, organizational learning, and overall perceptions of patient safety, staffing, and safety grade for the work area. Fig. 10.4 shows the variables in patient safety perceptions and their influence on the patient safety culture.

Based on the study findings, the researcher suggested that perceptions of patient safety culture may be more accurate when assessed in conjunction with measurement of safety organizing behaviors. The strength of safety in

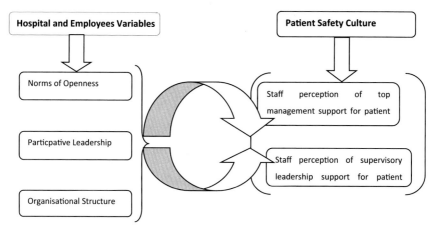

Figure 10.4 Patient safety perceptions.

developing behaviors in understanding patient safety can make it unique to existing research that focuses on hospital features and respondent characteristics. However, this field has limitations. Safety-setting encourages a hospital system that is convenient for patients and staff, too. Additionally, safety-organizing behavior may be accessed through self-reports. Using the 12 subdimensions of patient safety included by several reports, cross-national research clarifies the impact of long nurse working hours on patient safety. Evaluations of the impact of nurse working hours on patient safety outcome measures, patient safety grade, and number of events reported are based on odds ratios (ORs) calculated by a generalized linear mixed model. Most nurses working greater than or equal to 60 h per week have a significantly lower OR for patient safety grade than nurses working less than 40 h per week. In Japan, the United States, and Chinese Taiwan, significantly higher ORs were reported for the number of nurses working greater than or equal to 40 h per week. In most of the countries, the average "staffing" score was significantly lower for nurses working greater than 60 h per week than those in the less than 40 h per week group. In countries like Japan and Taiwan, the mean "teamwork within unit" score was significantly lower in the greater than or equal to 60 h group than in the less than 40 h group. The study concluded that long working hours were associated with deteriorations in patient safety grades and an increased number of reported events. Additionally, in all three countries, long working hours impacted "staffing" and "teamwork within units" among the 12 subdimensions of patient safety culture. A major strength of this study was that it was conducted across different countries, in which the researchers argued that common trends might be useful for improving patient safety culture in other countries. However, this study had some limitations. Objective indicators of "staffing" such as patient acuity or patient—nurse ratio were not collected; therefore, it was unclear how the actual workload or intensity impacted patient safety culture. Additionally, the response rate in the US was lower than in Japan and Taiwan, in which nonrespondent characteristics are not known, and the sample may not be representative of each entire country. Hospital patient safety culture across other countries, like the Netherlands, the US, and Taiwan are having similarities and differences in patient safety culture. Patient safety culture dimensions are the main outcome measures. Two of the 12 dimensions are similar across the three countries, with high scores on teamwork within units and low scores on handoffs and transitions. Significant differences between the three countries were found in the following patient safety culture dimensions: organizational learning—continuous improvement, management support for patient safety, communication openness, teamwork across units, and nonpunitive response to error. Fig. 10.5 shows core aspects of patient safety.

Additionally, differences were found among the frequency of event reporting, with US respondent scores significantly more positive than the other two countries. Overall, developed country respondents are more positive on the majority of safety culture 17 dimensions along with their higher overall safety grade than respondents in the other countries. However, responses between the developing country's hospitals are more consistent than in some developed countries, with more variation between hospitals. The large sample size across countries provides a broad picture of patient safety culture from many different perspectives and is a major strength of patient safety. Additionally, patient safety measure provides insight into different cultural backgrounds using a tool that is assessing culture itself. On the other hand, several limitations existed, including the

Figure 10.5 Core aspects of patient safety.

possibility of positive selection bias, variations in data collection methods between countries, differences in timeframes of survey administration, variation in sample size between countries, the potential for country-specific effects to influence the survey instrument, and limited verification of data accuracy against alternate assessment results. Overall, the researchers conducted a robust study with the following valuable concluding implications based on the research findings: Conducting comparisons on safety culture to identify opportunities for improvement is an important area for research with potentially useful implications for practice. The results have shown similarities and differences within and between the three countries. This means that within countries, hospitals with low scores on safety culture dimensions can learn from hospitals that have more developed safety cultures. Good examples can be found within each country, reducing the necessity to look over the borders when it comes to improving safety culture. However, for some dimensions with low scores nationally, countries can share best practices and learn from each other In countries like China and Hong Kong healthcare workers' attitudes and perceptions of patient safety culture are explored using a modified version of the Hospital Survey on Patient Safety Culture, which measures 10 patient safety culture dimensions. On five dimensions (teamwork within units, organization learning-continuous improvement, communication openness, nonpunitive response, and teamwork across units), the positive response rate is higher than for the others. Overall, a positive attitude toward patient safety culture within organizations is found among the surveyed healthcare workers in China as per some of the reported journals. Based on their findings, the researchers emphasized, the differences between China and the United States in patient safety culture suggest that cultural uniqueness should be considered whenever safety culture measurement tools are applied in different cultural settings. Several strengths and limitations are available. However, reported journals about the survey are modified, with deletion of 13 original items, potentially changing the framework of the original patient safety 19 culture surveys. Also, limited representation of hospital management in the sample may provide an incomplete picture of patient safety culture in China. Positive perceptions are found toward patient safety culture among good hospitals and their staff, in which percentages of positive response rates are highest among "teamwork within units" and lowest in the "staffing" dimension. Other differed in the following three dimensions: "Feedback and communication about error" "Communication openness," and "Frequency of event reporting"

Several strengths and weaknesses are identified in this. When compared with several databases, which included large samples in various healthcare organizations, the data has a lower internal consistency. Although both strengths and limitations of safety culture coverage are considered, the limitation of its use in a cultural setting is different from where it is developed. However, it is important to note that the application of this technique is found to be a good fit according to most of the confirmatory factor analysis indices. Based on research findings pointed out discrepancies between countries' data. Data suggest the cultural uniqueness should be considered whenever safety culture measurement tools are applied in different cultural settings. Twenty future research findings have been recommended to expand the survey worldwide to consider measurements to decipher individual and group perceptions and interactions related to patient safety culture.

Medical errors and adverse events

Improving the quality of health care and patient care has become an important goal over the past decade. The most common reports in the United States, such as parole and both, have made medical errors common and negatively affect patient outcomes. These publications are well aware of the shortcomings in the health sector. Healthcare professionals, governments, and medical associations around the world are encouraged to develop instruments to measure the quality of health in all medicines. Bug reporting agencies were established in 2000 in Australia and the United States, in 2003 in the United Kingdom, and in 2006 in France.

The concept of quality is a broader view of health from a process that is based on the doctor–patient relationship, the concept of efficiency, and an ethical approach to care. As for the quality of service, it is important to remember that safety is efficiency, safety, care, response to care, patience, and satisfaction. A series of important events from the 19th century laid the foundations for improving the quality of support. During the Crimean War of the 1850s, Florence Nightingale studied death in military hospitals. In 1912, Ernest Kodman developed a methodology for measuring the reversal of surgical procedures. In 1918, the American College of Surgery set the minimum standard, and hospitals must be accredited. In 1950, a medical audit system was introduced in the United States. The Joint Hospital Accreditation Commission was developed by Lambeck, and a year later, it created the accreditation of hospitals that perform standard quality measurements. In 1970, J.J. Williamson found a new methodology for discovering what can be achieved by a standard of care and patient care, but by examining patients. In 1992, Avidis Donabedian used an industrial model of the structure, process, and regression measures in the healthcare process. Finally, Palmer defined different dimensions of quality.

Quality assurance is a very complex task, and because of the complexity of their conditions, the need for emergency interventions, and the high workload in this area, patients in the intensive care unit are more likely to have more medical errors than other hospital patients. Therefore, the risk of medical errors should be taken into account when entering the intensive care unit. Safety must be established and measuring instruments installed. The overall control effect should be clearly defined. The impact of medical errors on patients and relatives should be examined. Prevention strategies need to be developed and evaluated. It is important to find the keys to developing a culture of patient safety in the intensive care unit. In this article, we will talk about these facts.

Measuring errors

Two methods can be used to assess and improve the quality of care. Improvement Model Identify problems, develop a problem-solving plan and evaluate the effectiveness and efficiency of the strategy. This approach is called PDAC by the Center for Health Improvement. The second way to measure safety is to use a monitoring system that identifies problems and evaluates them periodically using standard measures. The two methods are compatible and are often used in combination. Therefore, the management model can be seen as a tool for identifying opportunities for improvement through PDAC implementation.

Safety measures require a self-assessment process that determines what and how we set goals for improvement. Monitoring procedures require multiple identification methods to detect abnormalities and unintended consequences. These approaches apply not only to the regional level but also to the national level. The reporting process is done by government and government agencies. At the hospital level, the North American government and private agencies have been developing patient improvement databases and improvement programs and reporting adverse effects since 2005. The European Security Network promotes a culture of patient safety, provides a framework for safety education and training, develops a basic European approach to patient safety, and conducts reporting and learning processes. It was launched in Europe in 2008 to implement drug safety. Verification method. At the hospital level, there are various reporting systems for health professionals.

Medical examination. Review time that does not focus on the selected metric is time-consuming and depends on the information of the chart. The test can focus on selected indicators that can be evaluated using control data, travel summary, and death/disease data. Medical tests can be done manually or electronically using text words as a text search. Factors that may limit the use of clinical trial methods include a lack of electronic health records, a lack of diagnostic tools, and the use to determine side effects and spelling. Failure to enter words will make your search more difficult and increase the risk of incorrect answers. In addition, poor festival analysis requires good skills in data interpretation. Meta-analyzes comparing pharmacologists' rate of diagnosis and pharmacology showed that pharmacists had a higher rate of outcome identification.

Volunteer reporting is the most common way to identify health problems and adverse events. Restrictions include time constraints, lack of proper reporting system, fear of prosecution, refusal to report personal misconduct, lack of confidence in incidents and disputes, postreports, etc. However, this communication system is necessary to change the behavior, show the benefits of reporting bad things, and enable us to learn from mistakes. The presence of more than one security team simplifies the volunteer reporting process.

Medical abnormalities and serious cases can also be seen by direct monitoring near the bed. It is necessary to avoid this approach in order to solve problems. For example, the wrong medication errors can occur at any stage of the process (display, print, volunteers, advertising, and monitoring) and the description of the drug used in the study, how to evaluate drug therapy, and reporting system. The bedside pharmacist can add errors to the volunteer report that he did not come by mistake. The rate of error caused by medication errors ranged from 7.45/1000 days per patient to 560/1000 days of voluntary reporting. Also, in the IATROREF study, the presence of trained medical examiners collected medical errors, increasing the number of patients from 2.2/1000 to 597/1000 days.

Over the past few years, there has been a growing interest in the study and experience of Consumer Rating and Healthcare Providers in all countries with a long history. The United States and the United Kingdom. NHS. In 2007, the Organisation for Economic Co-operation and Development highlighted patient experience as a priority. Although many patients in the intensive care unit are very ill and cannot report their experiences, they can get information from their families.

Participating in these approaches is essential to ensure that medical disorders and adverse events are fully reported in a better understanding of critical care. The control system described above requires the use of accurate indicators. Ideally, each indicator should be presented as a numerical ratio (number of events, easily identifiable), an important indicator (nurse or vulnerable group), and a follow-up procedure. Eyes should be on the data collection table used by the coach. Data quality should be checked regularly (by checking and marking missing data). When the definitions are not the same for institutions or organizations, or when not all leaders agree with these definitions and cannot accurately describe the people at risk, it can be difficult to determine the extent of the disease.

According to Avedis Donabedian, there are three types of indicators that can be used: system indicators (what we have and what we have), system indicators (related to what we do and what we should do), and what we get and the benefits (as opposed to what we get). The organization has updated the quality standards of major hospitals since 2004. A list was compiled after examining online clinics using a variety of terms of "harmful," "iatrogenic," "clinics and severe," " medical disorder", and "epidemiology." This list contains 180 reports of misconduct. In July 2004, we sent a list of 180 cases to 30 professionals working in five intensive care unit (ICU) sites (cardiology, neurology, nephrology, psychology, and gastroenterology) and added 415 cases during this time with the impact of an accident, illness, or fatal event.

The choice of safety indicators depends on a number of factors, such as quality indicators previously studied in this unit, control methods, time to see additional signals, feedback to the organization, control actions initiated before they see results from these actions. Improving security requires time, organization, and resources. The goal is to achieve the best quality of resources. Both systems and performance indicators must be selected. Procedural guidance should be linked to complex outcomes. The results should be avoided, at least in part. Catheter infections associated with nosocomial infections manifest these symptoms. Other positive outcomes include random healing, pressure ulcer, reduction, hospitalization within 48 h and family satisfaction.

Comparing health errors with conditions of negative academic events can be difficult because of lack of variability and lack of interpretation. When clear definitions of harm are established prior to the study, the risk cannot be reduced. Two types of mismanagement and adverse events have been reported: (1) drug–related processes and (2) related processes or the ICU environment. The proper and effective administration of drugs is a challenge for nurses. The Critical Care Safety Study reported 80.5 daily medication errors in outpatients and 11 daily patients. In a recent SEE2 study in the world, the prevalence of parental drug errors was 745/1000 patient day. As the treatment progressed, there were 105/1000 patients per day. When bed rest was used to detect direct attention, one medical error was reported five times for a drug, and 23% of medical errors were anti-ulcer and anti-inflammatory. Vasopressors and mass KT, insulin, drug supply withholding, antiviral drugs, and storehouse often participate in medical malpractice. Insulin and anticoagulant drugs are associated with many nutritional or diagnostic problems. In recent years, evidence has supported insulin therapy and complex diabetes management, leading to insulin use in ICU patients. Clinical trials have shown that this strategy increases the incidence of hypoglycemic events. The IATROREF study found that the clinical error was estimated to be 757/1000 days for patients and 126 traumatic events/1000 days for patients taking insulin.

The ICU investigated a number of other medical errors and systemic complications. Respiration was associated with at least one event (0.004 per patient and 1 inhalation day) in 95/137 patients. Pneumothorax, a major complication of pneumonia and catheter insertion, was reported in 1.5 patients 5 days after the onset of asthma, and the risk of death was three times increased. All pumps, lines, and drains are used in the care of ICU patients. A French study of 14.5/100 patients per day and more than 22 clinics/1000 patients per day was erroneous. European Studies. Maintaining balance is very important, and electrical defects present to show poor care may occur during the life of the ICU and can lead to serious illness or death.

The risk factors for poor health habits and adverse events have been extensively studied. They come from the ICU or the patient. Modern medicine, the latest technology and diagnostic equipment used in the ICU are associated with an increased risk of infectious disease and adverse events. In the IATROREF study, risk factors for medical errors include airway obstruction, insulin use, intermediate catheters, and unexplained surgery. A study by the French Medical ICU identified people over the age of 65 and over 2 years of abnormalities when limbs are an independent risk factor for adverse events. In complex cases, several studies looking at Europe have found that there is none. Retirement type, reduced or reduced workload, and time exposure to risk factors independently mean traumatic events.

It is difficult to assess the effects of medical abuse or adverse events due to differences in Casmax, causes of death, and the presence of multiple events in one patient. Therefore, modern diagnostic methods must be used to assess the relationship between medical abuse or adverse events and patient outcomes. The IATROREF 1192 study demonstrated medically incorrect results in 1369 patients. Of the 128 patients (9.3%), 183 patients (15.4%) experienced adverse events that resulted in one or more health outcomes (n = 163) or the need for one or more interventions. After adjusting for exposure time to health problems, although frequent, had no effect on death. On the other hand, more than two adverse events were associated with a threefold increase in the risk of death.

Errors result from a combination of human factors and systems. People make mistakes, and the human error rate ranges from 30% to 80%. In the meantime, measures to increase attention and diligence are ineffective: people's mistakes are inevitable. But working conditions must be developed to minimize errors: As the reason puts it: "We cannot change the human condition. But we can change the way people work." For example, when two very similar drugs are stored in a presentation in the same area, a human-based approach is to train health professionals to pay greater attention to their needs. Similarly, to avoid mistakes. The systemic approach leads to the storage of the two drugs in different places. The key question in system policy is not to determine who is responsible for the error. But consider how this can go wrong. Therefore, without blaming the medical staff, the mechanism behind the error was identified. Organizational deficiencies can then be corrected to prevent errors.

Since the 1980s, the role of safety culture in the prevention of medical errors has been explored. Safety culture or safety environment (Sometimes the term is interchangeable, but "safety culture" is generally considered a more accepted term. "Safe weather") is a description of the security management flaws that led to this major disaster. Hence, the term was first used after the Chernobyl accident. Now the concept is designed to be applied at the individual level at the error level. The most widely used definition of a safety culture was developed by the UK Health and Safety Authority: A safety culture is a set of values, attitudes, cognitive abilities, and behaviors that define "loyalty, pattern, and health skills." For example, the six dimensions proposed are teamwork environment, job satisfaction, management awareness and management safe environment, working conditions and sense of stress, others are large corporate describes the scale of the ICU and the organization of hospitals is central to the concept of safety culture. The dimensions of the organization are both human and technological. Concerns about the high rate of drug-related errors have sparked interest in using technology to improve safety. New technologies introduced in recent years in electronic healthcare records. Computerized carriers support clinical decision-making with or without access to orders, barcode-coded drug administration, and smart infusion pumps, but these methods help reduce the number of errors. However, there is little evidence for joint mitigation, which suggests that new technologies produce new and harmful errors. Identify the clinical impact, and we need to address specific issues increasing their use in the ICU, as many errors are related to the less ideal organizational structures. For example, burnout syndrome affects almost half of the physicians and one-third of nurses in the French ICU. Burn syndrome can negatively affect the performance of healthcare workers who are at fault. Factors that increase the incidence of weakness include an increase in the number of patients, an increase in noise and light levels, long delays, changes in shift hours, and the occurrence of inconsistencies. The traditional internship work schedule includes shifts of more than 24 h and an average of 77—81 h of work per week compared with tables designed to reduce insomnia (abrupt, shift). (15 h, worked 60—63 h per week).) [61 Normal table 22% severe dosing error rate (193.2, $P < .001$ compared with 158.4/1000 patients per day) and 20.8% more severe drug error rate (85.7/1000 patients per day, 99.7, $P = .03$) and a 5.6-fold increase in primary diagnostic error (18.6 compared with the primary diagnosis) .3.3/1000 patients, $P < .001$) may be useful in intervention tests designed to increase work well-being and evaluate their impact on the medical rate.

Many goals were seen to have changed. The Agency for Research and Quality of Healthcare has identified five techniques that influence the practice of different physicians (research tools, proof-of-concept suggestions, reminders, thought-provoking subjects, and printable interventions). Various programs and contributions are more effective than they really were at the beginning of security, errors, and accidental radio removal methods. Also, communication systems reduce pneumonia. In 2003, the Michigan Patient Safety Program at Johns Hopkins University began removing the catheter: part, training, repair, and support. This describes effective ways of preventing pneumonia and reducing the risk of infection from the catheter.

Administrative errors

The majority of administrative anomalies are related to data. The exception is the other most common type of problem, which is negative. During the course, the injection site has been seen blocked many times during the course. These days, it includes patients who have taken the wrong drug or the wrong drug for a particular drug. The nurse distributed some amount of juice instead of the prescribed quantity, and the medicine of ceftriaxone should be given in different quantities. The patient did not prepare a pantoprazole dose again after discontinuing pantoprazole on time.

Transcriptional errors

The majority of errors are related to patients who require ions and are not registered with prescriptions or a sanatorium. As a result of these transcription errors, many drugs have been lost, and two are suspended while using drugs. Drugs such as valsartan, simvastatin, lansoprazole, calcium carbonate, and paracetamol have been indicated at withdrawal. Other errors mentioned include stopping the anion storage in the patient's history, but the anion storage table does not show this stop. Almost half of the drugs used for these types of defects are given to patients. These types of deficiencies include cefixime, paracetamol, pramipoxole, captopril, lactulose and laxadine (Galenium Pharmacia Laboratories, Indonesia, Semarang). Several results suggest that accuracy is needed in the transformation process to avoid errors in drug dosing and dispensing.

Prescription errors

The most common prescription errors are related to standard medications not available on patient procedure notes. This type of error is closely related to misdiagnosis, for example, 80 and 100 mg of aspirin are recommended. To evaluate the patient's progress note, the neurologist recommended 80 mg of aspirin, and the patient's progress note was 100 mg of aspirin daily. Alternatively, it is recommended to inject 50 mg of ranitidine twice daily instead of 50 mg 50 times daily as indicated in the patient's procedure notes. The third most common prescription drug is a history of misdiagnosis. For example, pharmacists (PIs) identify dexamethasone, spironolactone, and telmartine as the patient's most common drug interactions, but these are not found in patient progress notes.

Exchange errors

Exchange offenses detected during the study, 81 were related to cessation of distribution, of which 57 were related to drug administration failure, and six drug administration delays. The second most common type of error is a misdiagnosis of a patient's medication. Of the total, 24 error calculation problems were incorrect. In the district, such incidents were discovered at a pharmacy and the wrong medication was found. An example is a patient who received rampril 2.5 mg instead of the recommended rampril 5 mg. Drug Description "Prepare one tab of rampril 2.5 mg once a day."

Document errors and consequences

The accompanying drugs were classified as offenses in the identification document based on the classification system and chemotherapy. Gastrointestinal and metabolic drugs, the cardiovascular system, and the nervous system are the most common groups of drugs, but they are not listed in a patient's medication schedule. They are also the most common group of drugs that do not give the same drug but are distributed.

The potential consequences associated with document errors were further analyzed based on Lisby and other classifications. Errors in the rate of illegal documentation ($N = 513$) were also significant because patients were already receiving medication. As shown in Table 3, document errors were removed for various outcomes.

Stopping all medications can lead to negative consequences. The severity of these effects depends on the patient's medical condition and medications. Common antipyretics, painkillers, antibiotics, opioid painkillers, and laxatives are classified as necessary. This classification occurs because the absence of these drugs can lead to uncontrolled symptoms or worsening of the disease. 21 cases of drug withdrawal, including antibacterial drugs, were declared dangerous because cessation of antibiotic treatment would result in ineffective treatment and reinfection. Pesticides (such as fentanyl) are also dangerous because they need to be given temporarily to maintain effective blood levels to reduce the risk of further infection. Lack of anti-nutrient and anti-platelet agents can also lead to dangerous consequences 2. Therefore, lack of these drugs can be harmful.

Although drug intervention was not the primary focus of this study, the intervention was open. Recommendations for hospital treatment include timing of antibiotic use, adjustment or initiation of treatment, and dose adjustment. Its rating is lower than reports received from other studies, confirmed by 35 assessments. This low enrollment rate may reflect the general understanding of hospital clinical pharmacy services, especially between physicians and pharmacists, and the nature of healthcare provision in Indonesia.

Misdiagnosis is a surprising phenomenon in hospital and community ethics, and they are ridiculing the causes of more research errors and thinking of ways to prevent them. It is believed that 70,000 hospital deaths in the United States are due to misdiagnosis, and postoperative studies have shown that of patients are undiagnosed. The frequency of diagnostic errors in primary care has not been well studied. Please describe how they contributed. These factors include weaknesses and a history of physical examination, misdiagnosis by healthcare providers, especially routine and abnormal routine procedures, failure to perform proper testing, follow-up, and further evaluation of abnormal test results. The corresponding contribution of these errors varies from study to study, but everything is subtle and everything can be reported.

Additional testing, prescribing, or diagnostic procedures show symptoms that may be necessary individually or collectively. Further research into the primary care system is needed for emergency monitoring, symptoms, conditions, and identification of patients with red or orange symptoms. The role of basic "thinking" computer systems, including demographic and demographic data, historical and clinical data, and the ability to ask questions during counseling.

The study of early diagnosis of mental illness by Ahmed and colleagues can influence thinking, decision-making, and positive attitudes or clinical outcomes, providing clinical information, knowledge, training, and clinical experience. Patient capacity, Although experts say that the risk of misdiagnosis is higher and the treatment of people with mental illness is more negative, this study proved that medical knowledge is needed.

Simon and his colleagues made a brief remark on the promise of improving the "early diagnosis of schizophrenia" for early detection of schizophrenia in Bern, Switzerland. This randomized study has had a profound effect on physicians' knowledge of preventive schizophrenia. The A5 leaflet, distributed by the British Heart Foundation, raises awareness about the role of contraindications in the treatment of coronary heart disease.

William Osler praised the patient and said he would tell the patient about the test results. This phrase may be more accurate in the 21st century, but it reminds us of the importance of listening. One patient was so grateful that he did nothing else. The combination of our articles on QOF has always dealt with general practitioners—they deal with long-term diagnostic results and are in fact the most important practical diagnostic tasks, including the diagnosis and identification of serious diseases. I am here Of course, it has nothing to do with production. You need basic patient needs, disability, end of future patient care, level of competition for stress counseling, "suggestions" test results and methods to be checked by an independent consultant. Practical analysis of preparations, irregular schedule checks, cancer checks, etc. These are important factors in ensuring quality and patient safety.

To our knowledge, this is the first study to document the use and treatment of cases in Indonesian hospitals, and the study found that errors in drug administration were overlooked. This is due to the great work of the nurses. Most nurses are responsible for reporting patient education by prescribing and administering medication. It is a healthcare provider for information on pharmacies and homes. They want to fill the patient with information about entry and exit. This increase in performance is due to errors and poor handling.

This book is not what the chapter title is. Even two nurses need to pass a drug test. According to Werner and Aronson, there are 25 factors unrelated to mental illness. In addition, if used excessively, the document will err. The drug must be prescribed only after the government, not after preparation or administration. Therefore, these inaccuracies are related to errors during correction. Clear, accurate, complete and necessary for many purposes. Relevant documents show the strength of the medical services provided. Complete and complete information can be used to prevent inaccuracies from health professionals.

Inexpensive medications as well as over-the-counter or over-the-counter medications. These inaccuracies have a significant impact on hospital records. At present, you may prescribe medication to patients in three forms (medication, prescription, and over-the-counter instructions). The supervisor received an inaccurate report during the patient's medication change. In these cases, the procedure is on the table but does not contribute to the intelligence of the patient's growth. In this case, the doctor did not prescribe the medicine to the patient in their diet. This will give other healthcare providers (doctors or pharmacists) a lot of ideas about medication prescriptions for patients. When a detective finds a discrepancy between a patient's writing system and a prescription, he or she will ask the parent to provide him or her with accurate information. As the dean and others say. 15 Type this error in relation to the write process. Similar to the evidence discussed in this section, this type of error is related to inaccuracies in the wrong drug distribution by Werner and Aronson.

In a clinical study, three different pharmacies dispensed medicine to a church based on a patient's health insurance. This creates a big problem with the administration of the drug, especially when there is no help at the central pharmacy or when the patient's insurance does not cover the drug. Problems with drug distribution and health insurance resulted in 78 drug cases. These problems sometimes lead to other errors in the management system. This indicates that a negative attitude toward the drug distribution system or communication problems between the central pharmacy and the church is due to poor maintenance. It has been suggested that drug modifications are necessary to reduce drug inconsistencies in the distribution system.

This study shows that drug errors occur at any stage of the drug delivery process. Different health workers play different roles in this process, so they can prescribe the wrong type of medication. In addition, such failures in the distribution of medicines and participation in health insurance can lead to errors. The main limitation of this study is that there are only half-sections in the camp. Therefore, the level of overall results can be determined. In addition, having an early inspector in the department affects the behavior of other health workers, especially when there are differences and suggestions for resolving original and dangerous errors. Finally, the accuracy of some of the errors reported in this study depends on information received from nurses and patients because the researcher is unable to monitor all aspects of patient management.

Mistakes in drug use can be prevented. Therefore, through the implementation of a strategy to understand the nature of errors in the distribution of drugs, health services will be strengthened to prevent errors from occurring. However, there is no simple way to prevent drug errors during drug delivery. According to current research, a simple solution to a prescription or prescription drug that eliminates the need for prescription drugs seems to be a way to reduce deletion and duplication errors. Implementing a comprehensive computerized drug delivery system provides a comprehensive solution. In the late 1990s, research showed that the use of computer-aided drugs (CPOEs) reduced drug use. The maintenance provider has enough computer skills to influence the system. Therefore, the implementation of CPOE in educational institutions is not the right way out.

Mistakes in medications and materials can be minimized when pharmacists handle medications. Pharmaceutical manufacturers play an important role in reducing drug use by providing hospital-based pharmaceutical services such as more expensive drugs, 29 prescriptions, clinical trials, staff education, and patient counseling. Drug conciliation is an activity that ensures the continued use of inter-departmental drugs before being hospitalized or hospitalized. The 30-member panel said the drug collection was "a process of comparing drugs that patients take with some drugs." Although there are technical difficulties in assessing the impact of drug interventions on drug safety, the Health and Quality Research Organization 31 (IAEA) argues that this reduces the risk of drug interference. Drug incidents, especially preventive ones (e.g., misdiagnosis of drugs).

A good health system, health management and environment are required to ensure the safety of pharmacists. At present, the role of pharmacists in the safety of medicines in Indonesian hospitals is limited, as public pharmacies are involved in the distribution of other medicines in the provision of patient-related services. This study shows that pharmacies play an important role in diagnosing and preventing medical errors in addition to providing medical services in hospitals.

Communication and patient safety

Recent studies show that inefficient relationships between healthcare providers are the same. The main causes of medical errors—injuries of patients. The committee said more than 70% of media failures were ingrained youth center, when nurses are asked to identify risk factors for treating patients. One of the most important factors is contact with doctors. The source of the drug error was about half of the respondents. Drugs, they felt safe, but they felt safe Convey their concerns effectively. Focusing on security literature prevention is ineffective. Inadequate communication between team members is a possible cause of noncompliance. Accidents, when there is urgent treatment, communication problems can harm patients—prolong their lives. Use of housing resources as well as dissatisfaction from the fast caregiversare not recommended to participate in multidisciplinary intensive care research. Among the other special reasons, it facilitated communication between nurses and doctors. Patients at risk of regulated death—remain 1.8 times. Surgery communication events in the operating room, communication failure. About 30% of group exchanges pose a safety risk to some of these patients. Increase cognitive load, leg fracture I. Increase stress under operating conditions. The researchers found that relativity is simple щшпq divided into four categories:

(1) Lack of communication,

(2) Lack of communication

(3) Complete content issue

When he studied the results of the relationship he discovered that there was a connection with other researchers better communication between nurse and doctor collaboration more active results of the patient, viz. Low mortality rate, satisfaction. Low reading, effective communication between healthcare providers is a challenge for a number of reasons.

Related variable:

Healthcare services are complex and unpredictable from all walks of life at different times of the day different often affects more people. Modules usually make room for such features

• Therapists often have their own advice about what the patient needs.

• The health facilities had a historically hierarchical structure

Large energy distance between doctors and other healthcare providers. Instead of feeling, it often leads to a culture of self-control—a culture of self-control open communication secure communication (psychological well-being). Differences in training often lead to different results. Communication techniques that increase the complexity of the scene communication is inefficient. Despite the need for effective teamwork— effective communication for the care of patients, In most of the healthcare industry, education focuses primarily on personal technologies skills, teamwork, ignore communication skills.

Cultural barriers can be found in many organizations, and this quality can be relied upon the results lead to impeccable performance and disrupt professional training efforts. Special limitations described in anthropometry, In fact, human factors are like understanding excessive stress, fatigue, and disruption of action can lead to the breakdown of a personal relationship. Communication is known to deal with imperfect information, making wrong decisions. Health aviation benefits, such as complex health aviation errors fails knowing and understanding these problems leads to unrealistic expectations, a culture of guilt. Effective deviation from a team-based error management strategy. Lack of research based on interventions to improve cooperative relationships.

Patient safety culture in primary care

Initially, the focus is mostly on patient research research and policy recommendations hospital care over the past decade, this focus has expanded to include primary care. Primary care is an essential component of health care, and strong primary care is even better. Healthcare results in European countries strongest support gatekeeper system. More than 90% of primary care is delivered at the primary care level, which is an issue that is emphasized importance and relevance. In 2013, the number of general practitioners in a country estimated for example the Netherlands. At 11,075.15, primary care is easily accessible and GPS specialists are acting as a watchdog for the business. In addition to the general practice of care, primary care also covers various areas. Such as speech therapy, dental care, physical therapy, and midwifery. Compared with secondary care, primary care has relatively less exercise. Practice can use one hand or a group. Training, psycho or interdisciplinary, mostly management and corporate work this is done by a specialist, sometimes with the help of a nurse or administrator.

There is a big difference between primary and secondary care keep in mind that this type of error may be affected. Though there are risks serious damage is less commonly seen in primary care than in hospital accidents. Because of the large size, it is important to maintain patient safety in primary care. Number of patient interactions the presence of adverse events was studied in 16 studies. In Spain's primary care settings, 773 adverse events were found in seven instructions, seven cheeks. The team examined 1000 medical records in the Netherlands and found that 2.5% of patients call 18 study or equivalent note events in other major Dutch languages. Carers found less risk: 0.8%, 2.5% in dental care in the treatment practice, 19 connections of the midwife and 1.0% of the heart were identified. One of the main causes of accidents corporate security culture is defined as: "personal products and group value patterns, attitudes, perception, competence and behavior explain the commitment, pattern and health skills of the organization."

The security management characteristics of organizations with a positive safety culture are the importance of security is discussed through a general view of mutual trust and with confidence in the effectiveness of the warning. Classification of conservationist culture and cultural characteristics have reviewed seven subgroups: leadership, teamwork, evidence, communication, education, justice, It focuses on about 24 patients. An open and creative culture is seen as a facilitating factor for success. In addition to reporting the rate of adverse events, safety measures apply and suggested an improvement in preventive deaths, including the IOM report (Institute of medicine report). Similarly, the national patient protection authority states that a protective culture is to be developed creating a positive safety culture is the first phase of "Patient Protection from Seven Steps" In addition, a Dutch report by a former shell manager in the Netherlands mentions it is prohibited if there is a security hazard and a violation. It is also recommended that culture be positive for healthcare outcome.

The concept of the patient protective environment is also gaining concern for the patient security specialist in the first decade of the 21st century. This term is often used instead of patient care cultures, creating widespread confusion about how these two concepts are understood. There is an important difference between the two concepts. The Foundation has diverse climate and cultural research resources. Culture has been studied in anthropological studies and is largely a qualitative approach.

The emergence of the social psychology of the institutional environment and often occurs learning by quantitative methods. Relating to culture and climate. However, normal culture considered a deep root, it represents a climate that evolves over time exposure to more cultural surfaces. Culture comparision with personality and organizational environment in terms of its mood. It is considered to be more stable over time, on the other hand it is more susceptible to influence and change than the cultural climate. The concept of the patient safety environment and the patient safety culture contribute to confusion over Swedish guidelines for hospitals surveys related to the patient care culture The term "culture" often applies scholars will agree that it actually measures the climate beyond culture. According to the Swedish definition of patient protection culture the guidelines are "Interpersonal standards and attitudes among functional groups." It is one of the most important duty of a healthcare professional to protect the patient without making error. This definition differs from the general scientific understanding of defense culture as a collective and collective phenomenon.

Culture is often disputed as a concept and early anthropologists say that without human beings there would be no culture. It is important that noncivilized people do not exist. Culture groups are considered to be memory, and by applying memory to culture, culture seems to learn. There has been a lot of discussion about the differences in terminology culture and climate due to space and smallness significant changes. The climate includes work-related knowledge, related to the needs of facilities, operations, and small management. Housing culture is deep and inclusive prerequisites, confidential values and beliefs, Interpretation: It reflects the work related to the environment and high management organizations. However, there are many more the two terms are interconnected and must be treated differently. Chapters of different phenomena instead of two different classes phenomenon.

Company culture: The enterprise environment developed in the 70 and 80s of the last century corporate culture is like a corporate environment imagine how people feel and describe the content of their services and excerpts from it.

The concept of safe culture: The definition of a security culture is different. Brief and local concepts security culture is "how to do things here" share a protective attitude, belief, concept, and values private. A more detailed definition of protectionist culture is the result of value, attitudes, knowledge, skills, and actions are shared by everyone. All teams that recognize the organization's commitment and capabilities health and safety management. A deeper understanding of the differences identified by available for use in the safe environment. Supplement: Safety is related to religious beliefs and culture settings. The goal of knowing a safe environment can be considered as an indicator of the security culture in a given time or organization. Depending on the situation of the employees in the organization, they are all the same, but they look the same laws and procedures. So safety culture is a shared value the safety of the task force.

Patient safety culture is one aspect of social culture in the medical field protection. A cultural and dynamic healthy definition of patient safety introduced in 2006 by the European association for the quality of health think of a safety culture that takes steps to reduce the risk or harm to the patient feeding path. Definition: "Integrated model individual and organizational behavior based on common beliefs and values continuous efforts can reduce patient losses care distribution process."

A special feature of the patient safety culture has been identified recognized the environment as a leader risks and resilience in assignment of resources to frontline staff, a collaborative nature of the relationship between open, respectful, and respectful employees. A living environment in which all health workers can speak where patients and all staff can learn from mistakes improves their performance. Moreover, error-based training there are many system errors, but at the same time be responsible for people's actions. It is important to build a safe culture of open, honest patients improve patient safety. It means sharing of cultures, values, attitudes, laws, and beliefs. Practices, policies, and procedures on day-to-day security issues. It is to protect a cultural study was done to assess cultural health in health to date, the school of patient safety has been a major hospital. Some health services are available in health homes. Although injections affect patients the lower the number of first aid patients, the greater the number of patients and the greater the number of patients.

Several studies have examined the cultural safety of early responders to the disease. Some studies have been revised and confirmed with questions geared toward clinical practitioners and others. Basic care and treatment is available in the most of the countries the caretaker is watching the hospital. Many jobs involve large numbers of employee, the rules are different, but the test speed is low. After first aid note on patient safety, get acquainted (gain benefits) with current practices and other related professional organizations appreciate the equipment and enjoy the safety of the culture with patience. People were asked questions about their work industrial sectors are cooperative and highly cooperative in the health sector. Safety comes first for patients: first and foremost. Second, learning is about understanding cultural tenderness in a particular way.

Principles of clinical risk management

An important part of the recovery process is to involve network users in health development and safety plans. Attracting clients to investment. A risk and safety assessment plan is likely to be personalized. Beneficiary participation ensures a more accurate assessment of potential risk. Better understanding of the vulnerability, risk, and confidence of users of security services as well care programs are more closely aligned with care plans. Full integration is not always possible; employees may be asked to make decisions protection of service users, especially if they are in the early stages of recovery they traveled and became very ill. Having it now is weak and immeasurable unsatisfactory customer service. In the event of such an incident, it must be emphasized and all steps must be taken to ensure that the person enters into a joint conversation as soon as possible. This is a must support her in every way and continue to treat her with respect and dignity as a nurse informs and monitors opportunities for recovery and future participation. Studies show that many service users are unaware that risk can be identified. Health experts point out the level of risk when there is no language used to discuss potential risks to service users can be used as language. It is always influenced by other thoughts and actions. If the language can be dangerous. The principles apply to clinical and cultural policy, it is possible to eliminate potentially harmful language service users because they prefer to discuss personal safety with service methods and upgrades to protect them.

It is difficult to serve and refer customers. What effect does this answer have on us? Nurses can prevent the growth of communication and trust between service providers business relationship. Nurses represent the position of service providers at risk knowledge, skills and emotions drive his work and those experiences and results are diagnostic risk assessments. And the staff dealing with relationship problems with hope is a difficult and helpful, compassionate answer and how to plan. Recovery and risks take all unique and reflective experience ourselves and the world at every moment. The cause of the crisis each level of customer service is different and varied. Matters of communication and the nature of intimacy. Conversation talk, on the other hand, should be based on examples of respect and support. Set a default series, no question marks in the list.

Risk assessment and safety planning are therapies conversation involves human knowledge, not process. It is a process for nurses to talk to them in accordance with practical standards and best practices. Build the relationships you need, best wishes and respect. Each treatment relationship is unique and effective various factors. Every contact with a service user is an opportunity to do so develops nurse risk assessment and external safety plan establish communication and help minimize problems over time. Process these individuals may include their involvement in planning and assessing security risks the sisters feel connected. Create space for exploration and other topics provide an opportunity to understand conflict, engage in dialogue, and consider possible options. You also need to make positive decisions about risk. Develop a risk assessment and security plan customer service nurses call these people and ask them to show up ability to develop and communicate with people at risk at the same time. It is acknowledged that there is no security or guarantee.

Risk control and protection plans are inseparable bio–psycho–social assessment plan and rehabilitation/care plan. Security plan is created as a result of the risk assessment process.

Consider a number of issues related to human health and recovery. The relationship between risk and individual learning and recovery. Helps combine and promote effective risk and protection capabilities. The purpose of the collective risk assessment is to gather information for information take a general picture of that person and make sure that person continues to take risks. Promotes the safety of people and nurses and promotes efficiency and importance.

Service user. There are three main ways to assess clinical risk experiment: An unstructured answer based on the perception of its use evaluation. How to use health-proven instruments in measuring risk? "Dedicated to the second code method response from the clinical organization. Organized methods and procedures for clinical examination (equipment, Risk Assessment List or "Related memory" Risk assessment methods). It is interesting to note the structure of the clinic many people think of risk factors and safety perspectives like people, family/guardians and other coaches. Risks security must be considered at the user's discretion. Determining the cause of coercion is based on coercive methods.

Risk Assessment: Safety Plan, force effective power our resources are resources that you can use or support maintenance security plan. Security factors can come from within network location. Safety factors can be identified individually or together talk to a mental health team, family member or caregiver. A comprehensive assessment of human components or human capabilities take a good risk to move forward. Look for energy-based methods accident assessment safety plan reflects nursing past users discusses how successful strategy leadership strategies can best address this. Works with resource tax options with support system can be used at specific times. "Risk" is an option to use and measure productivity.

Many activities that identify potential problems risk assessment and plans and actions that reflect potential returns. A list of service user preferences (e.g., energy-based methods). Help achieve the desired results, including the use of available resources: Cost reduction dangerous products of the rehabilitation program. Health risk is a process that promotes government independence.

User freedom, user service for personal gain no operation is completely safe. Risks lead to natural disasters consumers understand potential risks, explaining the risks of self-regulation the purpose of life. "The goal is to eliminate all destructive threats destruction, deception. There can be no danger of falling reduce the number of people who develop the most appropriate leadership skills with recruitment. He is responsible for his actions. The role of nurses is to communicate and work together risks of timely equilibrium for personal development. Although its effects are somewhat less. If the healing process is based on improvement a person's ability to stand up for themselves, come forward reconnection.

The risk to mental health depends on the prevalence of the problem and the speaker. Dangerous species are considered acceptable because they are considered dangerous evaluation and control. Mental health partners found that the clients of the service are based on full risk others often monitor service users or threats to others. The medical services they use the latest version of the document implement policies and procedures to manage mental health and safety risks Irish treatment, suicide (suicide or suicide) and the threat to others attention to the level of knowledge of clinical nurses (Hilgin and others). Get a fair and complete estimate of the cost added to the price things to keep in mind: the risk of self-service to consumers. Accidents with service customers others; The creation of others by service users. And poses an "androgenic risk" or threat to services about the client's relationship with mental health services. There are a number of risk factors that affect the safety of people, including humans. Organizational and communication factors. Systematic exercises to assess the risks to mental health control of individual risk factors that harm organizational and communication factors. Careful research is useful in terms of each risk factor know the level of risk offered to customers. Individual factors focus on people and are often divided into static or energetic groups. Otherwise, the risk factors are constant irreversible risk factors are called fixed risk factors. Human life Demography, aspects of a person's personal history, personal characteristics. Mostly socioeconomic factors, risk factors and drug addiction unchanged elements of past or present human life exacerbate the danger look at the negative results. The dynamic factor is called the opposite. Understanding the diversity of factors, the human environment, and internal capabilities risk management. They can be different and can vary in weight and weight. These factors include local factors or direct social factors related to people's lives the mood gradually weakens and returns to its former "sexual" state perform treatment related to branches or medical services. Dynamic factors consider a high and broad context religious life, social relations, access to resources and past experience. Company or environmental factors are factors that depend on the context. They use a consistent or unbalanced approach to security management and planning.

Policies and practices that prevent people from collaborating and taking risks support and management capabilities, reception external services. Communication factors are based on the relationship between the client and the nurse. This can be determined by the following methods: lack of negative skills and knowledge of employees. An employee's or employee's behavior responds to fear, expectation, and inappropriate behavior start and define the procedure for deleting security plans.

Patient safety issues

There are several important things to keep in mind to help you adjust to security issues. It begins with the inclusion of a culture that is difficult to change, and the fact that many charities and related issues do not want to go beyond understanding or other powers. effect. Lack of awareness and inadequate education continue to create unpredictable "defects" in the immune system. Given the need for young people as an alternative to health care, another complexity of the "safety chart" may be relevant and useful. For example, before performing a complex test, it may help the patient to diagnose the anatomical condition, but the patient may be prescribed the wrong medication for high blood pressure.

In the case of the "privacy issues" mentioned above, group communication and the patient's "access" approach pose a significant risk of health errors. HOP or Regulation (HOP) refers to a policy that transfers patient healthcare services to a healthcare provider and includes the need for important information, treatment, and decision-making management. HOPs, also known as "focus" change systems, can have different levels and levels of transportation (e.g., return time, daily, weekly), making communication more effective. and not harsh. HOP occurs in hospitals and outpatient care systems during the transition between the two systems. HOPs vary widely and often depend on the employer's level of education, workload, maturity area, and reduced time spent in daily work. However, HOP is often overlooked as the underlying cause of the complications that cause the complications. In 2003 and 2011, the Institute for Advanced Health Studies ordered the principal to reduce working hours. This exercise routine replaces the short-term with other long-term questions. Therefore, there are many other HOPs about environmental spread. Supervisors can continue to schedule professional inspections hours, some of which provide a "safe place" Advanced communication systems and EMR are listed earlier. Given these new facts, healthcare providers should study in the nontraditional public health sector.

Colvin et al. She was born in the emergency department of HOP, where historical errors or failures can affect patients with a serious illness. The complexity of the situation due to the high connectivity provided by the ICU is very important for HOP and is closely related to several patient safety issues. Colvin et al. It also violates any transcript that the HOP (a) distributes related material, (b) inaccurate or contradictory information, (c) provides unsolicited or unsolicited information, and (d) consider what is expect. Problem or plan, (e) "ignorant or ignorant" and (f) unable to explain the night's answer. This demonstrated a lack of value and knowledge of HOP in all healthcare systems. Based on donations from 125 medical schools in the United States, a 2005 study of local pharmaceutical industry leaders found that 10% of companies teach students to develop HOPs in the regulatory framework. As there is no consensus between the above agencies and HOP agencies, it is important that this is addressed and immediate action is taken to ensure that HOP-related energy is trained and implemented.

Other challenges work together in the healthcare sector related to psychosocial activities and the organizational process in the work place. Communication barriers are studied in the context of managing different subjects. The success of the media is key to the performance of any team in the workplace. It turns out that "treatment" may be related to security. You may have little experience, such as medical students and minorities.

Human factors

In the field of patient safety, human factors appear to be directly or indirectly involved in secondary incidents. Many variables need to be considered including behavior, perception, sensitivity, and other factors that shape an individual's performance. Community efforts "Samin affects survival and it is difficult to see the overall impact on national statistics …" This was one of the reasons for imposing working time restrictions on residents in 2011; however, in 2017, the pendulum switched to a "mixed mode approach" because of clock limitations that were thought to help prevent fatigue -related errors that might not be explained by system errors in the manual.

Awareness of the importance of faults in equipment and systems has shifted its focus to "safety" for private healthcare providers, departments, and patient care organizations in general. A recent study clearly shows that most patient safety incidents associated with unintentional surgical procedures involve group or systemic faults and isolated human problems are rare. The complexity of the full review is illustrated by the fact that two or more safety exceptions were included in more than 52% of surgical cases in the same study. There are unknown aspects of patient safety, but perhaps the most interesting, which may affect all players over time as it happens, is self -harm in a general hospital. Hospital suicide was the second most common incident reported to the military (12% of all escort accidents) in the Joint Committee on Recognition of Health Organizations, but the study was very limited. As we read each chapter on patient safety machines, the importance of collaboration and organizational structure becomes clear to ensure our healthcare facilities and equipment are ready to succeed.

The main challenge in improving health security is the health culture itself. In a high -risk medical environment, quality audit activities can indicate a tendency to use a "culture of guilt" rather than "honest practice" or other collaborative models. Many health professionals are interested in corrective action and punishment on unwanted offenses. Anxiety failures can occur as a result of medical errors and therefore reduce the strength of registration and prevent change/correction of your personal future and that of your partner. Learn from the common sense mistakes of doctors. Misleading fears cannot be thought of as individual failures, but as a "shared responsibility" for future learning and improvement.

New vision problems have been reported and many changes in removal are taking place on a daily basis. However, we are hopeful that the increase in immunization risk will reduce the risk due to knowing and improving the process. We believe that the ultimate goal that can be achieved and achieved is "Half Media" But exercise is an opportunity to learn, think for yourself, and eventually improve. The difficulties of the healthcare industry, as well as the many specialized healthcare facilities that work with so many patients, are beyond that of any other company. Climate change has begun and, in addition, the end of current efforts to improve patient safety, including inadequate national care and attention, is increasingly in the region.

Based on these decisions and changes in disease management, our authors believe it is important to educate our readers about disease safety. Instead of going back to the "old" collection, we decided to stop our research in the last 5 years (2012−17) and provide information that could help restore patient safety around the world. The main points learned are clinical outcomes, tests, patient outcomes, other outcomes of individuals/staff, and hospitals.

Patient care policies

Investigating critical events in NHS organizations in critical situations the research in this case determines what lessons are to be learned evidence suggests that the NHS as a whole is not good at doing this on several issues not from only one country but from several countries. "Security comes first and the rate of change is very slow in framing new policies in the field of health care." A strong message in building patients is to strengthen leadership sure. Health care can be seen as risky from a statistical point of view in part Bungee jumping—and every year many people lose their lives on the plane and nuclear power has a very low risk. Thus the difference is considered important.

Much has changed since the study began: The NPSA has declined and we have seen the national development reporting and learning system (NRLS) with the participation of Royal Colleges. Working groups to protect patients are set up in collaboration with the national network among them. Introduced by those who for patient care programs. Some countries is working on a Safe Patient Initiative (with Health Foundation) in hospitals. In addition, NHS has carried out a number of programs the first patient care campaign was created with the aim of changing cultures among the NHS that give patients safety and the highest priority, everything is preventable death and unacceptable and harmful. There is also an increase focus on safety and quality and NHS for commission usage and application this is the agreement to ensure. Minimize erosive erosion and improve the security profile of healthcare providers remain at the forefront of radar policy locally and internationally. Recent developments include design the NPSA and its NRLS can teach lessons about conflict and patient safety. Its National Clinical Assessment Service for the Identification and Intervention of Physicians and Dentists. NPSA The report was launched in July 2001 by the organization Memory was started to recommend what the NHS could learn. From his experience. Safe NHS was developed for patients after an organization, the government plans on memory. One problem of planning—learning and learning lessons—related to education and research is needed to understand how the NHS can teach lessons and strategies for embedding lessons that improve patient safety and reduces risk.

Creating a Quality Circle: Patient Safety, accountability and openness and a beautiful culture where teachers are present (NHS) has been created encourage learning from adverse events. This is also suggested patient safety should be included in the curriculum at all levels and in the teacher qualifications. You need to learn important skills such as discussion, leadership, teamwork.

In 2006, NPSA began developing visionary lessons about the lessons of others. The industry is motivated to build training for art nurses and midwives and identify where they are vision training adapts to existing nursing education. Health Foundation recommends better training of health professionals through a report. Recommended by the Department of

Health Publications Protection the NHS works with the Royal College of Medicine for Innovation and Development other providers for education to ensure education training for patient safety support. It emphasizes the need for patient development certain courses that promote the required attitudes, behaviors, and skills. It is important that the NHS security is formally incorporated and connected with health and NHS health assessment programs. The World Health Organization has a global reputation for this problem (WHO) was responsible in 2004 for the World Alliance for the Protection of Patients World Health Organization to Implement World Health Council Resolution (2002) Member countries prioritize patient safety (WHO World Alliance) Patient Protection 2007. Meanwhile, the World Alliance is launching major campaigns includes the importance of intimate health care to minimize patient safety risks improve care security by activating infections in hospital surgical checklist and patient safety "Patient Agenda" emphasizes the potential focus of consumers on healthcare development.

Threats to patient safety

Several threats can cause patient's health in direct or indirect ways. This may include hazardous antiseptic, hospital acquired infections, dietary error, medical error etc. This can be broadly categorized into health (Does limitation and age), treatment (side effects of medication and limits), primary care (lack of efficient doctors in proportion to population, awareness), Gap and lack of communication (no feedback, unhelpful, rude, or disrespectful). Sometimes patient also become threat to their safety. These threats arise when a patient struggle to secure its own functional limitations. Some of them feel frustration from their problems and start to harm themselves from various ways like irregularity in taking medicine, noncooperation with hospital staff, unnecessary movement etc. This leads to create adverse condition for them in the form of injuries in case of falls, worsening of patient health due to proper medication and dieting and much more. Another factor that plays a very crucial threat in patient safety is under consideration and hiding of mild and moderate pain. This brings suppression of mild symptoms and attempts to normalize their health by their own home remedies. Fig. 10.6 shows threats to patient safety.

Globalization phase of this era is highly influenced by several pros and cons. Patient safety and healthcare policies are not untouched from the positive and negative effects of internationalization. Non government

Figure 10.6 Threats to patient safety.

organizations are playing important role in healthcare policy implementation and providing ground level services to the end users. But at the same
time, rapid industrialization, population growth, and increased human
activities are increasing global temperature leading to several types of new
diseases. Modern healthcare system is facing challenges associated with
globalization, population growth, and industrialization. These associated
challenges are unheard for small children either due to the lack of fund or
other reasons. Most of the poor and below poverty line people are unable
to keep away their children from noisy, dusty, and dreadful condition
because of their work location for example coal mine worker. Early care
and nutrition availability are important factors for a child in healthy growth
and development which is unavailable to extremely poor families that leads
to increase in disabled population and mortality. Some of world level organization like WHO and non government organizations like Medicines
sans Frontiers, World Federation of Public Health Associations, the International Association of National Institutes of Public Health, the International Union for Health Promotion and Education, and the Associations of
Schools of Public Health are working in organized manner as per WHO

guidelines. As per Alma-Ata declaration health is defined as human right. But global division on health status contradicts this declaration and creates a gross injustice among the people and countries too. Now a day's total world population and interconnected things leads to imbalance whole system due to imbalance in one thing. Higher rate of fertility increases poverty leading to disease burden on public health system and at the same time it increases out migration to developed countries.

Over the years, health care too has faced many errors in treatment of millions of patients. This has cause many deaths in the hospitals due to patient safety negligence. There is always a room for growth improvement and measurement to be taken toward zero patient harm. There are many concerns abiding with the healthcare safety throwing light on patient safety threats. Some of very common safety threats that are to be considered seriously are explained briefly in following sections.

1. **Workplace Safety:** Most of the experts argue that safety of the patients lie in the hands of healthcare workers. Unless the employees are provided with certain safety measures, hospital cannot proclaim about the safety of the patients. One of the most important duties of hospitals is to keep their patient safe. To keep patient safe, hospital staff should be well aware of their own safety measures and patient safety measures too. As we know, if doctor, nurses, and hospital staff will not be safe then how can they take care of patients with full alertness and energy. We generally think that hospitals are safe places that makes a sense about safe workforce can provide safe medication leading to safety and security of both patients and staffs. Safe and healthy workforce can use their full energy and alertness in rendering a good care to the patients. In the process of treatment, hospital staffs are in the pins and needles in taking care of patients but forgetting that indirectly they are endanger their lives from various threats on behalf of serving patients. It is very unfortunate for the workers that global community (except some countries) are still least bothered staff safety issues. In the recent years WHO has take some good steps toward safety and security of healthcare professionals for awareness and prevention of mis-happening with health professionals.

2. **Hospital Facility Safety**: Sometimes there are issue with hospital facilities that becomes threat to the patient's health and put safety at risk. Many times patient's health and safety are compromised because of on time building maintenance. Maintenance issues are very common in rural areas because of several issues. For example sewage, moisture, leak, and air ventilation comes in front line where people take these

things very carelessly. Another thing has to be taken care is sanitization of ward to maintain proper cleanliness. Infection control and risk assessment supervisors are very casual toward their job and responsibilities in taking care of hospital ambience neat and clean. In rural area, you can see live rats running within ward premises and also floor tiles, air conditions are never up to the mark.

So many diseases that can occur because of the hospital facilities like water supply, structure of building, ventilation etc. Keeping all these in mind government should make a suitable policy and strict regulation to implement this. For this purpose, hospitals should reevaluate their maintenance protocols to ensure patient safety.

3. **Medication Errors:** It is a common error that accounts around 5% of hospitalized inpatient are victim of adverse drug events yearly. Such types of evidence are seen in not only in general medication but it can be seen during surgeries too. Medication error accounts a significant number that occur from different kind of surgeries. It occurs because of incorrect dosages, labeling mistakes, negligence to tackle a problem got through various system and documentation error. These errors are prevalent under the category of medication error.

Medication error is having always room for improvement in reducing and preventing mortality rate. It is much needed to segregate for type of error occurring and what is frequency of their occurrence. So, we can develop appropriate strategies to mitigate such types of error.

4. **Diagnostic Error:** Medication error was much improved under the title "Improving diagnosis in healthcare." An annual report confirms that diagnostic error accounts around 5 to 15% of total hospital adverse events and in terms of patient deaths it accounts 10%. It shows that it is having ample space for improvement. In future hospitals should focus to improve diagnostic errors for betterment of patients. The possible solution to reduce diagnostic error may include better coordination among health providers.

Reprocessing issues

This issue provide a better opportunity for health care to work as a bridge increasing further scope and incidents. Infection rate in Corona period has increased up to a significant height because of non availability of Corona vaccine in the market till date. Hospitals are using protocols and procedure as provided by WHO to prevent the rate of infection at the same time most of the hospitals are sanitizing their wards on daily basis and reprocessing to check virus infection and prevention of the spread of Corona.

The Cyber insecurity of medical devices: A few medical devices availability in the market are not up to the mark and may have the patient health. One of the examples that you can take is Hospital Symbol of Infusion System, a computerized pump which is generally used for infusion therapy, is under threat of being hacked. Some of medical devices need internet connection for their smooth operation that gives a probe for their medical data to be hacked. On the same line of direction most of the devices are connected to the hospital network leading to chance of data insecurity if hospital network security is not strong. It can hurt to the patients emotionally as this informative data leads to social exploitation in rural areas where people think that if a person is having disease like AIDS, COVID-19, Chicken pox etc. then they used to start social isolation to the diseased person. The experts are hopeful to resolve the vulnerability of the data from the hackers as to protect the privacy of the patient.

Going transparent with quality data: Patients feedback play a crucial role or up gradation of healthcare system — good and bad experienced by the patients during their treatment brings the reality for the hospital administration and policy maker through patients feedback. But most of the rural hospitals are not having any kind of feedback system where as only few urban hospitals are providing online feedback facility to improve their quality and patients safety. Hospital physician and other medical staff should have a greater accountability to deliver a quality care. It can also help in rating a hospital in addition to provide better tools for individual assessment. From this we can develop a mechanism of salary hike and reviews and reviews promotion for health professional based on reviews and feedback received from patients.

Assessment of patient safety

Methodological aspects of safety assessment, along with their application in hospitals on patient safety are identified in a thematic review. The review included searches from electronic databases, patient safety organization websites, and reference list. Results showed that the review related to hospital measures of patient safety in the specified time period surrounded three main methodological areas: research approaches; survey tools for data collection; and levels of data aggregation. Based on this analysis, future researches are recommended to focus on clarification of core safety dimensions and identification of primary sources of safety variability. In addition, research using a mixed methods approach suggests allowing for in-depth research to identify the multiple components of safety. Although this may not directly use a safety assessment, it provides a comprehensive review and identified aspects and application of safety assessment, and offers a robust background to recommend future research. Because of the importance of patient safety assessments, a review of the literature about the development of patient safety among nursing staff is considered here. Patient safety should be recognized as a priority concern in healthcare environments, by incorporating leadership, teamwork, evidence-based care, communication, learning, just, patient-cantered care. Safety parameters are of complex nature and identified patient safety assessments parameters as a key factor in obtaining a comprehensive perspective on various strengths and weaknesses of patient safety to determine areas that require attention. As with other methods of research, this field is also having strengths and limitations. Combining ideas from different research provided a unique insight into patient safety culture assessment. However, selection and interpretation of research findings using different method of analysis are subject to research unbiased and must be considered as a limitation. The multilevel psychometric properties of the Hospital Survey on Patient Safety Culture are examined in a research. This topic analyzed survey data from different hospitals which included small hospital units and respondents to examine survey item and composite psychometric properties. Acceptable psychometric properties are found at all levels of analysis among all the dimensions. One exception is found in the staffing composite, which fell slightly lower than cut-offs in several areas, however it is conceptually crucial because of its effect on patient safety. Another exception is found for Supervisor/Manager Expectations and Actions Promoting Patient Safety, in which some hospital-level model fit indicator, is low. However, other psychometric properties related to this scale are considered good. Overall, the survey's items and dimensions are considered psychometrically sound among all levels of analysis: individual, unit, and hospital, and can be used to assess patient safety culture by researchers and hospitals. Both unit and hospital membership impact individuals' survey responses based on this study's multilevel psychometric results. Not only does the survey measure individual attitudes, but group culture at higher levels. Although this provides an in-depth analysis of the psychometric properties of the survey, it does not identify relationships among patient safety culture and outcomes, which is an area that requires further research. Because of the uneven distribution of positive and negative worded questions among the Hospital Survey on Patient Safety Culture dimensions, a research study is selected to examine the survey for acquiescence bias. Based on findings of several reports, it is concluded that the well-known Hospital Survey on Patient Safety Culture involves a risk of acquiescence bias which may lead to exaggerated reports of patient safety culture dimensions. The researchers suggested, "Balancing the number of positive and negative worded items in each composite could mitigate the mentioned bias and provide a more valid estimation of different elements of patient safety culture." Although this provides us new insight into potential acquiescence bias related to the Hospital Survey on Patient Safety Culture, it has limitations. The sample size of data should be adequate, but a larger sample could provide a better representation of the population. Also, the sample can be somewhat narrow in its focus, in which all participants are nurses, medical staff, and the majority are female. Another limitation of this is having individual differences among those completing the survey, which could be addressed by distributing both questionnaires to the same individual at different times. This method would provide stronger evidence of acquiescence bias. Event Reporting Practices among ICU Registered Nurses' perceptions of patient safety climate and potential predictors for patient safety perception and incident reporting are explored in a cross-sectional study. Among types of units and between hospitals, significant differences in RNs' perceptions of patient safety are found. Unit level variables are found to have significant impact on the outcome dimensions "overall perception of safety" and "frequency of incident reporting." However, among the outcome variables, differences are found in positive scores on "overall perception of safety" and "frequency of incident reporting." In all dimensions, concludes that patient safety climate is most positive among ICU RNs at the unit level, and areas for improvement included: "incident reporting, feedback and communication about errors, and organizational learning and continuous improvement." This identified several limitations which includes various healthcare professionals' limitations too. Additionally, generalizability is limited since the hospitals in this are small and within a limited area of the globe. Another limitation to this may have impacted the results are the known implementation of reorganization across units that are to occur after data collection.

Do's and don'ts of testing

The doctor's care toward a patient makes patient–doctor relationship and is considered as the center of primary medical care. However, this relationship cannot occur in a vacuum or vicinity of space. Every organization consists of human resource followed by certain, procedures, technologies, regulations, and organizational structure. The whole system that constitutes this organization has a significant impact on patient care. Scientist and policy maker have recently begun to start a more sustainable global perspective on primary health care and for the impact assessment of the larger system operation on the quality of patient care.

Most common and an important process in primary health care are testing. Tests performed in primary care may include laboratory, imaging, and special tests like ECG, EMG, EKG, EOG etc. The testing and their procedure can be defined in step by step manner that are conducted when a physician decides to order a test until the appropriate follow-up action is discussed with the patient and follow-though has occurred.

Less complex tests are performed within physician's offices, whereas complex and long procedural tests are sent to diagnostic lab facilities. Understanding of the steps that make up the testing process in primary care and delineated the steps in which physicians and their staff members perceive the most errors occurring. Although some authors have broken these actions down into "preanalytical, analytical and postanalytical" phases, we have expanded the pre- and postanalytical office-based actions into a series of steps, which taken together define the testing process.

- **Ordering:** Physician suggests going for a test and passes this information to the appropriate personnel.
- **Implementation:** The order is communicated to those performing the advised test or obtaining the specimen(s); the patient is prepared for the test or the specimen(s) are obtained.
- **Tracking:** The test order is monitored by technician internally (within the primary care practice) until the results are returned.
- **Return of results:** The results are forwarded to the concerned unit, office, or doctor from testing facilities or locations.
- **Response:** Based on medical test reports physician decides as to the about further action to be taken for the betterment of the patient and creates an action plan.

- **Documentation:** Hospital staff note all the entries in the medical record that the result has been reviewed; that the physician has responded to the result; and that the patient has been notified.
- **Notification:** Concerned patient is informed of his/her test result and the physician's recommendations for action.
- **Follow-up:** The process whereby abnormal results or results requiring action are monitored until such action is taken or the patient refuses the action.

Safety measures for diagnostic instruments

Accurate diagnosis and timely action is a key feature of good clinical practice. Getting optimal outcomes in terms of patient health is essential factor of treatment procedure. But we have seen that diagnostic errors are common in our health system. Eventhough efforts toward improving diagnostic error are not up to the mark and very less action has been taken to improve performance, quality, and safety. All these errors are of major concern when it comes to patient safety that can reduce several deaths and cost of treatment significantly. Several Technologies and tools are available in present day's diagnostic process. Information Technology (IT) plays an important role in most of the modern diagnostic processes. This technology is integral part of most of the diagnostic instrument with both positive and negative impacts.

Measurement and finding exact location of disease is crucial for better treatment in which diagnostic performance matters a lot.

Health policies

In most of the countries including India, the primary health care is given the highest priority which envisages the goal of healthy status for all. It focuses on early intervention, preventive, curative, and rehabilitation services. Global community has got varying health policies for their citizens but all in all aims to achieve an acceptable standard of health and nutrition for all of their population. The broad objective of health programs are advocated and prepared by expert committee for public health care, nutrition, sanitation, drinking water, and awareness about hygiene and cleaniness. It also addresses the malnutrition and nutritious food to all. More importance is to reduce disparities in healthcare services among regions and communities by providing affordable health to all, especially to economically weaker and underprivileged section of our population specifically women and children, the older persons, disabled, and tribal groups.

Conclusion: Purpose of healthcare system should reduce illness, injury, and disability with minimal side effects/errors and should improve people's health and normal health functioning of the people. Health care have a broader scope as it incorporate the global population dimension relating to public health and other social and political factors too. Thus, well-developed and error free health care encompass not only the spectrum of primary health care, hospital services, and all public health services but also the economy. WHO enumerates four key functions for health systems:

1. service provision
2. resource generation
3. financing
4. stewardship

Current issues that facing patient-related matters to our population and planet are numerous, vital, and real. After rigorous assessments and analyses starting from the late 1990s until today our health care and patient safety are not error free. Some of the undeveloped under developed and developing countries are using old methods, machines, and technology because of financial issues putting patient safety aside. This is the need of the time to change our old thinking of the 20th century, we are still concerned about diplomatic, and military power plays. We should face the real challenges collaborate like healthcare issues, patient safety, global warming, and the recognition of health as a basic human right with focus on implementation of the Alma-Ata Declaration on Primary Health Care along with improvement of global aid support and mechanisms, administrative strengthening of good governance. All these things could not be possible without active participation and involvement of the civil society. Although number of NGOs are increasing in recent years but their roles and responsibilities are limited up to certain extent. However, NGOs are not only responsible and accountable for this change, but government and society are also in general. Therefore, an important requirement of such type of NGOs that can reveal about the malpractices in clinics, hospitals, and medical institutions should report fearlessly. A code of conduct for NGOs is a first main recommendation and requirement for better implementation and coordination with different units and organizations. Poverty, the burden of disease, and violent conflux are interconnected. Basic needs like primary care, food, nutrition, provision of safe food and water, shelter, and education, and last but not least, social equality in daily life with healthy people and population, does not seem to be an extraordinary or unjustified one. This entire requirement makes a person to grow and educate himself to have a better understanding and responsible citizen. Basic requirements make human right a reality to be fulfilled in global world to a degree where people can feel safe.

Index

Note: 'Page numbers followed by "*f*" indicate figures and "*t*" indicate tables.'

Index

471

risks of medical imaging, 177—178
safety, 196
T1 and T2, 191
ultrasound, 197—199, 199f
ultrasound in anesthesiology, 200—203
PET scans, 200—203, 204f
urology, 199
X-ray radiography, 178—179
formation of radiography, 179
X-ray machine block diagram, 179f
Radiological imaging techniques, 178
Radiological recorders, 28—43
diagnostics, 31—33
magnetic resonance imaging, 28—33
Radiological techniques, 9
Radiological thickness, 178—179
Radiologist, 172
Radionuclides, 205—206
Radiopaque, 173
Radiotracers, 37—38, 69, 205—206
Rate-responsive pacemaker, 254, 257
Rayleigh scratching, 181
Readback values, 143
Real-time magnetic resonance imaging, 193
Recessed electrodes, 58—59
Rectifiers, 133—138. *See also* Amplifiers
bridge, 138
center-tap, 137—138
controlled, 135—136
full-wave, 136—137
half controlled, 136
half-wave, 136
uncontrolled, 134
Recurrent TMS, 353
"Redundant EDR", 76
Regional anesthesia, 316
Regular pacemaker tests, 262
Rehabilitation/care plan, 429
Relative pressure. *See* Gauge pressure
Relative sensor, 95
Remote alarm operation, 316
Remote patient monitoring (RPM), 80—82
Repetitive transcranial magnetic stimulation (rTMS), 353—355
mechanism, 354f

normal side effects of, 355
risks of, 355
serious side effects of, 355—356
Report module, 83
Reporting process, 412—413
Reproducibility, 5
Resistance temperature detectors (RTDs), 97—99
Resistive humidity sensor, 115
Resistive membrane transducer, 50
Resistive method, 49
Resolution, 5
Resonant pressure sensors, 106
Resource generation, 445
Respiration, mechanical ventilation for, 310—311
Respiratory
anesthesia, 284
flow waveform, 309f
function, 284
respiratory therapist-driven protocol, 311
Response
steps of testing process, 443
time, 92
Restrictions, 413
Reusable disk electrodes, 7
Reusable electrodes, 46
Revival, 284—285
RF. *See* Radio frequency (RF)
Rhabdomyolysis, 13
Rheobase, 324
Risk Assessment, 430
Robotic laparoscopic surgery, 397
Robotic surgery, 397
Rochelle salt, 50
Roller pump, 377
Rotating armature generators, 306
Rotating field generators, 306—307
Round soccer electrodes, 346
RPM. *See* Remote patient monitoring (RPM)
Rubidium-22, 205—207

S

SA node. *See* Sinoatrial node (SA node)
Safe culture concept, 427